The Development of
JET AND
TURBINE
Aero Engines

THE DEVELOPMENT OF
JET AND TURBINE
AERO ENGINES

4TH EDITION

BILL GUNSTON

First edition published in 1995
Reprinted in 1996
Second edition first published in 1997
Third edition first published in 2002
Fourth edition first published in 2006
Reprinted in 2010 (twice) and 2012

British Library Cataloguing in Publication Data
A catalogue record for this book is available from the British Library

ISBN 978 1 85260 618 3

Patrick Stephens is an imprint of
Haynes Publishing, Sparkford, Yeovil, Somerset, BA22 7JJ

Tel: 01963 440635 Fax: 01963 440001
Int. tel: +44 1963 440635 Fax: +44 1963 440001
E-mail: sales@haynes.co.uk
Website: www.haynes.co.uk

Haynes North America, Inc.
861 Lawrence Drive, Newbury Park,
California 91320 USA

Printed in the UK by MPG

Contents

Introduction

This book has a rather clumsy title because you can have jet aircraft without any gas turbines, and you can have gas-turbine aircraft without any propulsive jet. Examples of the former include rockets and ramjets, and the latter include many turboprop installations where the jetpipe is just bent round sideways, and most modern helicopters.

From the dawn of this century numerous inventors proposed methods of jet propulsion, while almost as large a number suggested that propellers might be driven by gas turbines. Just after the First World War a Frenchman, Maxime Guillaume, combined the two, conceiving the turbojet; but he lacked almost all of the knowledge and skills that could have led to anything being built and tested.

So it was left to a young RAF fighter pilot, No.364365 Fg Off Frank Whittle, to bash his head against a brick wall of disbelief for seven years and then have an even harder time trying to make the thing work with hardly any money. The wasted years 1929–36 meant that when Britain did at last run a turbojet there was a rival team in Germany, and the German team had all the resources of a large and prosperous aircraft company.

As in the case of a previous book for PSL, dealing with aircraft piston engines, I have divided this book into two parts. The first is intended to describe how the various parts of gas-turbine and jet engines work. As always, I have avoided mathematical formulae and tried to use language that will not frighten off readers who have no engineering background. If any casual reader does find it heavy going, skip it and read Part II. This simply tells the story of the different species of engine.

Soon after I left the RAF I went to the cinema and saw *Sound Barrier*, in which the young test pilot watches the supposedly supersonic fighter being run up in full afterburner and exclaims something like; 'It's the most exciting sound I ever heard!' We had a different outlook in those days. Supersonic jets blasting out glowing shock-diamonds, and the feel of the sound waves impacting on your chest, were a matter of national pride. And Britain had a lot to be proud of. For example, I also loved the synchronous whine of a Viscount, and am pleased that some will keep flying into the next century.

Today we are more environmentally conscious. I doubt if many people who campaign against aircraft disturbance will read this book, but if they do they will learn that, in the past 30 years, big passenger and cargo aircraft have multiplied in carrying capacity by 3, multiplied in range by 2 (the A340 goes anywhere you want non-stop), and in safety by roughly 30, while at the same time cutting out between 98 and 99 per cent of the noise. Offhand, I cannot think of any other human creation that can boast so many totally different improvements of similar magnitude. It is hard to believe that in another 30 years the aircraft of 1994 will appear crude and inefficient!

Bill Gunston
Haslemere
1994

PART I
HOW GAS TURBINES WORK

1 Principles

Every aeronautical propulsion system from the Wright brothers to spacecraft has had to be some form of jet engine. That contention would be harder to sustain if we had been able to emulate the birds and build successful ornithopters (flapping-wing machines), but it was the crucial decision to separate the lift, using a fixed wing, and the propulsion that made the aeroplane possible. To obtain forward thrust in flight it is necessary to accelerate a mass of air to the rear. At one extreme is the helicopter, which handles an enormous mass flow (mass handled per second) and in a complex way accelerates it downwards, to create lift, and to the rear, to create forward thrust. At the other extreme is the rocket, which, usually by a chemical reaction, takes materials initially at rest and accelerates them violently through a propulsive nozzle.

When I was a small boy, my father's large and erudite family had a great argument over the proposition that a rocket can work in a vacuum. A majority said; 'Nonsense, it must push against the air.' If any readers think this, I must disabuse them of the belief at once. The rocket is unique in that it starts off carrying all the materials that form the jet. It can therefore work in the vacuum of outer space; indeed, because there is then no atmospheric back-pressure trying to stop the jet emerging from the nozzle, it works much better. For example, the thrust at sea level of the LR99 rocket engine of the X–15 aircraft was nominally 50,000 lb, but at 45,000 ft this had increased to 57,000 lb, and it exceeded 60,000 lb above the sensible atmosphere.

All jet engines work by accelerating a flow of what engineers call 'working fluid'. In the case of rockets this is usually a gas (of various kinds) heated to incandescence. This book is more concerned with air-breathing engines, so called

because their power comes from burning a fuel in the oxygen of the atmosphere. With such engines, the working fluid is anything from 99 to 99.999 per cent air.

The simplest air-breathing engines are ramjets. When these were first produced someone called them 'flying stovepipes', because they are little more than a piece of pipe through which the air flows. Today, many aircraft would have ramjets but for the fact that these engines cannot start from rest, as explained in Chapter 12. Another simple jet engine is the pulsejet, described in the same chapter, but for all practical purposes today's aircraft use either piston engines (see the author's previous book) or gas-turbine engines. As explained later, a gas turbine is just a turbine (a modern form of windmill) turned at high speed by a flow of hot gas.

The first gas turbines were designed to turn a shaft, for example to generate electricity. By 1920 a few workers were considering using a gas turbine to turn an aircraft propeller. Such an engine is called a turboprop. Meanwhile, several other inventors were busy with jet engines in which, lacking any other kind of suitable power, the essential air compressor was driven by a piston engine. Some even tried to use the piston engine's exhaust as the jet. Whittle's great breakthrough was simply to connect a gas turbine to the compressor, to create the turbojet. This is the simplest possible form of gas-turbine engine. It comprises nothing but the three essentials: a compressor to compress the airflow, a combustion chamber to burn the fuel and create the flow of hot gas, and a turbine to drive the compressor. The hot gas downstream of the turbine is then expelled to atmosphere to create propulsive thrust, and to generate the maximum thrust it is necessary to take a little care with the

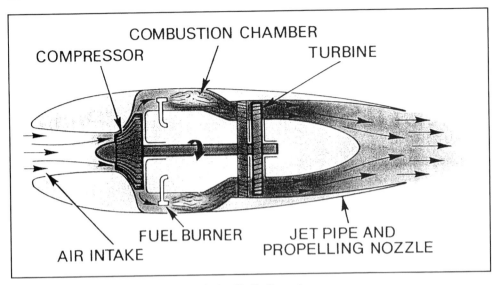

COMBUSTION CHAMBER
COMPRESSOR
TURBINE
FUEL BURNER
AIR INTAKE
JET PIPE AND
PROPELLING NOZZLE

Simplified diagram of a centrifugal turbojet (Rolls-Royce).

internal profile and area of the nozzle.

Compared with, for example, any kind of piston engine, a gas turbine can generate tremendous power. Today a turbine (turbine alone, not the whole engine) that could fit into a suitcase can generate 100,000 hp. If that is hard to imagine, it is roughly equivalent to 50 diesel locomotives or more than 1,000 family cars. The earliest turbojets, made largely from sheet metal, looked too light and flimsy even to contain their noise, let alone the power they generated, but they burned fuel at a daunting rate. Over the years their efficiency was transformed, until today they are among man's most efficient engines of any kind.

There are several kinds of efficiency, each of which will be discussed in turn. One is the basic thermal efficiency of any heat engine, defined as the useful power generated divided by the rate at which chemical energy is produced by burning the fuel. A second is the mechanical efficiency, which is the useful power delivered divided by the internal power generated, the difference being due to friction, and various 'windage' and pumping losses. A third is the propulsive efficiency, which is the work done in propelling the aircraft divided by the work done (in the same unit period of time) in accelerating the airflow to create thrust. Yet another is cycle efficiency, which is the ratio of useful work obtained divided by the useful work obtained (again in the same unit time, such as one second) from an engine with an ideal working cycle.

It is best to look at cycle efficiency first, because it is so fundamental. All heat engines have an operating cycle: they take in a working fluid (usually air), compress it, heat it, make it do useful work and then release it back to the atmosphere. Most readers will recall the Gas Laws, which describe how a theoretically perfect gas behaves. Boyle's Law states that, at any constant temperature, the volume of a given mass of gas is inversely proportional to its pressure; double the pressure and you halve the volume. Charles' Law states that, at any constant pressure, the volume of a given mass of gas is directly proportional to its absolute temperature. (What is that? See the Glossary.) Double the absolute temperature and you must double the volume to hold the pressure unchanged. These laws are universal.

Engineers naturally think in pictures, and every heat-engine thermodynamic cycle can be pictured by plotting the pressure at each point against either volume or temperature. This naturally gives a closed figure, because the working fluid has to be returned to the same atmosphere from which it came. In any PV (pressure/volume) diagram, any compression or expansion is a curve. Imagine we are using a bicycle pump: at the beginning we can push down a long way without any great increase in pressure, but when the air has been squeezed into a small space a small extra push will mean a big increase in pressure, and thus also in temperature (it follows from the Gas Laws), and this is a simplistic explanation for the fact that the plotted lines

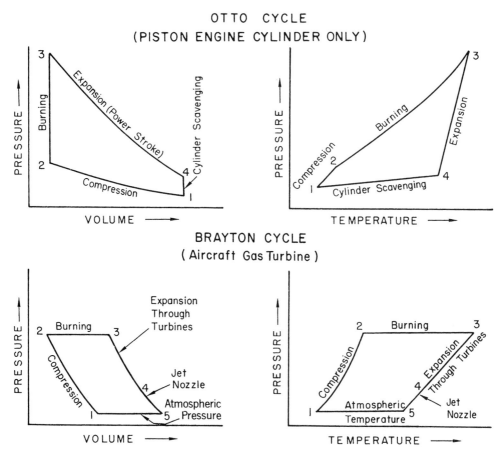

Comparison of the working cycles of a typical aircraft piston engine and gas turbine (Pratt & Whitney).

are not straight. But any change in state can be adiabatic or isothermal. An adiabatic change is one in which the fluid is perfectly insulated, so that no heat can pass to it or be lost from it. An isothermal change is one in which the temperature remains constant. In a real engine neither type of change is possible; the temperature always changes, and heat continuously passes through the engine structure from the hotter parts to the colder. Thus, in pumping up our tyre, we initially make the air (and thus the pump and tyre) hot, but everything gradually returns to what is called ambient temperature.

In a piston-engine cylinder the working fluid is dealt with intermittently in individual 'parcels'. Each parcel is taken into the cylinder, compressed into a small space (which makes it very hot), suddenly made much hotter still by combustion, and then allowed to expand by pushing a piston downwards. Another fundamental is that, in the common spark-ignition

(Otto-cycle) engine, the heat from burning the fuel is added at almost a constant volume, because it happens in a very small fraction of a second during which the piston is near the top of its stroke and moving relatively slowly.

A gas-turbine engine could hardly be more different. Instead of dealing with the working fluid in a rapid succession of parcels, it handles it in a smooth, continuous flow. Air is continuously sucked in at the inlet, compressed, heated by burning fuel and allowed to escape through the turbine and propulsive nozzle. This is called a Brayton cycle, and is totally unlike the Otto cycle of most piston engines. For example, because combustion does not take place in a sealed space, the heat from the burning fuel is added at constant pressure; indeed, in any real engine, because of aerodynamic drag the pressure actually falls slightly as the working fluid flows through the combustion chamber. Thus, the highest pressure in the engine is at the delivery from the compressor, before reaching the combustion chamber. Another difference is that, whereas all four parts of the Otto cycle take place inside the same part of the engine, each part of the Brayton cycle takes place in a different part.

Yet another difference is that, whereas once it is inside the cylinder the working fluid of a piston engine cannot move (apart from swirling turbulence), the working fluid in a gas turbine moves at high speed all the time. Aerodynamics thus plays a central part in the design of the engine. Before Whittle it was impossible to design aircraft to fly faster than about 805 km/h (500 mph), but the turbojet (and rocket) removed the previous barriers and made supersonic flight attainable. As explained in another PSL book (*Faster than Sound*), supersonic aerodynamics are in fundamental ways the opposite of the subsonic ones with which we are familiar.

Even the latter are basically the opposite of what we might expect. If we imagine a crowd trying to pass through a door, we should expect a convergent duct to slow down an airflow and greatly increase the pressure, to a maximum pressure at the narrowest point. In fact, as Bernoulli's Law tells us, the reverse happens. When air passes at subsonic speed (ie, slower than the speed of sound) through a convergent duct its velocity increases and its pressure decreases, reaching the greatest speed and lowest pressure at the narrowest point. This is called a venturi, and you still occasionally see one fixed outside the cockpit of lightplanes to provide suction to drive gyro instruments. Most people find it really hard to believe that the suc-

tion pipe is connected to the throat (the narrowest point).

Conversely, if air flows through a divergent duct at subsonic speed its velocity decreases while its pressure increases. As Bernoulli discovered, this is because the sum of all the energies (kinetic, thermal and potential) in any fluid flow must always be the same at all points in that flow, unless heat is added or taken away. Thus, as air flowing from a narrow pipe out through a large expanding cone must clearly slow down, so must its pressure increase, though at first glance this seems ridiculous. Such an expanding duct is called a diffuser.

Every part of a turbojet, or any other gas turbine, through which the working fluid flows is either a venturi or a diffuser. In some installed engines, such as those inside the wing roots of the RAF's 'V-bombers', it was necessary to add a long duct upstream of the compressor or downstream of the turbine, and such ducts can be simple pipes which merely cause friction and turbulence. But inside the engine the air and hot gas repeatedly finds itself being speeded up or slowed down and changed in pressure and direction by a succession of venturis and diffusers.

Supersonic aircraft are more complicated. For a start, they need a sharp-edged inlet whose shape is not only designed with great care but can also be altered to suit the flight Mach number. At Mach 0.8 (80 per cent of the local speed of sound) an ordinary subsonic inlet would still do, but as Mach number increases things get more difficult. At exactly Mach 1 (the local speed of sound) the Bernoulli laws are reversed.

In a turbojet the air flows at high speed through each part of the engine in turn, suffering tremendous changes in temperature, velocity and pressure (Rolls-Royce).

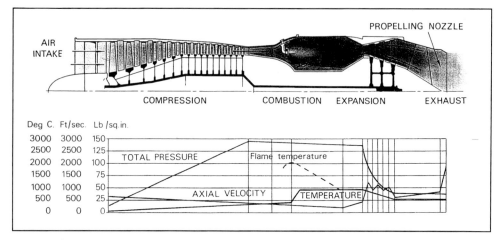

When a fluid flows at supersonic speed through a venturi it slows down and increases in pressure. When it flows through a diffuser it accelerates and falls in pressure.

That is why rocket nozzles end in divergent cones or bell-mouths. The rocket combustion chamber ends in a constriction called the throat, through which the hot gas escapes at the speed of sound. Downstream, the gas pressure decreases while the velocity increases, to increase the thrust. Of course, as in all jet engines, the pressure in the jet cannot fall below that of the surrounding atmosphere, but a rocket designed for use in space can have a very large expansion ratio (the area of the final nozzle can be over 200 times the cross-section at the throat). Thus, supersonic aeroplanes need a propelling nozzle which can be convergent at subsonic speeds and then converted to con/di (convergent to a throat followed by divergent) to produce the required supersonic jet. We will go into that in more detail later.

Every thermodynamic cycle operates between an upper limit of temperature, the heat source (which in a turbojet is the highest mean gas temperature at the exit from the combustion chamber) and a heat sink (invariably the atmosphere). Another pioneer thermodynamicist, Sadi-Carnot, gave his name to the ideal thermodynamic cycle, which is reversible. It comprises two perfect adiabatic changes linked by two perfect isothermal ones. I dwelt on this at some length in the *Piston Aero Engines* book. Suffice to say that in practice it is unattainable, and always will be. One reason for this is that, to achieve it, we would have to have a heat sink at 0°K (absolute zero, or –273°C), which is obviously impossible. We can note in passing that thermal efficiency is higher in Siberia, and better still at 60,000 ft. We cannot deliberately make the atmosphere colder, but we can work at the upper limit and keep trying to increase the maximum temperature that the hot parts of the engine (especially the first-stage turbine rotor blades) can stand. The first-stage turbine rotor blades have always been, and always will be, the critical parts which limit the power and efficiency a gas-turbine engine can achieve.

It goes without saying that engine designers ceaselessly strive for increased cycle efficiency. Higher efficiency means a smaller and lighter engine for a given thrust or power, and lower fuel consumption. Although we must forget about the Carnot cycle, we can still do a great deal. One fundamental is to increase the pressure ratio. Just as cycle efficiency varies with the difference between maximum and minimum

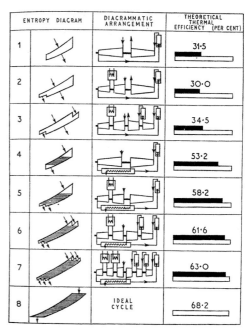

ENTROPY DIAGRAM	DIAGRAMMATIC ARRANGEMENT	THEORETICAL THERMAL EFFICIENCY (PER CENT)
1		31·5
2		30·0
3		34·5
4		53·2
5		58·2
6		61·6
7		63·0
8	IDEAL CYCLE	68·2

Nearly half a century ago the p.r. and TET of gas turbines were so poor that designers studied complex arrangements of heat exchangers. These thermal efficiencies assumed a p.r. of 4 and TET of 648°C (Sulzer).

temperature, so does it vary with the ratio of maximum over minimum (atmospheric) pressure. In 1950 most engines worked at a p.r. of about 4, and designers were striving against great difficulties to double this. Today we have reached 40!

Other obvious ways to increase efficiency are to cut down gas leakage (very hard at a p.r. of 40), reduce mechanical friction, improve aerodynamic design (not only to reduce internal drag but also to increase the operating efficiency of the compressor and turbine), minimize vibration and flutter, and try to reduce heat energy thrown away in lubricating oil and various cooling systems. In addition, we can make the cycle itself more efficient by adding extra items such as heat exchangers.

An accompanying diagram shows the theoretical gains from adding intercoolers between different stages of compression, a heat exchanger to transfer heat from the exhaust gas to the compressed air, and extra combustion chambers between different turbine stages. These diagrams were drawn (by Sulzer Bros of Zurich) in 1950, and reflect the technology of that era in assuming a compressor pressure ratio of 4.0 and a peak cycle temperature of 648°C (1,198°F, 921°K). The 1994 engineer would say; 'With such poor figures, no wonder they cast around for other ways of improving efficiency!' It needs no prior knowledge to see that an engine such as that proposed in (7) would be bulky, heavy, complicated, full of losses (such as gas leakage) which would partly nullify the theoretical gains, exceedingly expensive and, among other things, a pain to the maintenance engineers.

Such answers have achieved limited success in large stationary plant, but failed completely in aviation. Even the relatively simple addition of a heat exchanger to the Bristol Theseus was quickly abandoned. Instead, aero-engine designers have steadfastly increased the pressure ratio and the top cycle temperature. They have succeeded beyond the wildest dreams of the pioneer gas-turbine designers. I have vivid memories of the compressor designers at Rolls-Royce – among the best in the world then, as now – in deep trouble trying to achieve the very challenging pressure ratio of 8.8 in 1951, whilst at Mond Nickel and Henry Wiggin & Son the best turbine-blade metallurgists were striving to use internal blade cooling to work in a gas temperature of (whisper it!) 1,000°C.

What about the other kinds of efficiency? Perhaps the most basic is thermal efficiency. Like all efficiencies, it is some kind of output divided by an input. There is no problem with the input: it is the rate at which energy is supplied to the engine, measured by the fuel flow (gallons per minute, or pounds per second, or whatever) multiplied by the calorific value of the fuel (the actual quantity of heat energy liberated by completely burning unit mass of the fuel). But what is the output? In the case of a shaft-drive engine, such as in turboprops and helicopters, it has to be a power, measured in horsepower or kilowatts. But with any kind of jet engine, such as a turbojet or turbofan, the measure has to be of thrust. This is measured in pounds force or in Newtons (the internationally accepted unit of force, roughly equal to 0.1 kg or 0.22 lb). Thus, a typical engine of a wide-body transport may have a thrust during take-off of 245 kN (245,000 N) or 55,077 lb. We cannot directly compare horsepower and pounds thrust. If it is running on a test-stand (at zero forward speed, we hope!), the world's most powerful jet engine is still developing zero horsepower.

Fortunately it is possible to define thermal efficiency in terms of temperatures alone, in order to get simple numerical answers. If the highest temperature of the gas leaving the com-

bustion chamber is T_1 and the lowest temperature in the operating cycle is T_2, then thermal efficiency is numerically $(T_1-T_2)/T_1$. We need not worry too much about thermal efficiency or other 'internal' measures such as air-standard efficiency, except to note that the engine designer always wants to make T_1 as high as the first-stage turbine blades will stand. Within reason, he also wants to make compressor pressure ratio as high as possible. The optimum (best) p.r. varies with T_1; 40 years ago the optimum was not much more than 4, but today we can operate with the gas entering the first-stage turbine at white heat, hence our modern compressors with p.r. as high as 40. Taken together, these advances make today's wide-body airline engines the most efficient ever designed for any purpose.

Mechanical efficiency is relatively simple. It can never be 100 per cent because every machine containing moving parts must suffer losses due to friction, to windage (air drag) and even to the shearing action of adjacent layers of air, fuel and lubricating oil.

Propulsive efficiency is also straightforward; it is the percentage of the power produced by the engine that is actually put to use in moving the aircraft. For a turboprop, ignoring residual jet thrust, propulsive efficiency is the same as the efficiency of the propeller, which ought to be about 80 per cent from zero forward speed up to something over 640 km/h (400 mph); beyond that the blade tips become supersonic, causing a rapid fall-off in efficiency. For a jet engine it can most simply be expressed as $2V_a/(V_j+V_a)$, where V_a is the speed of the aircraft and V_j is the speed of the jet, in each case relative to the undisturbed atmosphere. Thus, efficiency for a typical subsonic jet will be about 60 per cent (speeds 850 and 2,000 ft/s), for a wide-body about 70 per cent (850 and 1,500 ft/s), and for Concorde about 90 per cent (2,000 and 2,450 ft/s).

Discussion of propulsive efficiency is central to why modern aircraft are so different. Clearly, we want to match the speed of the propulsive jet to the speed of the aircraft. Other things being equal, the faster the aircraft, the higher the propulsive efficiency. Thus, increasing the speed of the Spitfire from 350 mph in 1938 to 450 mph in 1944 made a substantial difference to efficiency, because the speed of the propeller slipstream did not increase significantly. This is despite the inevitable reduction in blade-tip efficiency as the rotating blades approach the speed of sound.

Thus, if we want to fly faster we have to

throw out the propeller and use a faster-moving jet. Fifty years ago designers had no choice but to use a turbojet, which generates a relatively small jet moving very fast indeed (say, 2,000 ft/s or 1,364 mph). As explained later, this means that it also generates intense noise. If our aeroplane has a maximum speed of 845 kp/h (525 mph) (typical of the first jet fighters), then propulsive efficiency will be only about 55 per cent. Today, aircraft designers have a wonderful choice of powerplants which fill the gap between propellers and rockets. An accompanying diagram shows how efficiency varies up to the speed of sound for each of these families of engine. Of course, there are side issues which complicate things. Whittle invented the turbofan in 1936, but nobody showed much interest. More than 20 years later British designers of both aircraft and engines looked at turbofans but timidly selected a BPR (bypass ratio, the ratio of the bypassed cool airflow from the fan, measured for example in pounds per second, to the hot gas flow through the core) of only about 0.3. This did little either to extend range or reduce noise. One factor was that the earliest of these 'bypass engines' had to fit inside a bomber's wing. Another was that when calculating the drag of an engine installed in an external pod, hung under a wing or on the rear fuselage, the aerodynamicists got it wrong and estimated the drag to be much greater than it actually proved to be. Today we can use BPRs of from 3 up to about 12, with 15 in prospect. This is doing wonders in making possible globe-girdling aircraft that are also quiet.

Later chapters also discuss the complicated

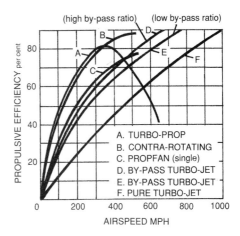

Broadbrush comparative propulsive efficiencies (Rolls-Royce).

inlets and nozzles needed by engines for highly supersonic aircraft, which must be able to change their shape and area in flight. It would be rash to claim, as did the Director of the US Patent Office in 1899, 'Everything that can be invented has been invented.' But we have come a long way in 60 years. Early gas turbines were barely self-supporting; if they did run, their fuel consumption was outrageous. Today's engines offer such power, reliability and fuel economy that thousands of engines originally designed for aircraft are powering ships and other fast marine vehicles, heavy battle tanks, and oil-pumping and electricity-generating stations all over the world.

2 Compressors

Every gas turbine incorporates some kind of compressor. This is encountered by the incoming air as soon as it enters the engine, and the air leaves the compressor at much greater pressure and also significantly higher temperature, but at unchanged, or even slightly reduced, velocity. Why have a compressor?

There are several reasons. Without it, the engine would be, like a ramjet, unable to start from rest. The combustion chamber would be rather like a bonfire, and airflow through the engine would be extremely sluggish. Referring back to the diagram of the working cycle, without compression the heating would have to take place at atmospheric pressure, on the bottom line, so the diagram would appear to be squashed almost flat. The power of the engine is broadly proportional to the area enclosed by the diagram, so the power output would be so small as to be useless. Not least, the cycle efficiency, and thus the specific fuel consumption (s.f.c., the rate at which fuel is burned for a given power or thrust), depends critically on the pressure ratio (p.r., the ratio of the air pressure at the delivery from the compressor to that at its entry). Higher p.r. means better fuel economy. In addition, almost all turbine-engined aircraft require various supplies of high-pressure air, as explained later.

Thus, the compressor is every bit as crucial a component as the turbine that drives it. Since the dawn of the jet age, designers have striven to combine greater compressor efficiency with higher operating p.r. Today the engine designers sit in front of screens on which appear precise and often beautiful pictures created by massive computing power, and nonchalantly design and redesign the compressor until it performs exactly as they wish. Then they build it and, lo and behold, it behaves exactly as the computer said it would. What a contrast with 1935, when Whittle was trying to design the first turbojet! He had no computer, and there was no established body of experience underlying the design of such an engine. He wanted to make a com-

A 'three-dimensional' plot showing the fantastic progress made in compressors: 1, RR Merlin piston-engine supercharger, 1937; 2, Whittle WU turbojet, 1937; 3, RR Avon Mk 532, 1960; 4, typical wide-body engine core, 1994.

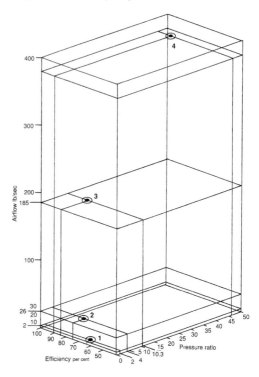

pressor with an airflow, p.r., and efficiency that today's designers would consider laughably inadequate. People also laughed at Whittle, but because they thought the demands he was making were not merely impossible but ridiculous. The advances made in the design of air compressors, driven by aviation, are without parallel in engineering. I have plotted typical figures on an accompanying 'three-dimensional' graph.

Compressors are most unusual in that the designer can choose from several totally different species. The reciprocating type uses a piston oscillating up and down a cylinder, as in a piston engine or bicycle pump. The variable-vane type uses a rotating drum mounted off-centre in a circular housing, round which it drives oscillating flat plates which scrub round the walls. The Roots type is one of several forms containing intermeshing rotors; at first glance they appear to operate 'the other way round', because the two rotors prevent any flow through the centre whilst carrying parcels of air round the outer edges. Lysholm was one of several designers who used intermeshing helical screws. The centrifugal type is a disc carrying radial vanes on either one face or on both sides. Air is drawn in near the centre and flung out between the vanes by centrifugal force, leaving with very high tangential velocity. The air then passes through a surrounding diffuser which converts the high speed into high pressure. The axial type is a drum on which are mounted a large number of radial blades or vanes, each resembling a small wing. Each ring of blades is called a stage, and between the rotor stages are arranged fixed vanes called stators, so that the air alternately passes between rotor blades and stator vanes. Without the stators the air would just be driven round and round by the moving blades, to no effect. Each stage makes the air a little more compressed and also a little hotter.

All these arrangements, and others, have been used to pump or compress air. Until Whittle came along, the air compressors with the highest performance were those used to supercharge the piston engines of racing or high-flying aircraft and racing cars. Some superchargers used the Roots principle, but in aircraft engines almost all were of the centrifugal type. A typical aero-engine supercharger of the 1930s handled an airflow of 2 lb/s and achieved a p.r. of about 2, with an efficiency of 60–65 per cent. Whittle wanted over 25 lb/s, a p.r. of 4+ and an efficiency close to 80 per cent. No wonder people thought he was living in cloud-cuckoo land. They would be dumbstruck by today's compressors, which can handle 400 lb/s with a p.r. of 45

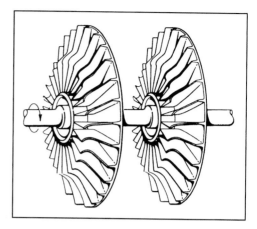

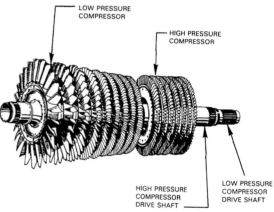

Two contrasting species of compressor: A, two-stage centrifugal; B, two-spool axial.

and with an efficiency of about 90 per cent. Of course, they are of the axial type.

Centrifugal compressors

Until the mid-1950s, axial compressors were bulky, heavy, costly, poor in performance, and prone to various operating difficulties. Wisely, the pioneers Whittle and von Ohain chose the centrifugal type. This was much more a known quantity, cheaper to make, equal in performance, much more robust, and more predictable in behaviour. It comprises the central rotating part called the impeller, the surrounding diffuser, and an enclosure called the manifold or casing.

The impeller is invariably machined from a single metal forging. The radial vanes are made as thin as possible, and the passages between each pair are naturally divergent in end view but convergent in the plane of the shaft axis. The impeller is mounted between bearings and rotat-

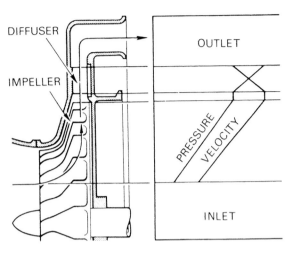

DIFFUSER

OUTLET

IMPELLER

PRESSURE

VELOCITY

INLET

Section through a centrifugal compressor, showing how the air velocity and pressure change from inlet to periphery and then through the diffuser (Rolls-Royce).

The airflow handled naturally depends on the size of the impeller and on its rotational speed. The p.r. achieved also depends on the rotational speed; the higher the speed at which the air is flung off the edge of the impeller, the greater is the kinetic energy which the diffuser can convert into pressure. In fact, p.r. is almost proportional to the square of the impeller's peripheral velocity. Thus – ignoring for the moment various complications – the designer wants to drive the impeller at the highest speed possible. Many domestic vacuum cleaners use centrifugal compressors, but these usually rotate at about 3,000 rpm (50 times a second) and handle a very small airflow at a p.r. of about 1.5. In Whittle's first engine the impeller was 19 in (483 mm) in diameter and designed to rotate at 17,750 rpm (296 times a second), giving a peripheral speed of about 1,500 ft/s, or at sea level Mach 1.36 (1.36 times the speed of sound).

ed at high speed by the turbine connected to the same shaft. The diffuser comprises a surrounding ring of slightly curved vanes, between which are divergent passages which slow down the initially supersonic airflow and increase its pressure. It is usual to design the compressor so that 55–65 per cent of the pressure rise occurs in the impeller itself and the rest in the diffuser. Of course, the diffuser vanes have to offer minimum drag to the high-speed compressed flow, and they must be set at precisely the correct angle.

Even this pioneer turbojet compressor of almost 60 years ago achieved a p.r. of 4, and heated the air some 150°C (just over 300°F, enough to roast a joint in the oven). This was still not too hot for the impeller to be made from an alloy of aluminium. Over the subsequent 25 years the design of centrifugal compressors did not change very much. To improve efficiency it became almost universal to add curved rotating guide vanes at the eye (entry), and in some engines curved fixed vanes were added upstream to impart an initial swirl to the air before it encountered the impeller. To eliminate leakage some piston-engine superchargers used the closed or shrouded type of impeller, in which the air passages between the radial vanes are enclosed by an outer wall on the entry side of the impeller. This has seldom been used in gas turbines, though Junkers even tried making an impeller with square-section radial passages like spokes. Virtually every modern impeller is open on the entry side, leakage being minimised by making the clearance between the impeller and the casing as small as possible, such as a thousandth of an inch.

Airflow through the diffuser of a centrifugal compressor (Rolls-Royce).

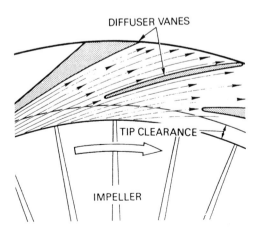

DIFFUSER VANES

TIP CLEARANCE

IMPELLER

At first glance it might seem obviously better to put the radial vanes on both sides of the impeller, because a double-sided impeller handles twice the airflow. The problem is that an engine with such a compressor has to be installed in a plenum chamber. This is an airtight box to which air is fed from the inlet of the aircraft. As the air travels at high speed along the inlet and is then slowed down in the plenum chamber its pressure rises, so the plenum is really a kind of diffuser. The drawback is that the chamber has to be larger in diameter than the engine, and this can increase the drag of the aircraft.

A single-sided impeller experiences considerable aerodynamic forces. The curved guide vanes pulling in one direction while changing the direction of the flow by about 90° usually imparts a force in the opposite direction. The net result can be a significant end-load on the shaft. If the compressor and turbine are rigidly connected, this force can either add to or help to balance the much greater axial load imparted to the shaft by the turbine. Usually, the turbine and compressor are on separate shafts joined by axially floating splines to transmit the drive, so the impeller thrust load has to be reacted by a large ball bearing. The double-sided impeller, being in effect two ordinary impellers back-to-back, cancels out the aerodynamic forces on each side of the central web, which can be an advantage.

Many engines have used two centrifugal compressors in series. Some of the earliest proposals for turbojets suggested as many as four impellers, the air being ducted inwards behind each impeller to reach the eye of the next one. This would incur considerable losses from scrubbing, turbulence and leakage, but using just two impellers has proved perfectly satisfactory in turboprops. There is no technical objection to using tandem centrifugal compressors in a turbojet, but no such engine has yet gone into production.

The reason for using two impellers in series is to achieve a higher overall p.r. In the Rolls-Royce Dart the p.r. achieved at maximum rpm was initially 5.5, rising during 20 years of development to 6.35. This was adequate for its day, but designers strove to run impellers faster and faster, to get greater p.r. and a bigger airflow. Smaller impellers can run at much higher rpm than large ones, because the limiting factors are the peripheral speed and the centrifugal stress in the impeller. In the 1960s the design of impellers quite suddenly made considerable progress, almost entirely because of a switch to titanium alloys. These stronger materials enabled designers to make the radial vanes curved, instead of straight, improving the airflow and increasing efficiency; aluminium impellers this shape would break. Even more important, titanium impellers can run much faster, and today superb centrifugal impellers machined from a single forging in titanium alloy are running at tip speeds around 1,800 ft/s, generating a p.r. exceeding 9 with a single stage.

Such compressors have been chosen for every modern turboprop or turboshaft of under 1,000 hp. They are relatively cheap, good at swallowing dirt without suffering damage, make for a compact and light engine, and can yield the lowest total weight of engine plus fuel for the typical mission. Once one gets into a higher power category, associated with longer flights, it pays to use either two compressors in series or, more often, to add one, two or three axial stages upstream. Although they make the engine longer, heavier, and more costly, the increased p.r. not only increases power but also reduces s.f.c., so that the engines-plus-fuel weight for the mission is actually reduced. With small engines it also makes sense to follow an axial compressor by a centrifugal impeller, to bring the air out to the diameter needed by the combustion chamber. With small airflows an axial compressor can have a diameter as small as 5 in (127 mm), and even today it would be difficult to make an efficient combustion chamber so slender. A centrifugal high-pressure (HP) stage brings the air out to a more manageable diameter, such as 15 in.

In the period after the Second World War, Frank Whittle was often criticised for picking the centrifugal compressor, which many observers considered obsolete, while the Germans were praised for preferring the axial. Even at the time this was a foolish judgement. Today, while nobody would use centrifugals in an engine of 80,000 lb thrust for 18 hr flights, nobody would use axial compressors in a turboprop or helicopter engine of 500 hp. The only question is: do you use one centrifugal impeller or two in series?

Axial compressors

The simplest axial compressor is the propeller. Such a compressor gets its name because the air flows in a more or less straight line in an axial direction parallel to the axis of rotation. Such a compressor is assembled from stages, each comprising a ring of moving rotor blades and a ring of stationary stator blades. The designer could in theory achieve any desired p.r. merely by picking the right number of stages, but in practice the design is – or at least was – exceedingly difficult. No less a firm than Rolls-Royce was almost defeated in the immediate post-war period trying to design a good axial compressor.

An axial-compressor rotor comprises a stack of rings or discs each carrying small rotor blades like projecting spokes. These blades are of aerofoil (wing-like) section, but they do not really resemble wings or even propeller blades because they are more or less rectangular, quite thin, and cambered (curved) so that while the back, corresponding to the upper surface of a wing, is convex, the pressure face is concave. Like a propeller, the blade must be twisted from root to tip. If the hub/tip ratio is about 0.9, as in the case of a small blade mounted on a large-diameter hub, the twist is not obvious. If the h/t ratio is about 0.3 (a long blade mounted on a small ring or disc) then it is most pronounced.

As the whole point of the compressor is to squeeze the air into a smaller space, then the annulus area, the cross-section area of the air-flow path, must get smaller and smaller from inlet to delivery. Depending on many factors, the designer can either keep the rotor hub dia-meter constant and taper the casing, or keep the casing diameter constant and make each succes-sive hub ring or disc larger than its predecessor, or take a middle course. Whichever is chosen, the first-stage LP (low-pressure) blades must always be the largest and the final HP (high-pressure) stage the smallest.

In early axial engines 50 years ago there was not much difference, because the pressure and temperature rise per stage was very small. Any attempt to aim higher resulted in complete breakdown of flow, so it was common to find as many as 15 stages being used to generate an overall p.r. of about 4. One can liken the work of a compressor to using a brush to push water up a sloping gutter, the degree of slope being a measure of the p.r. The centrifugal starts at the

bottom, and with one huge stroke pushes the water over the top. With the axial the brush takes many small strokes. At maximum rpm the brush makes rapid strokes and can push water up more efficiently than the centrifugal, but at lower speeds the water has a chance to run back and, especially if the p.r. (slope) is large, soon is flowing the wrong way. When this happens in an axial compressor it is called surg-ing, and the effect can be violent. Each blade is like a tuning fork, and any stalling or break-down of flow can cause vibration and flutter severe enough to snap blades off at their roots. The broken blades then hit their neighbours.

About 40 years ago several major aero-engine design teams were almost in despair over axial problems. Today the problems have been essen-tially consigned to history, despite the fact that we can now aim at a pressure/temperature rise per stage about 50 times greater. Thus, while the first-stage blades may be as tall as the pages of this book, the final blades may look more like fingernails. In consequence, while the LP blades may be aluminium, the material will have to change quite soon to titanium or steel, and the final HP blades are likely to be made of a high-nickel alloy – Nimonic or Inconel – very much like that used in the past for turbine blades!

Of course, if the air leaves the compressor at a temperature similar to that of the gas from the

The compressor of an Avon Mk 532 of 1960. On top, just over halfway along, can be seen a large bleed valve to assist starting, as described later (Rolls-Royce).

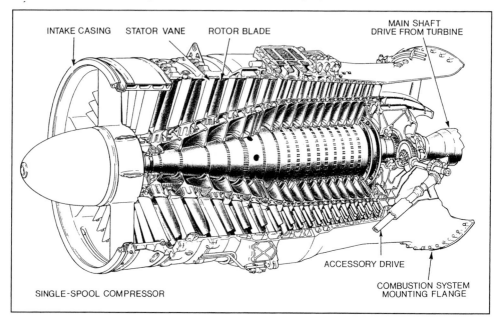

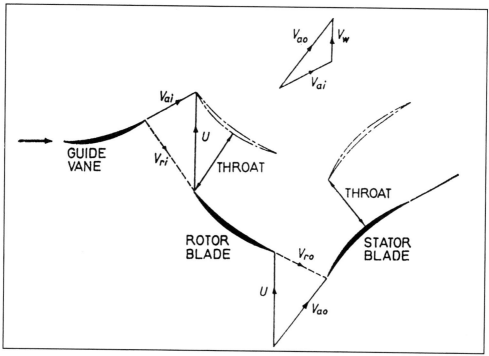

Velocities in an axial compressor: ai, absolute V (relative to engine) at inlet to 1st rotor; ao, absolute V at outlet from rotor stage; ri, relative to rotor blade at inlet; ro, relative to rotor blade at outlet; V_w, whirl V added to V_{ai} by speed of rotor blade U to give V_{ao}.

combustion chamber in turbojets of 50 years ago, one might think that not much fuel dare be burned without melting the turbine blades downstream. As power is proportional to the rate at which fuel is burned (other things being equal), this would lead to a rather useless engine. Fortunately, modern turbines can handle awesomely hot gas!

In early axial engines the inlet was usually a massive casting carrying the front bearing in the centre, with radial struts crossing the airflow and causing a loss of energy if not a turbulent wake. Next came a ring of IGVs (inlet guide vanes) which swirled the airflow in the direction of rotor rotation. Such guide vanes are useful in ensuring that the air meets the first-stage rotor blades at the correct angle of attack, preventing turbulence or even stalling of the rotor blades, which can stall (suffer sudden breakdown of the smooth airflow) just like a wing. The latest turbofans, however, have nothing in front of the first stage of blades, which are overhung ahead of the front bearing. With proper design this gives higher aerodynamic performance.

Whether IGVs are used or not, the ideal is that the air should meet the first rotor blades head-on; in other words that the angle of attack at the leading edge should be close to zero. This can be seen in a diagram in which the relative velocity between the air and the blades is called V_{ri}. This velocity is the resultant of the velocity of the blades U and that of the air relative to any fixed part of the engine V_{ai}. Without IGVs V_{ai} would be almost exactly at 90° to U, because the air would be coming straight in to the engine, so the rotor blades would have to be either set at a different angle (in propeller terminology, at a finer pitch) or else more sharply cambered, in order for the leading edge to slice into the air at precisely the correct angle. But this is a simplification; to describe an axial compressor in this chapter simplifications are unavoidable!

For example, this tidy diagram works at only one value of U, in other words at one particular rpm. Reducing this below the maximum (take-off power) value also reduces the speed at which the air enters the inlet, but there is no direct relationship and before long the angle at which the air meets the first row of rotor blades will be so 'off design' that flow can break down entirely. By far the worst case is when we enter

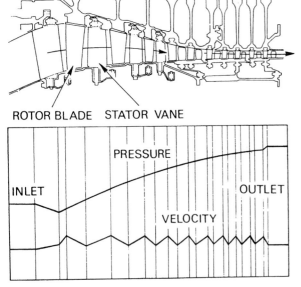

ROTOR BLADE   STATOR VANE

Each stage of rotor blades adds velocity; each stage of stator vanes diffuses this to increased pressure. Thus, V see-saws up and down while P builds continuously. This compressor has four variable stators, described later (Rolls-Royce).

the cockpit and press the starter button. The compressor starts rotating, with every blade far removed from the ideal design condition. The design pressure ratio might be, say, 20, but as the spool (the axial rotor) starts turning, the p.r. actually achieved might be about 3. This demands an exit area only slightly smaller than the inlet, but the actual outlet area (designed for p.r. of 20) is much less than one-tenth as great. So the first-stage blades start pumping air through the large inlet, oblivious of the fact that the air cannot get out at the other end of the compressor!

Thus, at low rpm the angle of attack at which the incoming air meets the first-stage blades is far too great, and the blades stall. Conversely, if we are able to get the engine running, then at full power the angle at which the air meets the final HP rotor blades could be much too low, causing a negative-angle stall. In either event, the result is complete flow breakdown and dangerous blade vibration. To get maximum performance the designer naturally wants every blade to work close to the limit, to pump the greatest airflow with the highest possible p.r. per stage. But if the p.r. demanded ever becomes higher

than any of the stages of blades can sustain, then the flow will – in about one-thousandth of a second – break down completely, resulting in a surge. The highly compressed air at the HP end and in the combustion chamber will be violently blown out at the inlet. There will be a loud bang, the flame in the combustion chamber will go out, and broken blades may be expelled from the inlet and/or from the jet nozzle.

It is the designer's job to ensure that, even in a violently manoeuvred fighter, the engine cannot be made to surge. The aerodynamic designers plot the compressor's actual p.r. against the airflow. We have already seen that flow through a diffuser is basically unstable. Although the progressive narrowing of the flow path makes an axial compressor look like a venturi, the actual flow between every pair of rotor or stator blades is in fact a diffuser, in which the reduction in velocity is converted to pressure. Because the flow through the compressor is basically diffusing it is highly unstable. Smooth flow can be sustained only by running almost on the proverbial knife-edge; any disturbance and the compressor can surge.

An accompanying set of what are called compressor characteristics shows how the surge line divides the undemanding stable area from the unstable region in which the engine not only ceases operating but may suffer damage. Designers naturally avoid the surge line, but are forced to approach it as close as they dare in order to get more power. Even quite recently, some design teams have found out the hard way that they have cut it too fine, a classic case being the TF30 in the F-111 and F-14A.

In the early post-war era some prototype axial engines refused even to start. The designers therefore were forced to remove some of the HP stages and down-grade the p.r. per stage until they got some kind of result, finding themselves

Simplified (theoretical) compressor characteristics (also called a compressor map) to show typical limits of stable airflow (Rolls-Royce).

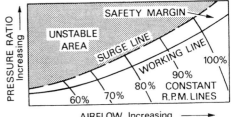

left with useless compressors far inferior to even the early centrifugals. Gradually they learned how to make axials work. Apart from correctly designing the rotor and stator blades of each stage – itself no mean achievement in the days before electronic computers – the two chief palliatives were to make some of the blades with variable incidence, and to cut holes at various places in the casing so that, by opening valves, some of the air coming in at the inlet could be allowed to escape.

We have already encountered VIGVs, variable inlet guide vanes, which ensure that the incoming air meets the first-stage rotor blades at the correct angle. Designers soon realised that they could solve some of their problems by using the same technique with subsequent stages of stator vanes. So far as the author knows, nobody has tried to make a variable-incidence rotor, which would be very difficult mechanically. Today most of the engines used in wide-body airliners or supersonic military aircraft have several stages of VSVs (variable stator vanes). The vanes of each stage are mounted in bearings and driven via short radial arms connected to a surrounding ring. These drive rings are linked together in such a way that, according to the position of an hydraulic ram – invariably driven by fuel pressure – the angles of anything

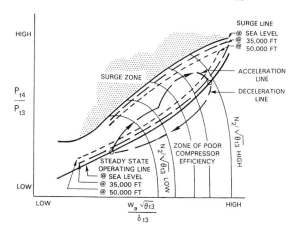

This is the same diagram, but drawn for the HP spool of a real engine; N_2 means HP rpm and W_a means airflow (Pratt & Whitney).

from about 70 to 500 VSVs are always kept precisely correct for peak compressor efficiency. On closing the throttle, as the engine 'spools down' (reduces rpm), the VSVs tend to close; on opening the throttle, as the engine 'spools up', the vanes open.

Whether the compressor has VSVs or not, the casing will today usually incorporate BVs

An engineer adjusting the VSVs of a J79, the first variable-stator engine to go into production. The operating ram can be seen at the top (General Electric).

(bleed valves). Their purpose is to release some of the airflow during starting and at low speeds, to prevent choking of the HP delivery end of the compressor and thus stalling at the inlet. Some are large apertures cut in the casing and sealed by a shutoff valve of either the sliding gate type or an outward-moving piston normally held against its seating by a spring. In some engines, part or all of the periphery right round the casing between two stator stages is perforated by quite large holes which are normally closed by a close-fitting ring slid axially towards the front or rear of the engine. The valve(s) can be opened when the engine is being started, and at any other necessary time, by rams driven by fuel pressure (or even by hydraulic oil, though the readily available fuel is more commonly used), or by the pressure of the compressor's own HP air, or even by electronically controlled solenoids whose magnetic coils are energized by the aircraft's low-voltage DC (direct current) supplies.

If possible, the air bled off is put to use cooling the turbines or balancing out thrusts inside the engine or, as described later, shrinking the compressor or turbine casing. Large bleed flows are in any case often extracted throughout each flight to drive the cabin ECS (environmental-control system), as explained in Chapter 7. Even greater bleed flows will be needed if the aircraft has blown flaps or BLC (boundary-layer control) slots. Jet-lift aircraft need extremely powerful bleed-air nozzles for flight control at low airspeeds.

Work has to be done by the turbines on compressing the air bled off, which then fails to pass through the combustor and turbine, so the result of a large bleed flow is an increase in turbine gas temperature and a reduction in thrust or output shaft power for any given rate of fuel consumption, so as far as possible air bleeds are minimised in cruising flight. During starting, however, the bleed flow can be frighteningly large, the compressed air shrieking out at high temperature from a large exhaust or grilled aperture in the side of a fighter fuselage or airline engine pod. The aperture is usually painted red, with a 'keep away' notice.

Another way of achieving higher p.r. without such severe problems is to use two compressors, an LP and an HP, each driven at its own best speed by a separate turbine. As already noted, many modern helicopter and turboprop engines use an axial compressor followed by a centrifugal, and these are often mounted on separate shafts, to rotate at different speeds. When this is done, the overall engine p.r. is that of the two compressors multiplied together. Supersonic military aircraft and wide-body airliners use tandem axial compressors, and so are called two-spool engines. A few, such as the big Rolls-Royce airline turbofans and the engines of the Tornado and Tu-160 'Blackjack' heavy bomber, are three-spool engines.

There is no technical reason why every axial stage should not be driven at its own best speed

In the 1940s designers began using two shafts, each rotating at their own best speed. By 1960 Rolls-Royce decided to split the work of compression into three in a triple-spool engine.

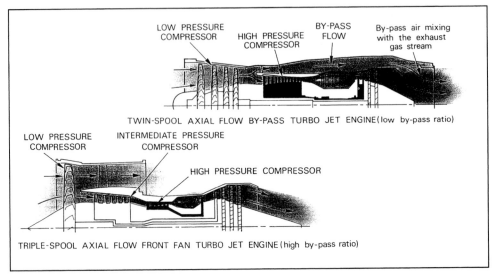

LOW PRESSURE COMPRESSOR   HIGH PRESSURE COMPRESSOR   BY-PASS FLOW   By-pass air mixing with the exhaust gas stream

TWIN-SPOOL AXIAL FLOW BY-PASS TURBO JET ENGINE(low by-pass ratio)

LOW PRESSURE COMPRESSOR   INTERMEDIATE PRESSURE COMPRESSOR

HIGH PRESSURE COMPRESSOR

TRIPLE-SPOOL AXIAL FLOW FRONT FAN TURBO JET ENGINE (high by-pass ratio)

by a separate turbine. Drawings of projected axial gas turbines of 50 and more years ago often showed about the same number of turbine stages as in the compressor, but it was soon realized that the stage loading of turbine blades can be far greater than that of an axial compressor, partly because of the much greater energy available in the gas flow. Accordingly, fewer stages are needed in the turbine, and, provided turbine efficiency is maintained, the fewer stages there are, the smaller the losses incurred. Accordingly, in a split-compressor engine, either two-spool or three-spool, the designer tries to divide up the work of compression in such a way that each spool can be driven by a single-stage turbine. The compressors in a two-shaft engine are called the LP and HP, or in the USA the low spool and the high spool. A three-shaft engine adds an IP (intermediate-pressure) compressor.

In practice, certain design features are common to virtually all gas turbines. Centrifugal compressors tend to rotate faster than axials for the same job. Accordingly they can be driven by a single-stage turbine, which will be smaller than that needed to drive the slower-running axial. If all the work of compression is done by a single axial compressor, the turbine will probably have to have two or more stages. In any split-compressor engine the drive shafts naturally have to be concentric, one running inside the other.

The faster- (or fastest-) running spool is always the HP. The actual speed may be around 3,500 rpm in a large engine, or 60,000 in a midget. The HP can be either an axial or a centrifugal compressor, but, except in some of the latest very-high-pressure turbofans, it almost always has a single-stage turbine, because a single stage can extract the required power from the very hot high-energy gas. The drive shaft has to be a tube, usually of several inches (perhaps 250 mm) diameter, running in bearings which are usually carried in webs (transverse discs) fixed to the casing. Down the centre runs the long LP shaft joining the LP turbine to the LP compressor and/or fan. It turns slower than the HP; if it drives a giant fan without a reduction gear it will turn very much slower. For this reason, and because the gas has already given up a lot of energy to the HP turbine, the LP turbine often needs several stages.

When running at high speed the combined turbine, shaft and compressor and/or fan form a giant gyroscope. Any attempt to rotate its axis results in enormous forces being imparted (at 90°) through the bearings to the aircraft. In the special case of a hovering jet-lift aircraft, especially an agile one such as the Harrier, this poses a control problem, and for this reason the main engine of such aircraft has LP and HP spools rotating in opposite directions. As far as possible this cancels out the gyroscopic couples of the two spools.

While designers are driven by the need to reduce fuel burn for a given power, which among other things means higher p.r. (and hence more stages of compression), they are also strongly motivated by the need for reduced costs. Costs can be broken down into the price of the engine and the ongoing costs of maintenance and parts-replacement, and both demand the lowest possible parts count. The net result is a strong demand for more work per stage, to achieve a higher p.r. with fewer stages. In the past 20 years fuel prices have tended to fall rather than rise, so engine manufacturers have moved away from demanding efficiency at almost any price, and instead have concentrated on making their engines from fewer parts. For example, while the PW4000 wide-body airline engine is based on the JT9D, its HP compressor has 27 per cent fewer blades, while the EJ200 fighter engine has 40 per cent fewer blades than in the previous-generation RB.199.

This in turn demands ever-better compressor blading. Most readers will not be engine designers, so we need not get too technical, but it is necessary to look slightly more deeply at how the blades actually work. To recap: the rotor blades add velocity, and the passages between the stators diffuse this to pressure. The profile of the blades is usually a compromise between two extremes. One is the free-vortex blade, which assumes that every particle of air spirals through the compressor in a perfect helix, meeting every stage at the same relative point; in other words, without any radial velocity inwards or outwards. In such flow the whirl velocity is greatest at the root and the pressure greatest at the tip, total energy being constant along the length of each blade. Such blades are sharply cambered at the root and almost flat at the tip. In contrast, the constant-reaction blade, though identical to the free-vortex blade at the mean height (the midpoint), retains similar camber along its entire length. Thus the plots of blade velocities would be symmetric at every point. Such a blade would generate higher work-per-stage than the free-vortex type, but at the cost of absorbing more power from the turbine and working closer to the stall. Real blades are based on intermediate values, always cognisant of the fact that the centrifugal acceleration imparted by the

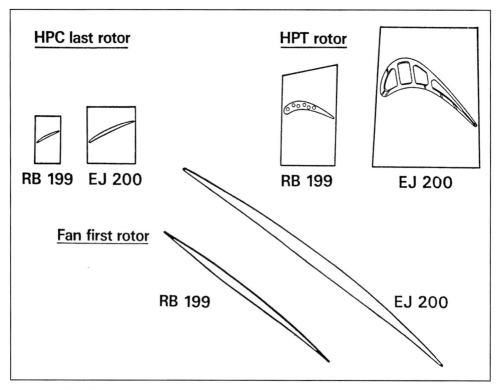

Since the RB.199 was designed, designers have achieved similar performance with fewer, but larger, blades. These diagrams show comparative cross-sections of the blades of the fan first rotor, HP compressor last rotor, and HP turbine of the RB.199 and today's EJ.200, engine of the Eurofighter.

rotor is balanced by increased pressure towards the tip of each blade.

To the uninitiated, a 1970s axial compressor might look very like its 1990s counterpart, but controlled-diffusion blading is actually quite different from traditional circular-arc profiles (Pratt & Whitney).

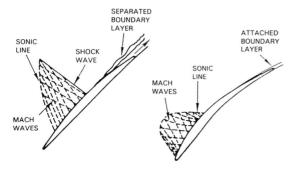

Aerodynamicists will note that it is impossible to eliminate the stagnant boundary layer of air along the inner and outer walls. Air bleeds do something to extract this sluggish air, but in recent years a little extra camber has been introduced to speed up the airflow at each end of the blade. Such blades are called 'end-bend', because of the bent-over corners at the root and tip, which look rather like a dog-eared book page.

During the past 15 years the aerodynamicists have tried harder, and instead of basing their blade design on traditional circular arcs they are making use of the enormous computer power now available in order to approach the ideal profile at each station along the blade. The result is more like a modern wing, with an increased leading-edge radius, a thicker front portion, greater pressure-face camber and a less-tapered aft section. Like the supercritical wing, this gives continuous acceleration across the upper (suction) face to the point of boundary-layer transition, whilst limiting peak Mach num-

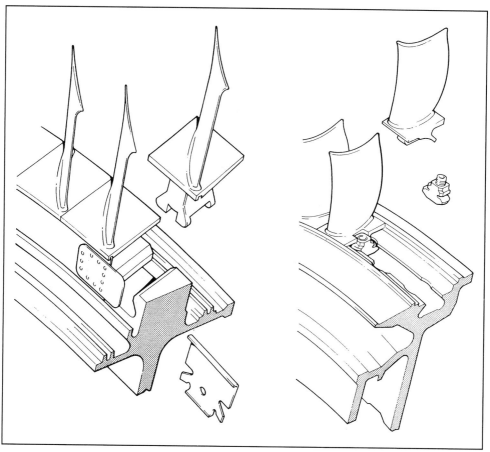

Typical construction of a modern HP spool, showing end-bend rotor blading.

ber to about 1.3. The region of supersonic flow is bounded at the rear by a mild sonic line instead of by a strong shockwave. This minimises transonic drag-rise, and diffusion (deceleration) downstream is carefully controlled to prevent separation of the airflow. Thus, the boundary layer remains attached, and skin friction is reduced. Such blades are called CDA (controlled-diffusion airfoil) in the USA and Codib (controlled-diffusion blade) in Britain. An incidental gain is that, while sharp leading edges quickly wear from erosion, blunter edges hardly alter in many thousands of hours.

Today, a range of engines called turbofans can completely fill the gap between turboprops (for aircraft slower than 450 mph) and turbojets (for supersonic aircraft). A key factor in the design of a turbofan is its BPR (bypass ratio). This is the ratio of the airflow discharged directly behind the fan as a propulsive jet to the airflow passing through the HP compressor, combustion chamber and turbine. The latter components, forming the very hot power-generating part of the engine, are today called the core.

The earliest turbofans had a modest BPR of less than 1; in other words, they were two-spool turbojets with a slightly oversized LP compressor, the excess air from which was sent round the rest of the engine in the bypass duct. For example, the first production Rolls-Royce Conway had a total LP airflow of 280 lb/s, of which about 200 lb/s went through the core and a little over 80 lb/s was discharged as the surrounding bypass jet. This did little except cool the outer casing, the BPR being about 0.3. Today's engines for supersonic fighters of necessity have to be slim, so they too have low BPRs. The F404 has a BPR of 0.27, and the EJ200 just under 0.4. GE describes the F404 as having a three-stage fan, while Eurojet says that the EJ200 has a three-stage LP compressor.

Whatever words are used, such engines are still just two-spool turbojets with an oversized LP compressor. All the LP bypass air is ducted round to mix with the core gas in the afterburner or augmentor, to form a single jet escaping through a single propulsive nozzle.

In contrast, with subsonic transports there is no particular restriction on engine diameter, apart from the fact that it may be difficult to fit large engines under a low-mounted wing. Gradually designers grew bolder, and they also discovered errors in their estimates of installed drag which made higher BPRs appear more attractive. Today a BPR of 9 is seen in the GE90, and since 1980 some design teams have been studying engines with a BPR around 15.

Some of these ultra-high-BPR projects have the fan(s) at the back, driven by independent turbines, but most of today's engines have the fan on the front. This fan is driven by the LP turbine. Most of its air is bypassed around the core, but in some engines the fan air is mixed with the core jet to obtain higher propulsive efficiency. Downstream of the fan, and rotating with it, there may well be from one to six stages of blading at the inlet to the core. These are called either the LP compressor or the core booster. It is not easy to achieve high efficiency with a giant fan and a small-diameter LP compressor rotating on the same shaft, because the ideal fan speed will be much slower than that for the compressor. Rolls-Royce does better by putting them on separate shafts, making three in all, each able to run at its own best speed.

In the latest engines with overall p.r. exceeding 30 it is so essential to minimize leakage that hot compressor bleed air is blown on to the casing to cool it slightly and shrink it so that there is almost no clearance between it and the high-speed tips of the rotor blades. In addition, or as an alternative, similar bleed air can be piped inwards to circulate inside the spool to control the dimensions of the discs or drum. Such thermal matching of the stator and rotor can reduce leakage very close to zero. Great care is needed to design the casing so that, both when cold and at 'red hot' temperature, it remains absolutely circular. Leakage can be further reduced by trenching; the tips of the moving blades run in shallow 'trenches' cut in the wall of the casing. These inner walls almost touch the rotor blade tips, so they are often coated with an abradable material so that if, during transients (when opening or closing the throttle), there should be a slight tip rub, it will not cause a fire or broken blades.

3 Combustion

It may seem simple just to burn a liquid fuel in a stream of air, but no part of the earliest gas turbines gave more trouble than the combustion chamber (in the USA often called the combustor). We have already seen how, when Whittle was designing the first turbojet, he wanted a compressor with far greater performance than any previously built, but in the case of the combustion chamber the intensity of combustion he wanted, measured in terms of fuel burned (heat released) per unit time per unit of chamber volume, was 'at least 20 times greater than had ever before been achieved'.

Yet no other part of an aircraft gas turbine has been so successfully reduced in size and weight. In Whittle's first engine a single enormous chamber was used, wrapped around and dwarfing the relatively tiny rotating machine in the centre. What a contrast with today's engines, where the combustion chamber is a relatively tiny unit in the heart of giant rotating machinery. Compared with the latest domestic kerosene heater or Calor-gas cooker, the combustion intensity is more than 5,000 times greater!

Modern gas turbines are able to generate tens of thousands of horsepower in a small space because of this awesome intensity of combustion. This in turn is made possible because of the enormous airflow delivered by the compressor, most of which is used to cool the flame and form a thin barrier shielding the chamber from the intensely hot, high-pressure gas which fills it. The primary design objective is to deliver to the turbine a flow of gas right at the limiting temperature that the turbine blades can withstand, with no part of the flow significantly cooler, and no part whatsoever even 1°C hotter.

In some respects the gas-turbine combustion problem is fundamentally simpler than that of piston engines. For one thing, instead of occurring in a rapid sequence of discrete events, it proceeds steadily and continuously. Again, the problem of detonation or 'pinking' is removed, so with proper design it is possible to run the engine on a wide range of liquid fuels, or even on gas. A third factor is that, though the pressures and temperatures inside the chamber(s) are the highest anywhere in the engine, the entire assembly can be made from thin and light material. At the same time, as explained in Chapter 8, the pressures and temperatures in the latest engines are so high that not only must the materials be very special, but they must be fashioned in new ways, replacing the welding of plain sheet.

Over the past 55 years the combustion process has been dramatically improved in aerodynamics, chemistry, physics and mechanical design. The aerodynamicists have designed the chamber(s) so that the airflow from the compressor is slowed down in the diffuser section immediately upstream as much as possible. This converts velocity into pressure, bringing down the speed so that the combustion flame is not perpetually being blown out. Air may leave the compressor at 500 to 600 ft/s (340–410 mph). The diffuser reduces this to perhaps 80 ft/s, giving a large increase in pressure. But typical liquid fuels burn at a flame-front speed of only a few feet per second, so the designer has to find a way of stabilizing the flow somewhere in the chamber at a very low speed. It is usual to design each fuel burner so that it is situated in the centre of a strong swirling vortex. The reduced pressure in the centre then tends to suck air back into it, causing a region where the flow is in the upstream direction, providing conditions for stable burning with the flame effectively anchored in space.

Though an increasing research effort is inves-

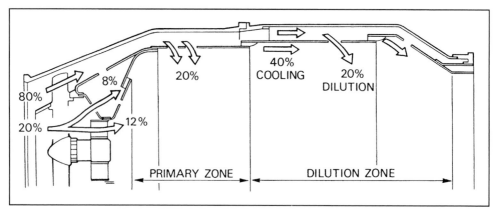

The combustion chamber receives hot high-pressure air from the compressor and divides it up along many different paths. A small proportion, 12 per cent in this case, is mixed with fuel around the burner (lower left); the rest is used for cooling and dilution (Rolls-Royce).

tigating the use of liquid natural gas, liquefied hydrogen and, for military engines, the same petrols (gasolines) and diesel fuels as used by surface vehicles, almost all of today's turbine aircraft use types of kerosenes (paraffin). Some overlap into the lighter distillates (petrols). Over the years the specifications (of which there are dozens) have become increasingly specific. Kerosenes are usually defined by a minimum flash point of 48.9°C (120°F) and a maximum end point of 300°C (572°F). Flash point is the temperature at which sufficient vapour forms to burn at the surface when a flame is applied. End point is the temperature at which all the liquid is distilled into vapour.

Jet fuels burn efficiently at an ideal air/fuel ratio (called the stoichiometric ratio) close to 15; in other words, 15 unit masses of air are needed for each unit mass of fuel. In terms of volume the ratio is nearer to 9,000. This seems a lot of air, but in fact many times more air even than this is needed to cool the flame to the level acceptable to the turbine. In typical gas-turbines the overall air/fuel ratio is from 40 to at least

Airflows in modern chambers are very complex. This velocity vector plot applies to a single plane in the combustor of a PW209; in a plane a few millimetres away the pattern would be different. Arrow lengths are proportional to velocity (P&W Canada).

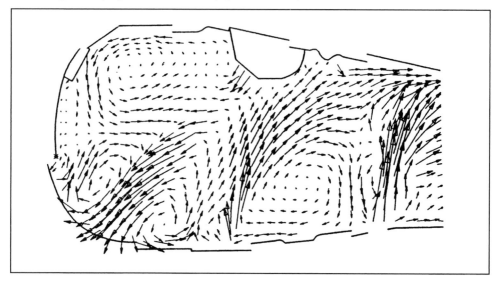

120. Typically, 15–20 per cent of the air is made to mix with the fuel in the burner. This primary airflow is then rapidly mixed with a surrounding secondary flow, probably also about 20 per cent of the total, which determines the aerodynamics that stabilize the flame.

The aerodynamicist also has to design the chamber(s) so that the rest of the air enters through various slots, holes or other gaps. This cooling airflow, which is typically 60 per cent of the total, has two functions. Some is directed into the heart of the hot combustion gas to reduce the mean temperature from a peak of something like 2,000°C to an acceptable TET (turbine entry temperature) of around 1,200–1,400°C. The rest is used to form a barrier between the hot gas and the inner walls of the chamber. Early chambers merely incorporated holes to admit the dilution and cooling air. Later, various forms of film-cooling were devised, including wiggle-strip and transpiration methods. In the biggest of all aero engines, the GE90, the entire surface of the combustor liner is perforated by hundreds of thousands of fine holes drilled by laser at an acute angle to spread a relatively small flow more uniformly over the metal.

Altogether, in a typical turbojet only about 25 per cent of the incoming oxygen is combined with fuel; thus 75 per cent is merely heated. In a turbofan the proportion of free oxygen available is clearly even greater, which makes such engines especially suited to thrust-boosting by duct burning or core afterburning.

The aerodynamicist is also ultimately responsible for ensuring that the losses in the combustion process are minimised, because these losses are mainly aerodynamic. Like almost everything in engineering design, the result has to be a compromise. Pressure drop could be minimised if the operating conditions never varied, or if nobody cared about harmful emissions. Unfortunately, the air/fuel ratio may be 40 on take-off but 130 (on a volume basis, roughly 80,000:1) during a steep high-airspeed letdown with engines at flight idle, and stable combustion must be preserved at all times. Pressure drop across a modern combustion chamber on take-off is unlikely to exceed 18 lb/in^2, and can be as low as 8 lb/in^2.

For any combustor there are upper (weak) and lower (rich) limits of air/fuel ratio, beyond which the flame is extinguished. The greater the air velocity through the chamber, the greater the risk of the flame being blown out, and on take-off these limits narrow to a single value, as shown by an accompanying graph called a sta-

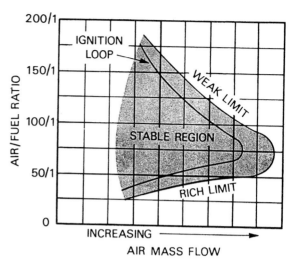

Combustion engineers plot mass flow against air/fuel ratio to see the limits of stable burning (Rolls-Royce).

bility loop. This also shows that during starting, with everything 'cold', the ignition loop is even more restrictive.

The actual combustion process is largely the domain of the chemists. They are not much bothered by combustion efficiency, because very close to 100 per cent of the heat energy released by burning the fuel is used to increase the energy of the gas flow to the turbine. Instead, their job is to get this 100 per cent result in an ever-smaller burning length. They have done pretty well in this regard. Whittle wanted to burn about 1,500 lb/h of fuel in a chamber with a volume of about 6 ft^3; a typical wide-body engine today needs an inner liner smaller than this in which to burn 20,000–24,000 lb/h to give nearly 100 times the thrust.

One might think that a fuel burner needed to be little more than a pipe to squirt the fuel into the airflow. As explained in Chapter 7, the engine fuel system has first to supply exactly the required (metered) flow of fuel to each burner. Usually the fuel is fed to a surrounding ring called a manifold, from which individual branches lead to the burners evenly spaced round the engine. The burners themselves can be of two basic types.

In a vaporizing burner the fuel is sprayed into the open end of a small tube which faces upstream and thus acts as a small ram air inlet. The fuel mixes with the air in the tube and is very rapidly heated because the tube is kept

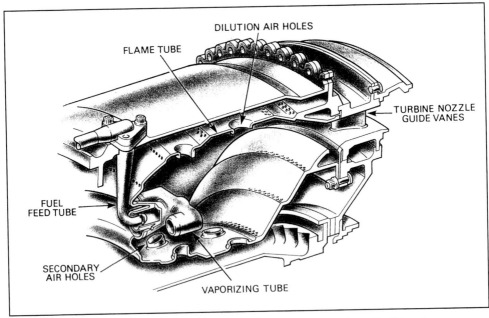

A vaporizing burner in an annular chamber. This example has a twin-outlet T-head.

extremely hot by being situated within the combustion flame. The tube bends round through 180°, either like a walking stick or by dividing into two. The flow of premixed fuel and air emerges from the tube as perfect vapour travelling upstream, where it encounters the flame and ignites and burns quickly and completely, making a second 180° turn and passing back around the vaporizing tube as incandescent gas. The open ends of the tube have to be sized to ensure that the combustible mixture flows in the correct direction.

Vaporizing burners give outstanding performance over a wide range of operating conditions, and in particular almost eliminate carbon formation. The main reason they are in a minority is that most manufacturers and specialist fuel-system suppliers were brought up on the alternative method, which is to use some kind of atomizing burner. In theory there is no problem with such a burner; the fuel is fed by the engine's HP pump at around 1,000 lb/in² to the individual burners at the upstream end of the chamber(s). Such a pressure is sufficient for a properly designed nozzle to squirt the fuel out in the form of a conical spray with such fine droplets that every last molecule is completely burned long before it passes out of the chamber.

Unfortunately, on our high-altitude letdown with the engine(s) at flight-idle the fuel flow will be less than one-tenth of the maximum. In the earliest turbojets the supply pressure was proportional to the square of the required fuel flow; it would therefore be reduced by the fuel control to only about 10 lb/in², a 100-fold reduction. At this low pressure satisfactory atomization could not be sustained, and each burner would splutter out small, medium and large droplets. The larger drops would need much longer to burn, and this could play havoc with a precisely calculated air/fuel ratio, clog the nozzle and liner with carbon deposits (and cause a visibly smoky jet), tend to cause flame-out whenever the pilot moved the throttle and, not least, could even punch holes in the first-stage turbine blades.

Lucas in England pioneered various types of burner to try to overcome this. They began in the pioneer Whittle days with the Simplex burner, which among other things blew air across the face of the atomizing nozzle to prevent carbon deposition, but could not handle very small fuel flows. The Duple and Duplex burners were a big advance, produced in many forms all incorporating a pilot nozzle and a main nozzle. At small fuel flows only the pilot nozzle was in use, and this was specially designed to give good atomization at low supply pressures. At around 100 lb/in² the incoming fuel pressure was sufficient to overcome a spring and open a valve in the burner to start feeding the main manifold or main flow plate. There are many

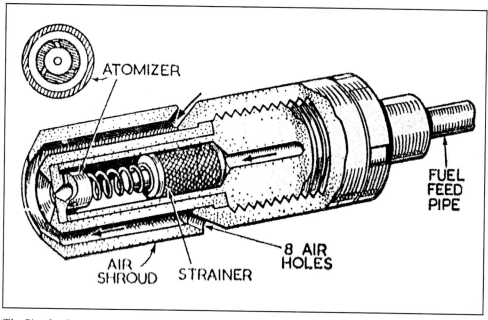

The Simplex burner was used in the first Whittle engines, of 1939–41. The fuel was rapidly swirled (rotated) before leaving via a fixed orifice.

variations on the pilot/main burner theme.

The Dowty alternative was the ingenious Spill Burner. In this, fuel was supplied at full pressure all the time, any excess over that needed for combustion being 'spilt' back from the burner along a return pipe. Fuel entered the atomizing element via one or more tangential ports, so that it found itself spinning round at high speed in a free vortex in a conical swirl chamber. As in the axial compressor, centrifugal effects mean that the pressure falls as the axis of rotation is approached, while the angular velocity must increase. The swirl chamber was so designed that it was possible for the air core of the vortex to be slightly larger in diameter than the discharge orifice, so all the fuel entering the chamber was spilt back to a circulating pump. Any restriction in the spill flow increased the pressure in the swirl chamber, reducing the diameter of the air core until a thin film of fuel was delivered from the orifice in the form of a hollow cone breaking into a finely atomized spray. As this was always formed at full pump pressure, atomization was complete; no fine orifices were needed, so dirty fuel could not cause a blockage, the continuous large flow through the burner kept down its temperature even when little fuel was needed by the engine, and eliminated coke formation, and construction was simplified. One of the few possible drawbacks

was that, at low power levels, continuously recirculating the same fuel (using only a little each time) could eventually make it hot and affect its quality.

Most of today's engines use some form of air-spray nozzle. Instead of requiring high fuel pressure, this atomizes the fuel by mixing it with high-velocity air. A proportion of the primary air passes at high speed through inner swirl vanes around the burner to mix with the fuel and give a rapidly rotating core of air/fuel mix. This in turn encounters the rest of the primary air which entered via the outer swirl vanes. It is impossible for any fuel-rich region to form, so carbon formation, and thus visible smoke, are virtually eliminated. A secondary advantage is that a lightweight pump can supply the fuel at the required modest pressure.

Indeed, in some conditions, such as high-altitude flight idle, the pressures in the manifolds surrounding the engine can be so low that the fuel supplied to each nozzle could be affected by the static head difference between the burners at the top and bottom of the engine. This is compensated by adding a distributor weight in the main manifold upstream of each burner. Burners under the engine have to push the spring-loaded weight upwards, while those on top merely have to help push the weight down against the spring.

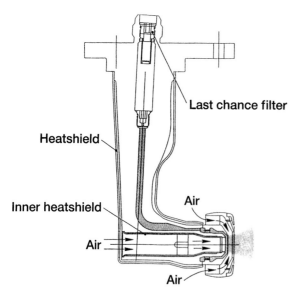

A modern airspray nozzle (from the IAE V2500) using high-velocity air at full compressor delivery pressure to atomise the fuel, which is supplied at quite low pressure (International Aero Engines).

More than 50 years ago Josef Szydlowski designed the first Turbomeca gas turbine with a startlingly simple form of fuel injection. He used a relatively cheap gear-type pump to feed fuel at about 45 (later 57) lb/in^2 to the interior of the hollow shaft joining the compressor and turbine. The fuel escaped along radial pipes terminating in plain holes spaced at intervals round a ring or thrower disc around the shaft. In general, the smaller the gas turbine, the higher its rotational speed, and a typical speed for early Turbomeca engines was 35,000 rpm. At this speed the fuel in the radial pipes was flung out at about 2,000 lb/in^2, sufficient for excellent atomization. No precisely calibrated burner or jet was needed, nor an HP fuel supply. A bypass return line was provided for use at low throttle openings. The technique has been retained in Turbomeca engines to this day.

This leads to the question of mechanical design, and the gas turbine is unusual in that there are many contrasting ways in which the combustion process can be arranged. In Whittle's first engine one giant curved chamber was used. This engine was then rebuilt with the compressor delivering air through ten pipes feeding a single reverse-flow chamber delivering through a reverse-flow turbine and a 180° bend to ten jetpipes. The objective was to

achieve significant heat transfer from the jet-pipes to the inlet air, to improve thermal efficiency.

Once Whittle had satisfied himself that, on starting the engine, multiple smaller chambers could all be made to light up without fail, and that all would keep burning, he then connected the ten air pipes to ten separate combustion chambers, each with air fed to the outer casing which carried a burner at the far end. Primary air entered round the burner and flowed through the chamber towards the front of the engine, about 80 per cent of the air entering through holes in the flame tube. The hot gas left at the front, curving down through 180° to escape through the turbine and single jetpipe.

Such a reverse-flow combustor has much to commend it, especially in shaft-drive engines. If a centrifugal compressor is used, it can be arranged to deliver through a diffuser ring at the full diameter of the engine, the folded combustor then turning the flow through two 180° bends to enter a small turbine in the centre. Another advantage is that the turbine and compressor can be brought close together, driving through a short rigid shaft.

On the other hand, there are inevitably aerodynamic losses in two sets of 180° bends, so it

In the Turbomeca system the fuel is piped (arrow) to a perforated ring on the main drive shaft.

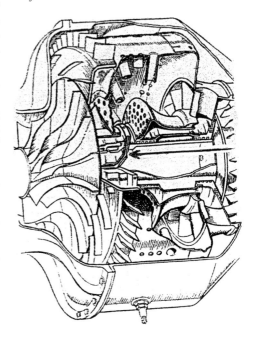

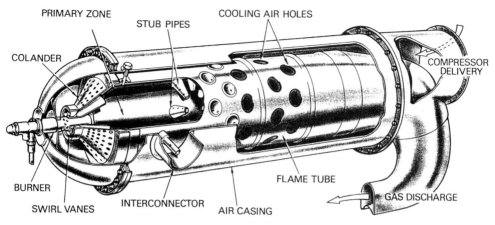

PRIMARY ZONE STUB PIPES COOLING AIR HOLES COLANDER COMPRESSOR DELIVERY BURNER SWIRL VANES INTERCONNECTOR AIR CASING FLAME TUBE GAS DISCHARGE

Whittle's first production engines used multiple tubular reverse-flow chambers. The discharge tubes led straight to the turbine nozzle ring (Rolls-Royce).

soon became more common for centrifugal turbojets to use multiple 'straight through' chambers. These were set pointing obliquely inwards, to link the large-diameter diffuser to the small-diameter turbine. In the Rolls-Royce Clyde and Dart turboprops the chambers were even arranged at a skewed angle to reduce the length of the engine and the angle through which the air had to be turned on leaving the diffuser.

In Germany, attention soon turned to the axial compressor, and this not only delivers through a small-diameter diffuser but also runs at lower rpm than the equivalent centrifugal, so it needs a larger turbine. Thus the combustion chamber had to link two rotating assemblies of similar diameter, and this led naturally to the design of an annular chamber making full use of the available volume. In the Jumo 004 there were really

Gas-turbine design occasionally appears simple, but is invariably the result of painstaking development. A good early example was the annular combustion chamber of the BMW 003A. Inner and outer flame tubes, flow from left to right.

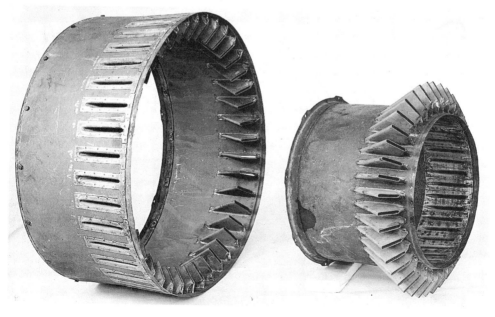

six separate chambers, each with an upstream burner in a simple flame tube, and this increased the engine diameter. The rival BMW 003 had one of the world's first truly annular chambers, made up of inner and outer flame-tube rings each incorporating 20 triangular secondary-air injectors, with 16 simple cup burners, and this kept the whole engine within a diameter of 690 mm (27.1 in).

The Jumo 004 had what is called a can-annular combustor, in Britain called a tubo-annular chamber, in which separate flame-tube cans are fitted inside an annular chamber. Like the use of multiple small chambers, this is easy to develop and improve because only one chamber need be tested, on a rig providing modest airflow. It is also easy to overhaul, because it is a simple task to undo the outer casing and remove the cans, whereas an annular chamber usually cannot be dismantled without removing the engine from the aircraft and then removing the complete turbine section.

When Pratt & Whitney designed the JT3 (J57) in 1948–49 it was finally decided to use a can-annular chamber of novel form. Tightly packed in the annular space were eight cans each made up from a stack of rings welded together. The unique feature was that a large hole in the centre of the flat can front admitted secondary air to a long tapering tube down the

The can-annular chamber of the JT3 (J57), designed in this form in 1949. In the centre are the LP and HP drive shafts (Pratt & Whitney).

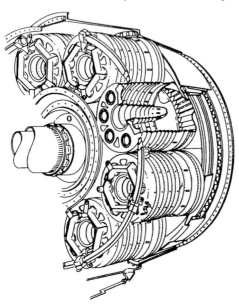

middle of the can. Thus secondary air could enter along both the outer radius and along the centre of each can, supporting increased combustion per unit length of can. Therefore, instead of having just one burner, each can had six, fed by a hexagonal sub-manifold which Pratt & Whitney called a cluster.

This classic combustor had many good features, and suffered a pressure drop at full power of only 9 lb/in^2, from 167 to 158. Yet the company designed the next-generation JT8D, the world's most widely used civil engine, with traditional cans, each with a single burner at the upstream end and with secondary air admitted only around the outer radius. The pressure drop (full power at sea level) increased to 13 lb/in^2, from 233 to 220.

Today's wide-body engines have to get so much power from a small-diameter core that there is no alternative to the fully annular combustor. Apart from probably being difficult to dismantle, as already noted, the annular chamber is in general superior to all alternatives. It obviously makes the best use of the space available between the compressor and turbine. For the same heat release, or power output, its length is only three-quarters that of a can-annular chamber of the same diameter, making a big difference to engine weight, rigidity and cost. An ideal arrangement of burners can be provided, with no need to worry that during the start the flame might fail to spread from the igniter(s) to all parts of the combustor. The wall area is considerably less than in any alternative chamber, and this reduces the required cooling airflow by at least 15 per cent. In turn this ensures virtually 100 per cent combustion, not only eliminating unburnt fuel but also oxidizing any CO (carbon monoxide) to non-toxic CO_2 (dioxide). Most design teams would also claim for the annular chamber improved exit temperature distribution, lower pressure loss, increased durability and reduced total burner weight. In the PW4000, maximum pressure loss is 18 lb/in^2 (from 440 lb/in^2 to 422), but this is a function of the high pressure at entry to the chamber. One might note that the famous steam locomotive *Flying Scotsman* was built with a maximum boiler pressure of 180 lb/in^2.

This will help explain why today's annular chambers are no longer fabricated by 'tin bashing': welding bits of plain sheet. Instead, they are precisely machined from centrifugal castings and extrusions in very special heat-resistant materials. The resulting assembly is thin-walled, remarkably light and dimensionally stable even when filled with white-hot gas at such great pressures. Even as little as 30 years ago limited buckling, cracking and other damage was not only tolerable but taken for granted. Today's chambers are made to resist damage or dimensional deformation.

Modern combustion chambers, especially for civil engines, also have to meet increasingly severe limitations on harmful or even visually offensive emissions. Designers are meeting the challenge by dividing the combustor into two parts, one of which operates all the time while the other is brought into use only on take-off and at other times when high power is needed in denser atmosphere; in other words, at high fuel flows. One of the latest airline engines is the IAE V2500, and in 1994 this entered service in a new version with a combustor specially designed to reduce NO_x. This is shorthand for the various oxides of nitrogen (NO, NO_2 and N_2O_4), all formed from the atmosphere and all sufficiently harmful to be the subject of aero-engine legislation (though out of every 1,000 parts of NO_x in today's atmosphere, 998 are produced by non-aviation activities).

In the natural atmosphere, of course, nitrogen (about 78 per cent) and oxygen (about 21 per cent) stay separate. The trouble starts when they meet in the engine combustion chamber, where the combination of extremely high temperature and high pressure causes the two types of gas atoms to start combining. The obvious answer is to get the atoms out of the chamber as quickly as possible; a shorter residence time means less formation of harmful oxides. Like most other airline turbofans, the V2500 has an HP turbine of greater diameter than the delivery diffuser from the HP compressor, so the combustor has an exterior more like part of a cone than a drum. This has been put to use in adding a second set of fuel burners further back around the larger mid-section of the chamber. Thus, there are 20 spray nozzles round the front of the chamber and 20 more further out round the mid-section.

Those at the front are called the pilot stage. They burn relatively little fuel, and are in use throughout each flight. Those mounted outboard and further back, and of necessity angled inwards, are the main stage and are brought into play during take-off and at other times when high fuel flows are required. When the main stage is working, the short length of combustion means that the gas has left the chamber before much combination of nitrogen and oxygen can take place. Tests have shown a reduction in oxide emissions exceeding 50 per cent.

An even newer engine is the huge GE90, tested at over 100,000 lb thrust. GE in effect gave it

Modern annular chambers are incredibly compact, in relation to the heat-release rate of the burning fuel. This shows a section through the combustor of the CF6 superimposed on a photograph of the liner (flame tube) (General Electric).

two complete combustors, one concentrically inside the other. On take-off the very hot air from the compressor, at a pressure of about 600 lb/in^2, encounters a split diffuser which diverts most of it inwards to pass through an inner annulus where, on take-off, most of the fuel is burned. The large airflow means that the burners in this inner combustor operate at very lean fuel/air ratios, and the high velocity minimises NO_x formation (though the path length is greater than in the V2500 main stage). The rest of the air passes through the outer (pilot) annulus, burning in similar airspray nozzles.

As the engine climbs into thinner air, the electronic control system, sensing all operating conditions, gradually reduces and then shuts down the inner annulus, leaving only the pilot annulus burning. As the inner annulus is not needed in cruising flight, it is designed specifically for minimum NO_x emissions. The outer annulus is tuned to minimise low-power emissions,

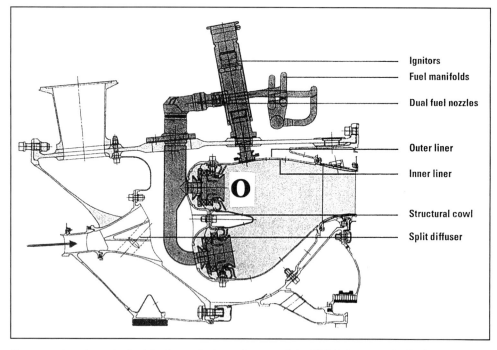

Ignitors

Fuel manifolds

Dual fuel nozzles

Outer liner

Inner liner

Structural cowl

Split diffuser

Cross-section through the 'dual-annular' combustor of the GE90, largest of all of today's aero engines. The highly compressed air enters at left (arrow). In cruising flight fuel is fed to the outer annulus (O) only (General Electric).

notably CO and unburned HC (hydrocarbons), and to operate with good starting, altitude relight and approach-power flameout character-istics. GE claims that the overall result is a reduction in HC and CO better than 50 per cent, and in NO_x greater than 35 per cent.

4 Turbines

In any gas turbine, the turbine has the task of extracting energy from the hot gas leaving the combustion chamber(s) and converting this into shaft power to drive the compressor. In the case of a turbojet, only just enough energy is extracted to accomplish this, because as much as possible must be left in the gas flow to form the propulsive jet. But in a high-BPR turbofan the thrust comes almost entirely from the fan. The core jet is relatively unimportant, and far more energy is extracted to drive the compressor and fan. In a turboprop or turboshaft engine the objective is to extract all the energy from the gas, and convert it into power supplied to the output gearbox driving the propeller or rotors.

The turbine is the key element, and no matter how well it is designed and constructed it is always the component which limits the power that can be achieved. This is because the blades on the turbine rotor each have to transmit a large sideways bending load – the force imparted by the gas which turns the rotor – whilst under extremely severe centrifugal stress caused by the high-speed rotation, continuously surrounded by the gas at a temperature of up to 1,700°C travelling at up to 2,600 ft/s (1,773 mph). The tips of the blades are likely to be travelling at over 1,500 ft/s, because of the rotation, while it might be noted that nickel-chrome stainless steels *melt* at about 1,400°C!

The power extracted from the dense high-temperature gas flow is astonishing. Each blade in the HP turbine of a modern wide-body engine, smaller than a credit card, puts over 500 hp into the drive shaft. But the pressure on designers to increase TET is relentless, because hotter engines are more powerful, for a given size and weight, and have higher thermodynamic efficiency.

About 50 years ago critics of jet propulsion pointed to the rate at which seemingly small turbojets could consume fuel. They overlooked the power generated. Very approximately, the compressor drive shaft must transmit about 100 hp for every 1 lb/s of air delivered. Thus, the Rolls-Royce Nene, a 1944 engine with a maximum airflow of 39.4 lb/s, needed a turbine developing about 4,000 hp, or twice as much as the most powerful piston engines then in use. Today, the engines of the Tu-160 bomber have turbines which on take-off each generate about 80,000 hp. The industrial version of the Rolls-Royce Trent is conservatively rated at 50 megawatts, or about 66,000 hp, but this is the surplus power available at the output, not including that needed to drive the compressor!

The first turbojet to fly, a Heinkel HeS 3b, had an inward-radial turbine, operating very

Every gas-turbine designer strives to achieve the highest turbine gas temperature (TGT) possible. We have already come a long way, because with today's aircooled blades the TGT can be much higher (indicated by crosses) than the limiting metal temperatures shown here. Construction is described in Chapter 8.

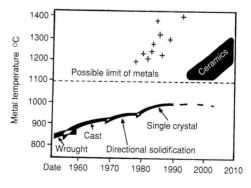

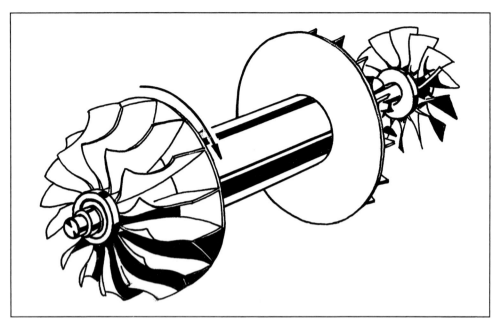

The first turbojet to fly, the HeS 3b, had an inward-radial turbine. A closely similar turbine was used on the next Heinkel engine, the HeS 8A. This turbine is shown here, with the axial/centrifugal compressor at the far end.

roughly like a centrifugal compressor in reverse. Instead of a diffuser there were convergent nozzles which accelerated the gas so that it impinged at high speed on the vanes of the rotor, travelled inwards between these and then added a further reaction in leaving between the curved exit vanes. Such a turbine is extremely limited in power and efficiency, and every practical gas turbine in today's aircraft uses a turbine of the axial-flow type, consisting of one or more stages each comprising a ring of NGVs (nozzle guide vanes) which direct the gas on to the ring of blades mounted on the rotor.

Most early turbojets could drive their compressor with a single-stage turbine. The Adour, used in the Jaguar and Hawk, the Olympus used in Concorde, and the new EJ 200 designed for the Eurofighter all have LP and HP compressors each driven by a single-stage turbine. At the other extreme, the GE90 turbofan has a two-stage HP turbine and six-stage LP, most of the latter's power being needed to drive the enormous fan. In general the designer likes to minimise the number of stages his turbine must have, but in the latest airline engines LP turbines with five or six stages have been accepted in order to extract the required power at the relatively low speed of the fan, having regard to the limit on the diameter of the turbine. A few tur-

bofan engines, such as the Lycomings used in the BAe 146 and RJ family, match a small turbine to a large fan by interposing a reduction gearbox.

Like an axial compressor, an axial turbine can be designed according to two quite different principles, impulse and reaction. In an impulse turbine the NGVs are sharply cambered, so that they start wide apart and finish with their trailing edges close together and pointing in the direction of rotation. Thus they form curving convergent passages for the hot gas, and these impart to the flow the whole increase in velocity and the whole drop in pressure across the turbine. The gas leaves the NGVs at about 2,500 ft/s (at the high temperature this is about the speed of sound) with very rapid rotary whirl. In this free-vortex flow it encounters the turbine rotor blades, and finds that the spaces between them are sharply curving (in the opposite sense) but of uniform width. Thus, as the gas flows through the rotor, it changes in direction but, ignoring slight boundary-layer friction effects, without any change in speed or pressure. The rotor blades are pushed round purely by the reaction to the change in flow direction.

A reaction turbine is the exact opposite. The NGVs form passages of constant width (they may well be circular arcs in profile) and thus do

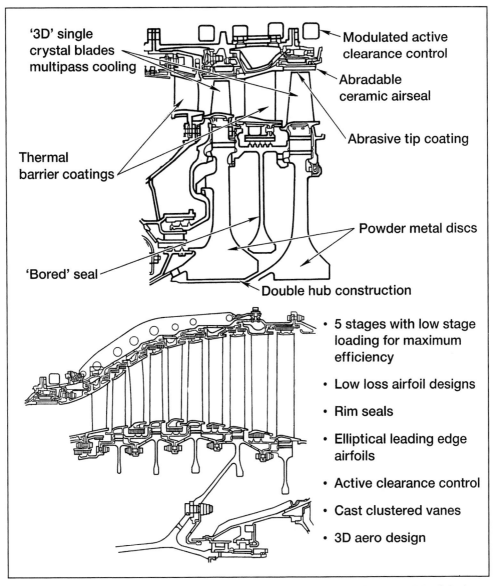

'3D' single crystal blades multipass cooling

Modulated active clearance control

Abradable ceramic airseal

Abrasive tip coating

Thermal barrier coatings

Powder metal discs

'Bored' seal

Double hub construction

- 5 stages with low stage loading for maximum efficiency

- Low loss airfoil designs

- Rim seals

- Elliptical leading edge airfoils

- Active clearance control

- Cast clustered vanes

- 3D aero design

In sharpest contrast, the IAE V2500 has a two-stage HP turbine and five-stage LP of the latest axial design. Note the progressive increase in blade and vane length, reflecting expansion of the gas. The modulated clearance control comprises surrounding square-section pipes blowing air on the casing (International Aero Engines).

nothing to the flow except change its direction, imparting whirl in the direction of rotation. In contrast, the rotor blades have their leading edges wide apart but curve round so that their trailing edges are much closer together, forming convergent passages which accelerate the flow and reduce its pressure. The rotor is turned by the aerodynamic force caused by the accelera-

tion and expansion of the flow between the blades.

Pure reaction turbines are almost unknown, though pure impulse designs are common in such things as engine starters and piston-engine turbochargers. It is safe to say that every turbine used in modern aircraft propulsion has blading of so-called impulse-reaction type, getting the

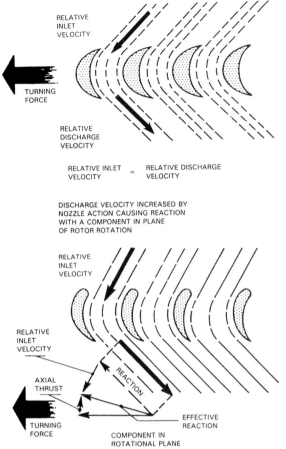

RELATIVE
INLET
VELOCITY

TURNING
FORCE

RELATIVE
DISCHARGE
VELOCITY

RELATIVE INLET = RELATIVE DISCHARGE
VELOCITY = VELOCITY

DISCHARGE VELOCITY INCREASED BY
NOZZLE ACTION CAUSING REACTION
WITH A COMPONENT IN PLANE
OF ROTOR ROTATION

RELATIVE
INLET
VELOCITY

RELATIVE
INLET
VELOCITY

AXIAL
THRUST

REACTION

TURNING
FORCE

EFFECTIVE
REACTION

COMPONENT IN
ROTATIONAL PLANE

*In an impulse turbine (top) the gas is changed
in direction but not in pressure as it passes
through the rotor. In the pure reaction type
(above) it is the NGVs that alter direction only,
the converging passages between the rotor
blades exchanging pressure for velocity (Pratt
& Whitney).*

best results from both techniques. The reaction
level has a profound effect on the shape and performance of the blades. A low degree of reaction, say 15–20 per cent, reduces relative blade
temperature; it also means that at the root the
stagger angle (the angle of incidence) of the
blade can be very small, so the attachment can
be almost directly across the disc, which makes
construction easier. On the other hand, most of
the convex suction face experiences an adverse
pressure gradient, promoting flow separation,
and in any off-design condition performance
deteriorates rapidly. In any case, the different
conditions along the blade mean that the degree

of reaction should change continuously from
root to tip.

At mid-length the degree of reaction of most
blades is close to 50 per cent. At the root the
stagger angle of each rotor blade will still be a
minimum, but the camber will be considerable,
giving a result much closer to the impulse type.
We saw in the axial compressor that in free-vortex flow the whirl velocity is a maximum at the
root, whereas the blade velocity is a maximum
at the tip. Thus the blade must be twisted, with
greater stagger at the tip, so that the gas does
equal work at all stations along the length of the
blade.

As with the axial compressor, the design of
turbine blades has progressed from a 'dark art'
to become an almost exact science. Forty years
ago designers had to use crude empirical formulae, using slide-rules and tables, to make up the
aerofoil profile only at mid-length out of circular arcs. Today's colossal computer power
enables the designer to achieve a near-perfect
visualization of the 3-D flow around every
blade, choosing 10 to 15 stations from root to
tip and even including boundary-layer effects.

A further consideration is that the gas should
leave the turbine with all the whirl removed.
Any residual whirl does nothing for jet thrust or
shaft power but merely increases scrubbing on
the jetpipe, increasing losses and causing vibration which can lead to fatigue failure of the
pipe, or of the struts supporting the exhaust-
cone fairing downstream of the turbine disc. In
some engines the cone struts are cambered, to
remove residual whirl, but with modern turbines
the gas downstream should be almost perfectly
axial anyway. This ideal is attainable at only
one running condition; in a long-range transport
this regime is usually high-altitude cruise, while
for a fighter it might be medium-altitude combat
power. In other conditions the jet is to some
degree a spiral, though the resulting losses
should not be significant.

In turbojets it is also desirable for the gas
downstream of the turbine to have roughly the
same axial velocity everywhere, and this objective can exert an additional influence on blade
design. This factor is not so important in turboprops and high-BPR turbofans where the gas in
the jetpipe has lost most of its energy.

Immediately upstream of the turbine the gas
has higher energy than anywhere else in the
engine, with a combination of high velocity,
high pressure and exceedingly high temperature. Thus, its acceleration through the NGVs
and rotor blades will be very rapid indeed. Even
though the speed of sound in the incandescent

An as-cast high-pressure turbine (HPT) blade for a GE90. Although still unfinished, it already combines several 'leading edge' technologies, including 3D computer aerodynamic design, complex internal cooling, and exotic alloys (General Electric).

gas may be as high as 2,500 ft/s, this flow velocity will certainly be approached, and may be exceeded at the roots of the first-stage rotor blades, subsequently spreading slightly out along the blades. Shockwaves will form, and the flow will be choked. This does not imply any kind of barrier; what it does mean is that the flow between the blades cannot be increased by raising the pressure upstream. If, for example in an emergency combat situation, the fuel flow to the burners were to be increased, the only result would be an increase in pressure in the combustion chamber. This would instantly move the compressor regime closer to the surge line, and if it went over it there would be a violent surge, with compressor stall, flameout and possibly damage. Choked turbines are therefore to be avoided.

While in any single-shaft engine the designer would naturally like the simplicity and low cost of a single-stage turbine, the shaft power needed may demand the addition of a second, or even a third stage. In Rolls-Royce's first axial turbojet,

which became the Avon, a large single-stage turbine was used. Even though results were poor, there was resistance to adding a second stage. But when a completely new two-stage turbine was finally designed it was not only smaller in diameter but even considerably lighter than the original, besides having much higher efficiency. Later, with a compressor of improved performance, it was found worth adding a third stage.

Of course, in any two-shaft engine there have to be two mechanically independent turbines, and in a three-spool engine there must be three. Each consists of a ring of NGVs directing the flow on to the turbine rotor blades downstream. With a multi-stage turbine, or a series of separate turbines, the NGVs of the later stages have to be designed to accept the flow from the last turbine rotor upstream, which of course varies according to the operating condition. Usually all spools rotate in the same direction, but where gyroscopic factors are important they can spin in opposite directions. In the wartime Daimler-Benz ZTL 007 complicated spools rotated in opposition, while in the Metrovick F.3 the two fan stages were driven by two two-stage turbines which intermeshed and rotated in opposite directions, without any NGVs or stators, as explained at the end of this chapter. This attractive idea has now been repeated in the GE UnDucted Fan and the engines of the F-22 fighter.

In the first turboprops a single turbine was used, driving the compressor and, via a reduction gear, the propeller. In the Armstrong Siddeley Python only a single large turbine disc was used, with a forked periphery carrying two rows of blades. This remains almost unique, and most shaft-drive turbines have several stages of blading in order to extract as much energy as possible. Even with a variable-pitch propeller it is almost certainly impossible to match the ideal rotational speeds of the propeller and compressor with a fixed drive gear under all flight conditions. This inevitably led to the so-called free-turbine engine, in which one turbine drives the compressor and a different turbine drives the output shaft connected to the propeller or helicopter gearbox. The only connection between the compressor turbine and the power turbine is the mainstream of gas, so each can run at its own best speed.

On 4 February 1954 the second prototype Bristol Britannia was lost solely because the main input pinion (gearwheel) on the end of the free-turbine shaft of one of its engines stripped its teeth. With a single-shaft engine, stripping

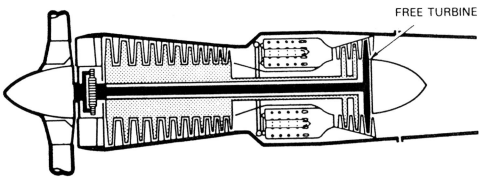

FREE TURBINE

In a free-turbine engine the power turbine, in this case driving a geared propeller, is not connected to the gas generator, except by the flow of gas. Thus, if the main gearbox input pinion were to strip its teeth, the turbine would instantly overspeed dangerously, unless prevented (Pratt & Whitney).

the teeth of the pinion driving the propeller gearbox would merely be a nuisance, because the turbine would still have to drive the compressor. With a free-turbine engine it was disastrous, because, freed instantly from all load, the power turbine overspeeded so violently that in a split second the complete rotor broke free from the shaft. It passed through the oil tank, setting the oil on fire. The result was not only an urgent redesign of the reduction gear, but also the addition of an extremely reliable and sensitive device which would immediately cut off the fuel if free-turbine shaft torque suddenly fell almost to zero. It is remarkable that today many free-turbine engines, most with plain spur gears, have no overspeed protection.

Overspeed of this kind inevitably means disintegration of the turbine, but when one considers the task facing each turbine blade it is remarkable that they can survive a single hour of normal operation. In contrast, modern blades can make over 4,000 Atlantic crossings with no maintenance and little deterioration. The three principal factors that limit the life of a blade are fatigue, creep and corrosion.

Fatigue is the progressive weakening of a metal part subjected to loads which vary in magnitude and direction. The popular example is breaking the lid off a tin can; few of us are strong enough to do this by a direct pull, but if we bend the lid to and fro it will eventually snap off by itself. The severe centrifugal load on the blade acts always in the same direction, but the lateral gas load which turns the rotor varies greatly between take-off and high-altitude flight-idle, and the effect of these loads is enormously magnified by the variation in temperature. Most airline engines go through just one

operating cycle on each flight, consisting of start, taxi, take-off, climb, cruise, letdown, reverse, taxi, shutdown. In contrast, a fighter engine may suffer a dozen cycles between cruise, loiter and combat in the course of a single mission, and such engines accrue fatigue damage far more rapidly.

A related problem is caused by sudden variation in the gas flow, which can occur as a high-frequency pulsation, or thermal shock as is suffered when combustion begins during the starting cycle or if for any reason the flame is extinguished in flight. Localised thermal shock is caused by hot-streak ignition of an afterburner.

Creep is the gradual elongation of a part subjected to a severe sustained tensile load. The high rotational speed of gas turbines imposes an enormous centrifugal load on the turbine rotor blades and on the disc or ring that carries them. Typically, a blade weighing 4 oz (113 grammes) will suffer a longitudinal stress at the root due to its own mass of four tons. Moreover, while the designer is forever trying to make the rotational speed higher, the centrifugal load goes up as the square of the rpm. Thus, to get a higher pressure ratio, Pratt & Whitney designed the core of the PW4000 to turn 25 per cent faster than that of the JT9D, but in so doing they increased the centrifugal stresses by over 56 per cent.

Even this stress would not be such a problem were it not combined with high temperature. Typically, a part under severe tensile load goes through three phases of creep. For a short time it elongates quite rapidly (though we are only talking about a tiny fraction of its own length) in what is called the primary phase. Then, one hopes for thousands of hours, very slow creep is

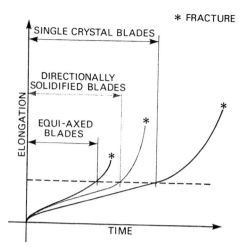

* FRACTURE

SINGLE CRYSTAL BLADES

DIRECTIONALLY
SOLIDIFIED BLADES

ELONGATION

EQUI-AXED
BLADES

TIME

Heavily loaded turbine rotor blades 'creep'; over a period they gradually stretch. These curves show non-quantitatively how new methods of construction (see later) have enabled blades to last much longer, or run much hotter. The broken line shows acceptable elongation (Rolls-Royce).

experienced at a fairly constant and thus predictable rate, proportional to the service operating time. At the end of this secondary phase the material begins to say 'I've had enough' and elongation ceases to be proportional to time but proceeds ever more rapidly until failure occurs. Thus, creep strength of a material must be expressed in terms of both temperature and time. Chapter 8 outlines the enormous subject of how to manufacture blades to make them stronger, and how to cool them so that they are kept at temperatures far below that of the hurricane of surrounding gas, thermally insulated by a sheath of air.

The third problem, corrosion, is also tackled by choosing the right materials and then adding a protective coating. This too is a matter for Chapter 8.

Clearly there must be a small clearance between the tips of the rotor blades and the surrounding casing, and this clearance must allow for future creep. At the transonic relative speeds between the fixed casing and moving blades even a light rub would cause instantaneous rapid heating and probable failure within seconds. It is hardly surprising that Whittle's very first engine was badly damaged by a turbine rub against the NGVs. Nine blades quickly failed, and the broken blades and severe out-of-balance forces caused further secondary damage. Yet

everything possible must be done to minimise gas leakage, as explained shortly.

Typically, about 3.5 per cent of the gas energy is lost in passing between the rotor blades and about 1.5 per cent each due to the NGVs, gas escape round the rotor-blade tips and losses in the exhaust system. Thus, the overall efficiency of a good turbine is around 92 per cent, which is commendable. Sometimes designers fall far short of this; in the prototype RB.211 the HP turbine was found to have an efficiency of about 65 per cent!

Apart from polishing the surface, not much can be done to reduce drag and boundary-layer losses in the passages between either the NGVs or rotor blades, but many ways have been tried to minimise leakages. Almost always the NGVs are provided with platforms at both inner and outer ends which abut with their neighbours to form a continuous ring. The roots of the rotor blades are sealed by being mounted in a continuous ring or disc, though gas might still escape inwards on each side of the rim. It is the tips that present the main problem. Many designers have added tip platforms, called shrouds, which like the NGVs form a continuous ring which almost eliminates leakage. In some cases they have made possible a thinner and more efficient blade by eliminating blade vibration, and they can also act rather like a wingtip fence in increasing aerodynamic efficiency (they increase the 'lift', which is the force causing rotation).

On the other hand, tip shrouds not only add weight, and thus increase the centrifugal stress, but also increase the manufacturing cost. It would be nice to leave them out, and in the past 15 years designers of the most advanced engines have introduced a technique which is dramatically improving turbine performance. In ACC (active clearance control) the rotor blades are made so that there is a very small clearance inside the casing. The latter is surrounded with rings supplied with cooling air. This air may be simply fan air from the bypass duct rammed in through inlets, or it may be air bled off from the compressor and supplied via pipes. The air is sprayed radially inwards through small holes spaced uniformly all round the casing, cooling it and causing it to shrink. The engine control system continuously monitors the clearance between the blades and casing and keeps it at about one-thousandth of an inch. Thus the blades need no tip shrouds, and so can run hotter and at higher speeds but with minimum tip loss.

Although materials and construction are dealt

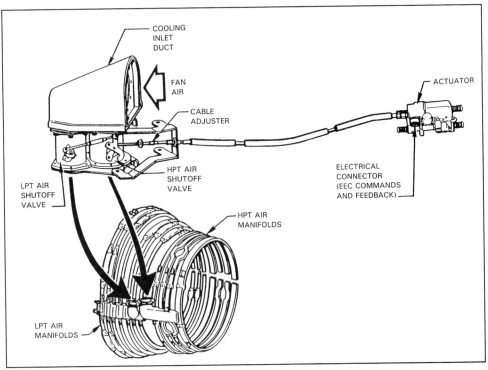

The latest high-power engines have active clearance control, in which cooling air is sprayed round the casings of the compressors and/or turbines. In the PW4000 the turbine casings are shrunk against the rotating blades by air from the fan stream (Pratt & Whitney).

with in Chapter 8, turbine rotor blades are so special that they need further description here. No other engineering components have to meet such severe demands whilst operating with virtually perfect reliability for tens of thousands of hours. It is also self-evident that such blades are likely to be difficult to make, because they have to resist deformation or damage even at temperatures far above the melting point of most metals.

The blades of Whittle's first engine were laboriously machined and hand-finished from individual forgings in a heat-resistant steel, while during the Second World War the several German companies did remarkably well in fabricating hollow blades which could be cooled by centrifugally induced air discharged at the tip. These blades were variously held by pins, by brazing, and by swollen 'bulb' roots inserted into matching slots in the disc. The latter method had been in use for many years on de Laval steam turbines, but the much higher temperatures encountered with gas turbines proved too severe with this type of fixing and there were many disc failures.

By 1940 the 'fir-tree' root had solved the problem, and today this type of fixing is almost universal on all engines other than those in which the blisk (blades plus disc) is a single piece of material. Early examples typically had five or six serrations on each side, though some modern engines have fewer and the HP turbine of the GE90 has just two. The blade is slid into an exactly corresponding fir-tree aperture broached across the edge of the disc, and held by various methods against lateral movement. At rest the blade is free in the disc, with room for thermal expansion, but as the engine speeds up the serrations bear the enormous centrifugal load and the joint becomes rigid. The root and disc are made with great precision to ensure that all the mating faces share the load.

Whittle's first engine had a turbine disc cooled by a flow of water, but he soon found that a properly controlled airflow was adequate. As explained in Chapter 7, various air supplies are fed to the turbine to flow across the front and rear faces of the disc(s) and then if possible to cool the blades. The hollow German blades with their internal airflow were able to use inferior materials, and the relentless quest for turbines able to run in hotter gas led by the early 1950s to experiments with passages for cooling air in otherwise solid blades.

This was done first with the NGVs, because they are situated in the hottest gas anywhere in the engine, and also because, as they do not move, arranging for air to flow through them was relatively simple. The air can flow through the NGVs in either direction, but eventually it must be discharged into the main gas flow. Any cooling air tapped from the compressor will be heated by compression, and it will then be made much hotter by removing heat from the turbine, but it will still be much colder than the main gas flow. It was eventually discovered that this discharge of relatively cool air into the mainstream can cause severe aerodynamic losses. A flow of cool air at 90° (right angles) to the mainstream at the roots of the blades, where flow may be supersonic, causes a momentum loss far greater than had been expected. This loss is now minimised by ensuring that all cooling air is ejected into the mainstream in the downstream direction, and at the highest possible velocity.

In early engines it was common to add a ring of vanes on each side of the turbine disc(s) to pump cooling air outwards, or alternatively to add a small centrifugal impeller on the main shaft purely to provide cooling air. Today the internal airflows are usually quite complicated, with air being bled off at various compressor stages and led through every part of the turbines and their drive shafts. In some cases these secondary flows are used not to cool but to heat particular places. For example, very hot HP air can be fed first to the compressor rotor bore, and then via the drive shaft to the turbine, to even out thermal expansion, improve matching of the rotor and case, and facilitate tip clearance control.

As noted earlier, over 50 years ago Metropolitan-Vickers tested a remarkably advanced turbofan in which the fan comprised two sets of blades carried on the outside of freely-running contrarotating turbines. Sadly, nothing was done to develop this at the time, but in recent years several companies – notably GE with a propfan – have tested intermeshing contrarotating turbines. Such turbines are complicated, expensive and difficult to dismantle, but they can generate enormous power for their size since, except for the initial ring of NGVs, every stage of blading extracts energy from the gas. Each ring of blades serves as the NGV for the next stage immediately behind. Of course, one half of the turbine has conventional blades carried on inner rings or discs, while the other half has blades carried by their outer ends in large rings surrounding the gas flow. An illustration

A simplified indication of the airflows in a modern airline engine is given by this part-section through a GE90. Air downstream of the core booster can be seen piped (shown schematically by single lines) to the active-clearance spray rings of the HP and LP turbines (General Electric).

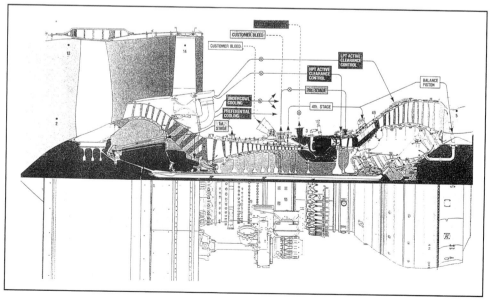

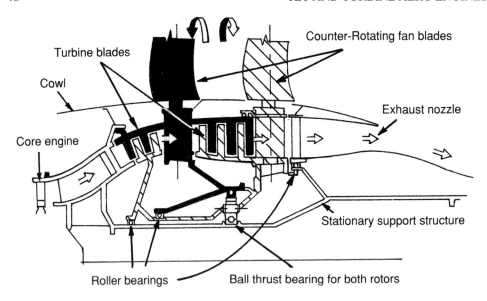

Turbine blades

Cowl

Core engine

Counter-Rotating fan blades

Exhaust nozzle

Stationary support structure

Roller bearings   Ball thrust bearing for both rotors

This cross-section shows how the turbine blades of an integrated contra-rotating propfan intermesh. The diagram depicts the General Electric UDF but the Rolls-Royce project on page 69 is similar

on page 69 shows such a turbine driving prop-fan blades, but it would be equally suited to driving a propeller or helicopter rotor.

The simplest arrangement of all has been achieved in the latest fighter engines, such as the F119. A single ring of NGVs is followed by the single-stage HP rotor turning in one direction and the single-stage LP rotor turning in opposition, with no LP stators. This is obviously going to catch on.

5 Jetpipes

In the simplest form of gas turbine, the turbojet, the hot gas leaving the turbine is simply allowed to escape through a plain tube, called the jetpipe, to the propelling nozzle. Few things in aircraft propulsion could be less complicated. But in 1941, when told how simple the turbojet was, the General Manager of Rolls-Royce, later the Lord Hives, replied; 'Don't worry, we'll soon design the simplicity out of it!' Thus, this chapter has quite a lot to say.

In a turbojet the turbine extracts as little energy from the gas as possible. Downstream, the gas is still travelling at high speed, typically 1,000–1,200 ft/s, and is still hot, probably 700–850°C. The jetpipe is designed so that as much of the gas energy as possible is converted into velocity in the propelling nozzle, because the net thrust imparted to the aircraft depends on the mass flow and on the increase in velocity (more strictly, in the momentum) of the flow as it passes from inlet to nozzle. The thrust depends on the *square* of the velocity.

Immediately downstream of the turbine is the exhaust cone. This fixed fairing starts with a diameter matching that of the turbine disc (in a multistage turbine, with the final LP disc). Usually it tapers either to a point or to a small hole to let out cooling air, but in some modern engines it is not so much a cone as a rounded hemisphere. It reduces aerodynamic drag, but an equally important role is to prevent the hot gas from eddying across the rear face of the disc and overheating it. Another function is that the annular duct for the jet formed between the cone and the outer wall increases in cross-sectional area, forming a diffuser. This slows down the gas, typically to about 900–950 ft/s, which at the prevailing temperature is around Mach 0.5 (50 per cent of the speed of sound). This reduces scrubbing on the pipe wall, with consequent energy loss. The cone is held by thin but long-chord struts, if necessary cambered to remove any residual whirl in the flow.

Losses will be minimised if the jetpipe is made as short as possible. As explained in Chapter 9, many aircraft have to have long pipes, adding weight as well as increasing wall friction loss. The entire structure of the exhaust cone, pipe and nozzle has to be made of heat-resistant material, able over long periods to resist any distortion, buckling or other deformation, and especially to avoid any tendency to crack. The requirement is not simple, because the whole hot end of the engine contains a powerful flow of transonic hot gas whose energy can generate severe high-frequency pulsation, leading to fatigue failure from prolonged vibration. It is also almost always necessary to lag the entire pipe with a heat-insulating blanket (Chapter 8).

In a simple turbojet the pipe terminates in a convergent nozzle. The nozzle area and degree of convergence depend on the upstream gas conditions (temperature, pressure, and velocity, all of which the nozzle itself affects) and the design maximum speed of the aircraft. Normally the nozzle is designed to accelerate the jet as nearly as possible back to atmospheric pressure, and thus reach the highest speed possible. In a simple convergent nozzle the jet velocity in most operating conditions will be close to Mach 1, the speed of sound, which in the hot gas means about 2,000 ft/s (the muzzle velocity of a typical rifle bullet). At exactly Mach 1 a clear limit on jet expansion and acceleration is reached, because a normal (90°) shockwave will inevitably form across the nozzle, causing it to be choked.

We have already seen, with the passages between the blades of a compressor or turbine,

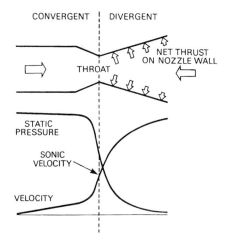

No matter whether the engine is a turbojet, turbofan, ramjet or rocket, the fastest aircraft must be propelled by accelerating a jet through a con/di nozzle (Rolls-Royce).

that the flow through a choked duct cannot be increased by pressure alone. Any attempt to increase pressure upstream of a choked nozzle merely pushes the compressor towards the surge line, though the increased static pressure of the gas passing through the nozzle does add what is called pressure thrust, equal to the extra pressure multiplied by the nozzle area. This is in addition to the basic net thrust due to the change in momentum of the flow through the engine. Pressure thrust is an inefficient answer; it is much better to continue trying to reduce the jet pressure to atmospheric, and thus convert all the excess energy to velocity.

There is one way to increase flow through a choked nozzle; if the upstream temperature is increased, the velocity can be raised to almost any desired value (the limit with a chemically fuelled rocket is about 13,200 ft/s or 8,980 mph). To make use of this possibility it is necessary to add a divergent section to the nozzle downstream of the throat in which to continue the expansion and acceleration of the flow beyond Mach 1. A convergent and then divergent nozzle is called C-D in the USA and con/di in Britain.

Such a nozzle is essential for highly supersonic aircraft, and the increased gas temperature is provided by burning additional fuel downstream of the turbine, as described later. A con/di nozzle is bulky, heavy and, because it must be made of heat-resistant material, is also expensive. Moreover, no matter how its area may be made variable, the total base drag can be

excessive in subsonic flight. Such nozzles are normally seen only in a fully variable form which can be matched to any flight condition, and only on aircraft capable of flying at high supersonic Mach numbers when the pressure ratio across the nozzle is greater than in subsonic flight, typically much higher than 2.

Even in subsonic aircraft a con/di nozzle is preferable if the jetpipe pressure is high enough, because for the greatest thrust we always want to accelerate the flow in the nozzle to the highest value possible. If the compressor delivers air at 400–600 lb/sq in, which is the case in some of the latest airline engines at take-off, there is no way a plain convergent nozzle can expand the gas downstream of the turbine back to atmospheric pressure, so a lot of energy will be wasted. Worse, this lost energy will appear downstream of the nozzle as noise.

Nozzles can thus be seen as much more than plain holes. To make things harder, a simple nozzle is right for only one condition of flight. If the nozzle is the correct area for the mass flow, pressure, and temperature in one flight condition, such as high-altitude cruise, it is going to be too small for maximum thrust on take-off; internal pressure will rise and the compressor will surge. If the nozzle is made larger, then in other flight conditions the pressure, temperature, and velocity will be too low, and much less thrust will be generated. In general, jet velocity increases as the square root of the absolute temperature, whereas gas volume is directly proportional to absolute temperature. Nozzle area varies directly with gas volume and inversely with jet velocity, so an increase in velocity requires an increase in nozzle area in the same ratio.

Thus, it is remarkable that almost all modern jet engines, except those burning extra fuel in the jetpipe as described later, have nozzles of fixed area. But in the pioneer German turbojets of 50 years ago the nozzle area was varied over a wide range by a central 'bullet' sliding in and out, positioned by an actuator controlled by a unit sensing throttle position, burner fuel pressure, and ram pressure. For starting, the bullet was fully retracted, leaving the nozzle with maximum area. As rpm climbed above 30 per cent the bullet began to move to the rear, almost reaching minimum nozzle area at full power for take-off. At full power at high altitude the bullet would be fully aft, giving minimum area for peak performance.

In Allied engines no such variation was provided. All that was done was to decide on an optimum nozzle area, sized either for take-off or for some high-altitude combat condition, and then trim the nozzles of individual engines in service until the operating conditions (such as j.p.t., jetpipe temperature) were at the specified values. This trimming was done by inserting small streamlined obstructions into guide slots inside the nozzle. Called 'mice', these half-round pressings from sheet were not really part of the engine; the jetpipe is usually regarded as part of the airframe. They are seldom seen today, neither are permanent methods for varying nozzle area such as that fitted to the RA.29/3 and 29/6 airline turbojets as described later in this chapter.

Today turbojets are rare, and virtually all jet aircraft are powered by turbofans. The largest transports have engines of high BPR, in which almost all the thrust is provided by the fan. This pumps air along the bypass duct, bounded by the fan case on the outside and the core cowling on the inside. The duct terminates at a point decided by conditions of installed performance, weight, engine accessibility and the design of the reverser. In many of the latest engines the duct is continued beyond the core nozzle. Often, no attempt is made to mix the fan and core streams forcibly, but the limited mixing in a so-called 'integrated nozzle' increases propulsive efficiency and can reduce noise.

The core usually has a plain nozzle, with no reverser or special low-noise shape, and the nozzle is sized for peak efficiency having regard to the fact that the jet does not emerge into the atmosphere but into the fan jet. As noted, if the overall engine p.r. is high enough it may be worth accelerating the jet to a choked throat and then making the remaining section slightly divergent to achieve the highest possible velocity.

Early turbofans had a BPR of 1 or less, yet in the pioneer JT3D Pratt & Whitney chose to discharge the fan air direct to atmosphere from a very short duct. In one of its first installations, the B-52H, the use of two engines in each nacelle meant that the fan air from each engine had to be discharged through a 'banana duct' (the name describing the nozzle shape) on the outboard side of the nacelle, which did nothing for installed efficiency. In contrast, since 1980 Rolls-Royce has fitted lower-BPR engines with special jetpipe mixers as far upstream as possible so that the fan air is forcibly injected into the core gas and thoroughly mixed before the combined jet passes out of the nozzle.

It was noted earlier that variable-area and con/di nozzles are essential if thrust is boosted by burning additional fuel in the jetpipe. This is

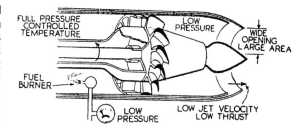

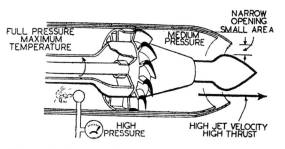

Schematic diagram of a German turbojet of the Second World War, showing the relationship between internal pressures, fuel supply pressure and thrust, according to position of the nozzle bullet.

An early mixer, typical of those on 'bypass jet engines' (Rolls-Royce).

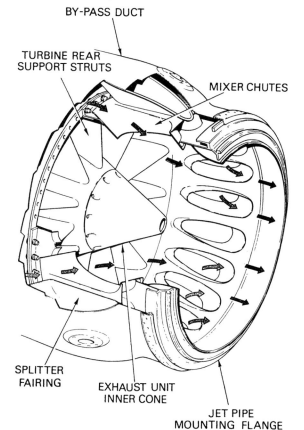

called afterburning, or in Britain reheat, so that the jetpipe is called an afterburner or reheat jetpipe. The abbreviation a/b is often used for both the afterburner and the process. The percentage increase in thrust is called augmentation, and this word is often particularly applied to a reheat turbofan, the jetpipe then being termed an augmenter. Another variation is a duct heater, which provides augmentation by burning fuel in the fresh air in the fan duct but not in the flow from the core. In highly supersonic aircraft, such as the SR-71, at the maximum flight Mach number virtually all the thrust comes from the afterburner, the engine machinery merely getting in the way. Such an engine can be termed a turbo-ramjet.

Whatever words are used, afterburning always results in a tremendous increase in fuel consumption, so that the s.f.c. (specific fuel consumption) may well be more than doubled. For example, the engine of the MiG-29 has an s.f.c. of 0.77 lb/h/lb in the maximum dry (unaugmented) condition at sea level, but lighting the afterburner raises s.f.c. to 2.1! Accordingly, a/b is normally used only for take-off, or for a brief dash at maximum speed by a supersonic aircraft, or to boost thrust at any speed in close combat. Any attempt to use it for prolonged periods, especially at low level, would mean such rapid fuel burn that mission radius would be so short as to be useless. In contrast, changing the engine of the F-14 Tomcat to the F110 has enabled take-offs to be made without augmentation, increasing mission radius by no less than 62 per cent. At the other extreme, the Tu-144 transport was rendered uneconomic by the fact that, with its original engines, supersonic cruise required sustained use of a/b.

Afterburning is possible for two main reasons: there are no turbine blades to limit temperature of the gas flow, and there is still an abundance of oxygen in the gas with which to combine the additional fuel. We saw earlier that, even in a turbojet, about 75 per cent of the oxygen molecules entering the combustion chamber escape from it uncombined with fuel. In a turbofan the proportion of available oxygen is even greater. Indeed, as the flow from the bypass duct is pure air, in some engines a/b fuel is burned in the core and bypass flows separately, the bypass fuel being proportionally higher in order for both flows to reach the same temperature at the nozzle. Alternatively, the bypass air can be mixed with the core gas upstream of the afterburner, in which case the pressures in the two streams must be properly matched before mixing. In either event, the objective is to achieve a uniform gas temperature upstream of the nozzle of 1,500–1,900°C.

The RRTI Adour which powers the Jaguar has a remarkably short afterburner. In this case the fully open nozzle does not have a divergent profile.

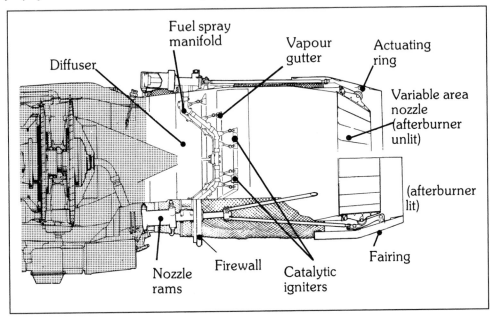

This temperature greatly increases the energy of the jet, so that its velocity and thrust can be increased by anything up to about 90 per cent. It also increases the gas volume for a given pressure, so the reheat jetpipe has to be of greater diameter than the optimum pipe without a/b. Considerable length also has to be provided for complete combustion of the large flow of a/b fuel. Not least, a variable-area nozzle is essential, the area being increased to handle the expanded jet volume. If this were not done, the jetpipe pressure would be greatly increased, feeding back through the turbine to stall the compressor. The objective is to leave the basic engine unaware whether or not a/b is taking place.

The earliest a/b jetpipes were relatively simple on/off units. The pilot could move his power lever (throttle) all the way forward to obtain maximum non-a/b thrust, in the USA called MIL (military) power. If necessary he could then rock the lever outboard and move it a short way further forward. The a/b would light with a muffled explosion, the nozzle would suddenly open fully, and the pilot would experience what an F-100 pilot called 'a kick from a well-fed mule'. Soon designers did better, the British inventing PTR (part-throttle reheat) and US designers 'modulated afterburning'. This gives a smooth increase in thrust all the way from flight-idle to maximum reheat, with nozzle area and profile being modulated along with the fuel flow to suit the requirements of each throttle setting.

An afterburner needs such large fuel flows that it is usual to feed it from a completely separate pump, run up to speed only upon selection of a/b operation. This feeds the same fuel as used in the main engine to burners arranged in concentric rings as far upstream from the nozzle as possible. Provision must be made for stable operation over a very wide range of gas conditions, from the aircraft stationary on the runway to Mach 2+ or subsonic loiter at 60,000 ft, and over a wide range of a/b mixture strengths. The explosive ignition of the early systems inevitably sent pressure waves upstream which could disturb compressor behaviour and result in a surge. Today it has been replaced by 'soft light' ignition at minimum a/b fuel flow, from where further movement of the throttle lever will progressively bring in additional burners, giving a smooth increase in augmentation all the way to the maximum. In combat this must be achievable extremely quickly; banging the throttle wide open should give maximum a/b thrust within about one second (because the basic engine will already be running at or close to maximum rpm).

Surprisingly, even in a 1970s engine, the Pratt & Whitney F100, pilots found that going into a/b could initiate a chain of events that required a lot of sorting out. Cruising at high altitude, you encounter a bad guy and instantly go into a/b. The a/b lights up, but in the thin air some or all of the flames blow out, leaving a cloud of raw fuel. The hot core jet ignites this cloud with a powerful 'whoomph', sending a pressure pulse back through the fan duct. The fan blades stall, which in turn stalls the HP compressor. The large volume of HP air in the diffuser and combustor 'belches' out forwards, clearing the HP stall and allowing the engine to spool-up again. Pressure in the a/b rises, the a/b relights with a second muffled explosion and the cycle repeats, seven times per second (it shakes your eyeballs out).

There is worse. If thrust drops below a critical level the HP spool remains stalled, and unburned fuel burns among the guide vanes and blades of the turbines, causing severe overheat. This 'stall stagnation' continues unless the pilot can get down below flight idle and restart the engine, hoping that the turbine has not been damaged. Pilots do not like shutting down and restarting in combat, so the F100s had the fan-duct splitter (leading edge) extended and were given a modified a/b fuel and nozzle control system.

With a hurricane of gas coming into the jetpipe from the engine, the task of maintaining stable combustion, and preventing the flame from being blown out, or blown out of the nozzle, is at least as severe as in the engine combustion chamber. The high temperature of the core gas is no help; it is still essential to provide regions where the gas is so turbulent that the mean velocity is only a few feet per second, with local areas where the gas reverses direction and flows upstream. This inevitably involves inserting baffles or gutters into the flow, and their presence causes pressure drop and loss of thrust. Nobody has yet succeeded in making them retractable, so the maximum non-a/b thrust is less than that for the same engine with a plain jetpipe.

These baffles are called gutters, vapour gutters, flame stabilizers, flameholders, or various other names. Their usual form is a series of concentric rings each with a cross-section like < or C with the open side facing downstream. As the high-velocity gas flows past these rings it breaks into violent local turbulence immediately downstream. The struts which carry the rings

Removal of the jetpipe from a Tumansky R-11F2S-300 two-spool turbojet reveals three afterburner gutter rings.

may themselves form radial stabilizers, and in any case radial flameholders must link the rings to ensure that the flame spreads from the point(s) of ignition to all rings, so that at maximum a/b power burning fuel fills the a/b with flame to achieve the greatest temperature possible. Thrust augmentation is equal to the square root of the ratio of the absolute temperatures. Thus, if the gas coming from the turbine is at 700°C (973°K) and the mean temperature at the nozzle is 1,700°C (1,973°K), then the augmentation will be $\sqrt{1,973/973} = \sqrt{2.028} = 1.424 = 42.4$ per cent. The augmentation achieved is greatly magnified at high supersonic flight Mach numbers.

Most a/bs have a separate fuel and control system, though a FADEC (Chapter 7) can handle all engine functions. Fuel is usually supplied by a special centrifugal pump, often driven by a high-power bleed-air turbine at a pressure from 50 to 500 lb/in^2, via a metering valve and pressure regulators to manifold rings on which are numerous burners. These can be immediately upstream of the flameholder rings, or at their leading edges or even inside the flameholders,

but in each case they spray fuel into the turbulent zone created by the flameholder gutter. Some designs spray upstream. The burners are spaced so that, when all are spraying the maximum fuel flow, the flames fill the a/b but do not overlap.

The fuel manifolds do not have to be in the same lateral plane, any more than do the flameholders; sometimes the aerodynamics dictate quite large axial separations. When the pilot selects a/b operation, fuel is initially supplied to one manifold only, usually the innermost or the next adjacent. Some form of reliable positive ignition is then needed, operating before any unburnt fuel can build up in the gas flow. Many engines have a catalytic igniter at the centre of the flameholders, fed with its own fuel. This is sprayed on to a platinum-based element which causes an instant chemical reaction, without itself being affected. Another popular choice is a spark igniter very like an enlarged version of those used in the main engine. It has to have powerful capacitor-discharge energy to produce a sufficiently large and intensely hot spark to ensure ignition even at 60,000 ft. In the USA many engines have used 'hot-shot' or 'hot-streak' ignition, in which a brief stream of un-atomized fuel is squirted into one of the burners of the main engine combustion chamber. This sends a hot flame back through the turbine, aimed to hit the turbulence zone of the first a/b flameholder to receive fuel vapour. Its presence should be too brief to damage the turbine.

Today a/b lightup is virtually foolproof, but for over 30 years – say, 1944–74 – it was a common occurrence for pilots to have problems even at sea level, so that twin-engined fighters were seen taking off with one big nozzle full of flame and the other closed down and streaming fuel vapour. Modern military engines have a lightoff detector. Usually this is an electrical circuit triggered by a sensor responding to the UV (ultra-violet) light in the a/b flame.

We have already seen that early afterburners were simple on/off systems. They could be used only when the basic engine had already reached full power, and then they were cut in at maximum fuel flow, the nozzle suddenly opening fully and ignition causing a muffled explosion. With such an arrangement the nozzle could be equally crude, and the most common was the eyelid type. This comprised left and right halves, each pivoted at top and bottom. When closed they formed a nozzle of oval shape, matched to the basic engine. As afterburning was selected the two halves were quickly pulled open by hydraulic rams to give a nozzle of cir-

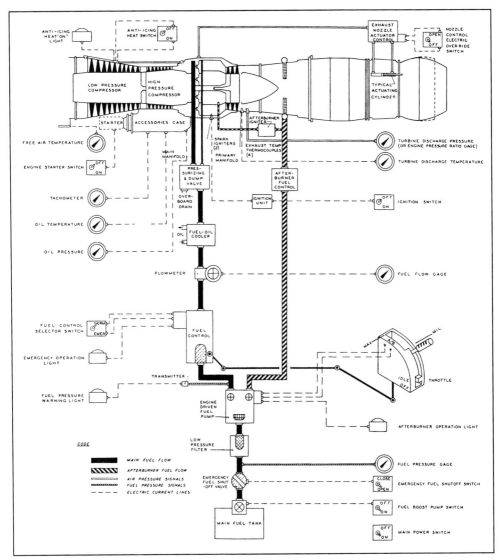

This diagram of the control system for a two-spool turbojet with afterburner does not explain the nozzle control. Usually it senses the difference between the pressure at the entry to the combustion chamber and that immediately downstream of the turbine (Pratt & Whitney).

cular shape matched to afterburning operation. Such engines were used in aircraft of the 1950s, such as the F-86D Sabre, Cougar, Swift, Lansen, MiG-17 and Mirage III.

The next stage, seen in such aircraft as the Javelin, Lightning, Super Sabre and MiG-19, was to fit the nozzle with a ring of flaps which either opened to give maximum area or simply closed to reduce the diameter. Although they had no properly designed divergent portion, because of the greatly increased gas temperature

the jet velocity was highly supersonic, as shown by the shock diamonds visible within it. These glowing regions are formed by the interaction of the shockwaves which start at the nozzle throat, where flow becomes supersonic, and are then repeatedly reflected from the boundary between the jet and the surrounding atmosphere. The shockwaves, almost transverse (straight across the jet) at Mach 1, lean back more and more acutely as jet Mach number increases. The angle is actually $\sin^{-1} 1/M$, where M is the Mach number. Next time you watch a fighter take off, study the jet; as a/b fuel flow builds to the maximum, not only does the flame downstream of the nozzle grow larger and longer, but the brilliant shock diamonds inside it become more and more elongated and move further apart, almost as if they were made of elastic and being pulled out of the nozzle. From a good side-on photograph you can calculate the Mach number in the jet.

Nozzles for highly supersonic aircraft are inevitably more complex. Some are a single unit assembled from long flaps or petals, the upstream portions of which are hinged to the jetpipe and driven either by a surrounding ring or by individual actuators interlinked via a synchronizing ring to ensure uniform movement. The downstream portion may be rigidly attached to form a constant-diameter tube when closed and a divergent nozzle when open. More often, the aft part of the nozzle is made of hinged portions called tail feathers which have no actuating linkage. As they are freely floating, they open out to an angle determined by the pressure of the surrounding ram air on the outside and the exhaust gas on the inside.

Most modern con/di nozzles incorporate auxiliary arrangements to reduce the petal hinge moment and stabilize the flow. Hinge moment is a measure of the force needed to open or close each petal or flap, and if it is large the weight and hydraulic power of the drive system can become excessive. Pratt & Whitney calls some

of its nozzles 'balance-beam' designs because each petal is hinged to an upstream section which forms a counterbalance to the gas pressure acting on the main segment. The huge nozzle of the Russian D-30F6 used in the MiG-31 has flow in the supersonic regime stabilized by perforations in the divergent segments which at lower speeds are covered by auxiliary valve plates.

For obvious reasons, variable supersonic nozzles must be made of special refractory (heat-resistant) materials. They are large, and so they are not only expensive but also heavy, and this weight comes at the greatest possible distance from the aircraft CG (centre of gravity).

For flight speeds around Mach 3 it is common to use an inner nozzle on the end of the jetpipe exhausting into a completely separate outer nozzle which often forms part of the airframe. At low speeds a hurricane of air flows inwards through the gap between the nozzles, entrained by the jet. This fills the aft (ejector) nozzle and adds to the thrust. At maximum speed – attainable only at high altitude – the gap is sealed, and the ejector forms a large divergent final nozzle capable of expanding the jet back to the very low pressure of the surrounding atmosphere. The pressure ratio across the nozzle can be far greater than 2, and the jet velocity some 3,000 mph.

In some engines a large divergent nozzle can be opened at low flight speeds, or when the aircraft is on the ground, to reduce thrust. One rea-

The F110-129, a fighter engine of the 1990s, has a fan duct/core mixer, a blunt turbine tailcone, three flameholder gutters (flame area black), a corrugated augmenter liner, and a con/di nozzle (General Electric).

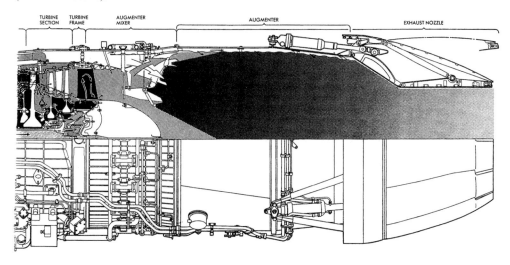

TURBINE SECTION   TURBINE FRAME   AUGMENTER MIXER   AUGMENTER   EXHAUST NOZZLE

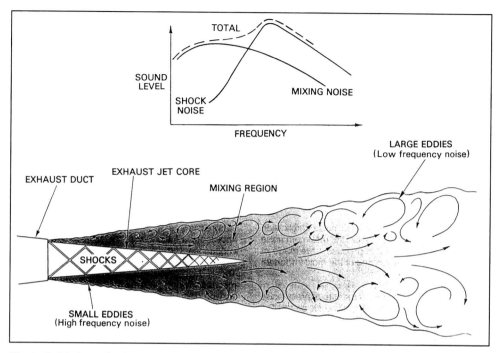

The jet behind an afterburning engine is an intense noise source.

son for doing this is that it allows idle engine speed (rpm) to be set much higher for any given thrust level, so that any subsequent spool-up (overshooting from an approach to a carrier, for example) will be faster. Another reason is that, without this method of reducing idle thrust, some aircraft would taxi too fast without frequent application of the wheel brakes, which would be highly undesirable from the viewpoints of brake life and heat dissipation.

In almost every afterburner the heat-release from the burning fuel is so great that a heatshield inside the jetpipe is necessary. This is a separate inner wall which, though attached at many places to the jetpipe, does not have to carry structural loads. It is made from refractory material, and is separated from the jetpipe by a gap through which passes relatively cool air in which no combustion takes place. Often the heatshield is made from numerous short rings welded together by corrugations through which the cooling air can reach the heatshield's inner wall. In some engines the problems of expansion/contraction and vibration are so severe that the rear part of the shield is made from numerous tiles, each held to the jetpipe by rivets.

A further reason for having a separate liner inside the jetpipe is to reduce or eliminate combustion instability. In early afterburners it was found extremely difficult to achieve stable and steady combustion. The fuel/air ratio tended to fluctuate rapidly, one moment richer and the next leaner, sometimes forming a regular cycle at audible frequencies. This would inevitably result in significant pressure fluctations, and because of the enormous energies involved these would be very violent. Such instability was called rumble or screech. The noise could be sufficiently intense to cause rapid cracking and eventual destruction of the afterburner. A common cure was to add a heatshield perforated with small holes. This absorbs the violent pulsation, preventing it from building up into large-amplitude screech and keeping residual vibration below the danger level.

Of course, any afterburning engine is still going to be extremely noisy. The fundamental equation of the noise produced by a jet issuing into the atmosphere shows that the noise is proportional to the eighth power of the jet velocity. Thus, if switching in an afterburner increases jet velocity from 2,000 to 3,000 ft/s, noise will be multiplied by nearly 26 times. Not much can be done about this, and in any case afterburners are almost confined to combat aircraft where noise is not a primary consideration. Fortunately, the noise from a non-afterburning jet can be reduced by various methods.

Almost all of the noise from a jet is caused by the violent turbulent mixing at the boundary between the jet and the atmosphere. Initially the edge of the jet is clearly defined and there is thus a high-speed shearing action between static and rapidly moving molecules of air and gas. This quickly breaks down into small eddies which cause high-frequency noise of great intensity. The eddies progressively mix the jet with the surrounding atmosphere, blurring the jet boundary. Continued mixing makes the eddies larger and larger, progressively reducing the eddy frequency and thus the pitch of the noise, and also reducing the jet velocity. Eventually, at a distance of perhaps 200 ft, the jet is no longer clearly definable, but is replaced by a much larger mass of eddies characterised by the fact that, in general, the air in line with the jet axis is moving away from the aircraft. By this time the writhing air mass is also distorted by the natural wind, so that it can constitute a hazard to light aircraft beside an airport runway in a crosswind.

Once the gas has left the nozzle, nothing can be done to quieten its noise, but noise can be significantly reduced by the design of the engine exhaust system. Remembering that awesome '8th-power of V_j' effect, everything possible must be done to minimize jet velocity. By far the best answer, for subsonic aircraft, is to replace a turbojet by a high-BPR turbofan. In any turbofan, provided there are no severe penalties in engine thrust and weight, it pays to mix the fan airflow with the core jet inside a full-length jetpipe. This can be done within the available length either with a chute-type mixer or by making the inner wall of the bypass duct (ie, the core jetpipe) progressively change from a circular section to deep corrugations.

Jet noise became an important environmental factor with the introduction of commercial transports powered by turbojets in the mid 1950s. These engines were noisy by any standard, and whereas the pioneer Comet 1 and Tu-104 had plain jetpipes, the Comet 4, 707 and DC-8 were all provided from the outset with nozzles which promoted more rapid mixing of the jet with the atmosphere. The objective was to maximize the length of the nozzle periphery. In the Avon engines of the Comet and Caravelle the nozzle (named for F. B. Greatrex of Rolls-Royce) had an expanding diameter so that, whilst keeping cross-section area constant, it could incorporate six large inward-pointing chutes bringing fresh air to mix with the jet. In the DC-8 and Conway-engined 707 the jet was diverted outwards by internal vanes to escape

through eight radial lobes. The effect with this corrugated-perimeter nozzle was the same as with the Greatrex but even more pronounced. With the Pratt & Whitney-engined 707 the nozzle was terminated in 20 separate circular tubes.

In the Avon 531 of some Caravelle VIs the nozzle was fitted with a hinged flap between two of the lobes to trim area to different flight conditions. Concordes have unique nozzles, the inner primary petals having retractable 'spades' to break up the periphery and the outer (reverser) eyelids in effect squashing the jet on take-off to reduce noise.

All of these nozzles had little effect on the high-frequency noise from immediately downstream (in fact, by lengthening the periphery they mostly increased it), but this noise is rapidly attenuated by the atmosphere, and some that may reach the listener is beyond the audible range of frequencies. Their main achievement was to reduce the more enduring lower-frequency noise. The fact that they were heavy and also degraded engine performance was obvious, and by modern standards the aircraft were still unacceptably noisy. The breakthrough came when BPR (turbofan bypass ratio) was raised from below unity up to 5 or more. This has resulted in engines which obtain their thrust by accelerating a far greater mass of air more gently, and their noise is of a much less-obtrusive character. The fan jet is slower and cooler than that from a turbojet, and the annular nozzle already has a large periphery. The hot core jet is relatively small, and has had most of its energy removed by the multiple turbine stages, and again presents no significant problem. Methods of reducing noise are further discussed in Chapter 9.

The last factor to be considered in this chapter is how to deflect the direction in which the jet is sent out. This may be done to provide a lift force for STOL (short take-off and landing), VTOL (vertical) or STOVL, or to enhance manoeuvrability in combat, or to provide a braking force for a CTOL (conventional) aircraft after landing. It is obviously easier to alter the jet direction with a jet engine than with traditional propellers, though (for example) tilt-rotor aircraft feature on later pages.

The earliest 'jet deflection' schemes involved tilting the whole aircraft by standing it on its tail for a VTO. Next, in the early 1950s, came various crude schemes in which a large valve box was inserted in the jetpipe. In normal operation the gas from the turbojet passed straight through horizontally, but for STOL the pilot could rotate a giant butterfly valve in the box which switched the jet to a short jetpipe pointing

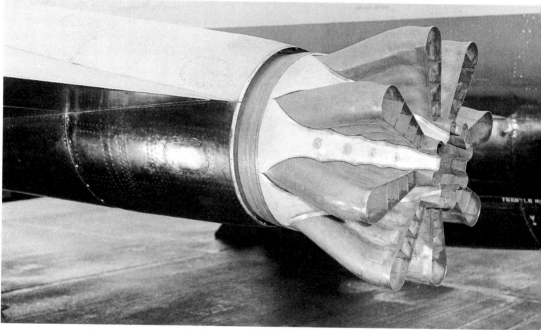

In 1957 Canberras were used to test noise-suppressing nozzles for the 707: top, as used by the JT3C; above, for the Conway (Rolls-Royce).

sharply downwards. A more effective arrangement is to make the jetpipe(s) swivel to an angle selected by the pilot. The engine can have from one to four nozzles, the thrust lines being positioned so that the resultant of their forces passes as nearly as possible through the aircraft CG.

Thrust-vectoring for jet lift can be done in several ways. Switch-in deflectors resemble the valve box just described, but instead of diverting the jet to a downward-pointing pipe they switch it to a circular nozzle on the side, across which are numerous parallel curved deflector vanes (called a cascade). The outer ring of this nozzle can be rotated to make the emergent jet blow down, for lift, or ahead for braking. In the pioneer Pegasus engine the fan air is discharged through left and right side front nozzles and the core jet likewise through left and right rear nozzles. The four nozzles are all able to be rotated by the pilot through an angular range of some 100°, all being mechanically linked to ensure that they move in unison. Other engines have a single jetpipe incorporating rotating wedge sections so that the nozzle can point aft or downwards. Jet-lift schemes are discussed in Chapter 14.

Today, thrust reversers are almost universal in jet transports and increasingly common in fighters. In turbojets they spoil or partly reverse the thrust of the entire jet; in today's high-BPR turbofans they are usually fitted in the fan stream only, the core jet being relatively unimportant. They are invariably made as part of the airframe. Obviously, they add to the engine's installed weight and cost, but their advantages can be considerable. On every landing they can significantly reduce the ground run and/or save on the use of wheelbrakes, which can pay off in prolonging the life of brakes and possibly tyres. On a contaminated runway – one covered with water, snow or ice – they can make the difference between a normal landing and a dangerous overrun. But in fact, the first reversers or thrust-spoilers were developed so that naval pilots could land on carriers with the engine(s) at high rpm, ready for a quick application of power if they missed the arrester wires.

Just as today we wonder why the designers of early retractable landing gears made them so complicated and unattractive, so do we wonder why they took so long to arrive at elegant designs of reverser. One of the many patentees of early schemes was ONERA of France, which suggested that, after landing, the pilot should actuate a toroidal deflector – a deflector ring

One of the first jet reversers to be flight-tested was this arrangement by SNECMA.

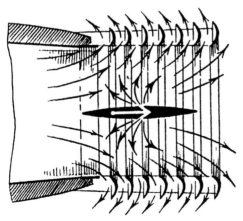

able to move to front or rear – so that from 50 to 80 per cent of the airflow from the compressor should be diverted forwards into a completely separate combustion chamber to produce a powerful jet ejected through a forward-facing annular nozzle. Apart from its complexity, cost, bulk and weight, this scheme guaranteed that a high proportion of the jet would at once be reingested by the engine inlet. Hot-gas reingestion has always been a factor to be watched in jet-deflection schemes, and is the reason why most of today's airline pilots cancel reverse thrust at something over 30 mph.

A better French reverser was tested by the national engine maker SNECMA. The end of the jetpipe was formed by a cascade of channel-section rings. In normal operation the gas passed straight through, hardly any being diverted out through the gaps between the rings. To provide a net braking thrust some means had to be found to divert most of the jet outwards between the rings. In the first prototype the jet-pipe upstream was fitted with a ring of pivoted flat vanes. Normally edge-on, these caused little drag, but if they were rotated some 30° they caused the jet to rotate rapidly so that much of the gas swirled into the vanes under centrifugal force. Later this idea was replaced by fitting a streamlined strut across the reverser from which a large flow of compressor-bleed air could be discharged, positively blowing the gas into the rings. The main drawbacks were that there were always obstructions in the jet, and that nothing was done to stop some of the gas escaping rearwards in the normal way.

These shortcomings are rectified in today's reversers, which positively shut off the normal nozzle. Turbojets and mixed-flow turbofans use either of two types. The clamshell-door reverser incorporates two curved doors which are normally recessed flush on opposite sides of the jet-pipe, where they seal off large apertures in the wall containing transverse cascade vanes. When the pilot selects reverse thrust he closes the throttles and then pulls up an auxiliary reverse selector lever mounted on the front of each throttle lever. On each engine this releases a locking mechanism, supplies high-pressure bleed air to rams which rotate the clamshell doors, and then spools up the engine to high power. The twin doors meet and close off the jetpipe, simultaneously opening the side apertures through which the jet gas escapes. The cascades turn the gas diagonally forwards to give reverse thrust.

The alternative scheme is the target-type reverser. Here, the outer fairing of the jet nozzle

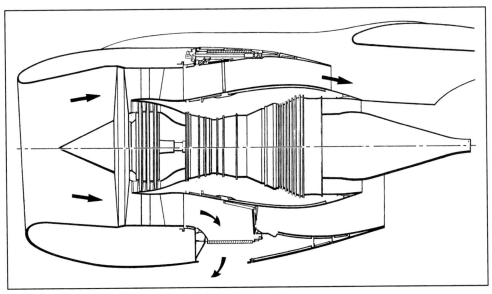

Most modern airline turbofans use the translating-cowl type of reverser; as the aft section moves back it shuts off the fan duct, diverting the jet through cascade vanes.

is formed by two segments, each mounted on a pair of pivoted arms and also connected to a pneumatic ram. When reverse is selected, the system is unlocked and air pressure is applied to the rams, pulling each segment round so that they meet in the centre, forming a pair of large 'buckets' downstream of the nozzle, blocking the path of the jet. Sometimes each bucket incorporates an inner curved vane which deflects the gas through a greater angle than the bucket itself.

High-BPR turbofans use either of two types of cold-stream (fan duct) reverser. In one, the outer wall of the fan duct incorporates a large number of panels hinged at their rear edges and connected by pivoted links to a translating (ie, moving to front or rear) sleeve which can form the complete rear section of the fan duct. To reverse thrust, bleed air is supplied to a pneumatic motor which, via flexible shafts and gearboxes, drives screwjacks which translate the sleeve to the rear. This not only pulls all the hinged blocker doors inwards to seal off the duct, but also opens a ring of cascade vanes all round the duct. In some designs the blocker doors comprise front and rear portions hinged together, so that, as the outer cowl moves aft, the doors are folded inwards to block the duct.

In the alternative type of fan reverser the outer wall of the fan duct, complete with the adjacent section of pod cowl, is made of pivoted sections rather like those of the bucket-type jet reverser. Each section covers from 75° to 90° of the periphery, so that four or five pivoted doors suffice to block off the duct. Each door is pivoted just behind its mid-point to minimise the load on the drive actuator. In the reverse position each door acts as both the blocker and the deflector.

It will be appreciated that it is never possible to achieve perfect deflection, by turning the fan air or the jet through 180°. In practice, deflection angles are usually 130°–150°, so for any given throttle opening the reversed thrust is only about 70 per cent as great as in the forward direction. Exceedingly reliable checks and safeguards are necessary to ensure that inadvertent operation is impossible (as a Boeing 767 of Lauda Air showed, such operation, even at high altitude, can be non-survivable). Any failure of the drive system should automatically put the reverser in forward thrust. Reverser operation is obvious to the pilot, but he also has cockpit indications confirming reverser selection, unlocking, and operation.

High-BPR engines are usually hung in a pod under the wing, where the reversed fan air can be discharged all round, except at the very top where the pod is hung on the pylon strut. Engines hung on or in the rear fuselage cannot have such all-round reverse. On a side-mounted engine the cascade sections, or reverser buckets, have to deflect the jets above and below, whereas a jetpipe close under the wing trailing edge,

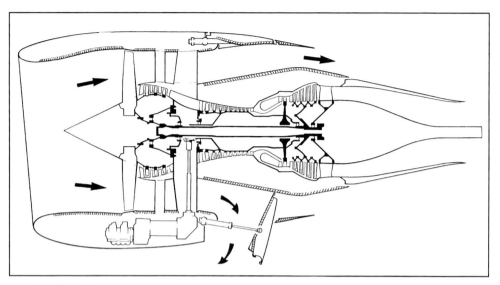

In A319/320/321 aircraft the CFM56 has a four-door reverser. The doors are actually set round the pod at 45°, but are depicted here above (in cruise) and below (in reverse).

as in the early models of Boeing 737, has to incorporate a reverser discharging to left and right. The single-engined Viggen reverses its jet through three apertures on the sides and underneath, while the Tornado has upper and lower buckets slightly canted to avoid excessive scrubbing on the fin.

In general, reversers and their drive systems have a harsh life. Those in fan ducts are generally part of the airframe. The cascade and blocker-door ring is frequently hung direct on the pylon, and remains in place when the engine is removed. It can carry the translating sleeve or cowl and also help share loads around the cowl. In most engines it also provides maintenance access to the core. Reversers in or around the jetpipe are seldom required to provide access inside the pipe, but they have to be cycled routinely during ground testing. Such reversers,

complete with their immediate operating system, have to be capable of reliable operation, without lubrication, at an ambient temperature of about 600°C (1,112°F). Among the other requirements are rapid and positive action, and minimum cost, bulk, and weight, despite the enormous loads imposed by the hot gas, and an absence of any significant effect on normal engine operation.

Today it is at last being appreciated that, especially on fighters where the installed engine thrust exceeds the total weight of the aircraft, it is a good idea to be able to vector that thrust in different directions. Even if the objective is not to lift the aircraft on jet thrust alone (Chapter 14), a vectoring nozzle can exert powerful control of the flight attitude or trajectory, reduce cruise and trim drag, and bring the aircraft quickly to rest after a slow landing (Chapter 9).

6 Propellers and fans

Until 1929 many workers had tried to produce jet engines, while others had devoted their energies to gas turbines driving propellers (in those days often called 'airscrews'). Whittle's breakthrough was using a gas turbine solely to produce a propulsive jet, which at a stroke released aircraft from the speed limitation imposed by the piston engine driving a propeller. Meanwhile, the 'airscrew-turbine engine', soon given the neater title of turboprop, became popular for aircraft flying at speeds up to about 725 km/h (450 mph). In the Soviet Union very powerful turboprops were even developed for aircraft capable of 900 km/h (Chapter 17).

Various designers, again led by Whittle, considered ways of bridging the seemingly large gap between the turbojet and the turboprop by inventing different forms of ducted-fan and bypass jet engines. These took a surprisingly long time to be accepted, which is odd because today, again given a neater title – turbofan – they have virtually replaced the turbojet. Gradually designers became bolder in their choice of BPR (bypass ratio), until in today's large high-BPR engines a single-stage fan produces virtually all of the thrust. At the same time, propeller designers have developed thinner and sweptback blades to produce engines called propfans. We have reached the stage where the fan can be considered a multibladed shrouded propeller, while propfans can be thought of as turbofans of exceptionally high BPR (with or without a surrounding shroud).

Thus there is no way fans and propellers can any longer be considered as different species. Indeed, in an increasing number of engines the designer's answer is to use a reduction gearbox to drive a relatively large fan by a small but fast-running turbine. In passing, the versatility of the gas turbine is such that a few designers have even considered driving propellers by a turbine of large diameter powered by a number of small gas turbines arranged around it, each blowing gas through a small part of its periphery. This seems to have little to commend it, though the tip-drive concept has been seen in large tip-drive lift fans.

Chronologically, the first partner to the simple turbojet was the turboprop. Indeed, the first turboprop to fly was merely a turbojet to which was added a reduction gear to drive a propeller. Today there are an increasing number of standardized core engines (compressor, combustion chamber, and turbine) which form the basis for a turbojet, various turbofans, a turboprop, and a turboshaft engine for helicopters. Provided the extra parts are properly matched, this idea is eminently sensible, because it increases core maturity and reduces costs all round.

In the early days of flying, with the notable exception of the Wright brothers, it was common to drive the propeller directly from the crankshaft. Soon Renault decided to drive it at half-speed from the camshaft, and by the 1920s it was realised that the ponderous weight and extra cost of a reduction gear might be well worthwhile, especially in the most powerful engines. With gas turbines there was never any doubt; the only problem was how to design the reduction gear in order to make an output shaft turning at perhaps 10,000 rpm drive a propeller at about 1,500. Today drive ratios of 15 or even 20 are not uncommon, and gearboxes are also featured in many propfans.

Propeller and fan gearboxes are always of the speed-reduction type. The simplest form is the intermeshing spur gear, in which a small toothed wheel drives a larger one alongside. Most propeller and fan gearboxes are of a dif-

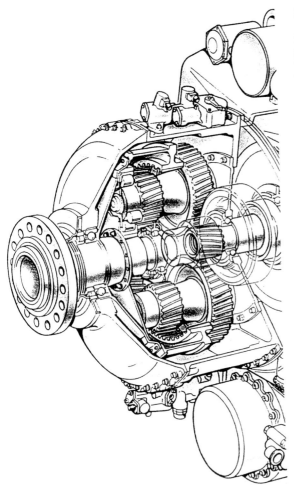

A typical propeller gearbox of the axi-centred type. All power comes in at a single helical-tooth pinion which meshes with three layshafts giving two-stage reduction to an outer annulus ring gear.

PSL book, *Piston Aero Engines.*

To convert greater power into thrust without increasing propeller diameter, it is sometimes necessary to use two propellers, one close behind the other, turning in opposite directions. Where these form a single unit, mechanically connected via the gearbox so that anyone trying to push one unit round will also have to turn the other in the opposite direction, the result is called a contraprop, short for contrarotating propeller. Where each half is independent and can be turned separately, the combination is called a pair of coaxial propellers. Some turboprops have been developed with two independent gas turbines, called power sections, each driving one coaxial propeller. The advantage of such an arrangement is that the aircraft can take off on both engines and then cruise on one, burning less fuel than with both units operating at a low power setting, and avoiding asymmetric problems.

Contraprops and coaxial propellers also have other advantages. They virtually eliminate the large gyroscopic effects of a single propeller, and they create a propulsive jet with very little rotation. The slipstream behind a normal propeller has powerful 'whirl'; it rotates rapidly, and this has a significant adverse effect on whatever fixed or movable aerodynamic surfaces might be in the wake, such as the tail. They also eliminate drive torque, which in a single-engined aircraft can try to dig one landing gear into the ground on take-off and leave the other almost skimming the surface, causing a large drag on one side only.

All of these effects were magnified by the greater power of turboprops, but because of their simplicity and lower noise single-rotation propellers are today almost universal on modern turboprop aircraft. A more significant effect of the greater powers was that the reduction gears were more highly loaded, and even greater care had to be taken in the design and precision manufacture of the gears themselves. Not only did the intermeshing teeth run at higher speed, but their mechanical loading (stress) was greater. Among other things, this called for the development of new so-called 'synthetic' lubricating oils, which instead of being merely the thicker distillates from petroleum are based on large-molecule chemicals called esters. Large oil flows at high pressure were required to lubricate the heavily loaded teeth, take away the heat, and also provide power for rapidly changing the pitch of the propeller blades. A separate oil supply may also be needed to serve a torquemeter (Chapter 7).

ferent type, called the planetary or Farman configuration, in which a small input pinion (gearwheel) on the high-speed turbine shaft engages with three or four larger gears arranged around it. These can rotate around the inside of a surrounding toothed ring or, to reduce speed still further, can each drive a smaller gear on the same shaft so that these smaller gears can run round a surrounding annulus ring more slowly. The orbiting planetary gears are carried on a cage or set of radial arms called a spider mounted on the rear end of the propeller shaft. Different types of reduction gear are depicted on page 82 of another

Without getting too deep into a slightly peripheral subject, the form of the teeth can be important. Common spur gears, in which the teeth are parallel to the axis of rotation, are the cheapest to make, but they engage across their length with an impulsive force, and when the meshing frequency – number of teeth multiplied by rpm – is the same as the natural frequency of vibration of any part of the gear-train then dangerous vibration will probably build up. In contrast, helical gears, with teeth cut diagonally, engage progressively from one end of each tooth to the other in a kind of shearing action. Thus the meshing frequency changes from a rapid succession of impacts to a drive almost as smooth as that from a belt. Many designers even go to the expense of using double-helical gears, often called 'herringbone' from their appearance. In the Lynx helicopter the main drive gearbox uses W–N (Wildhaber/Novikov) conformal gears with even smoother-meshing teeth having the profile of convex or concave circular arcs. Tomorrow's propfans and high-BPR turbofans are likely to need gearboxes transmitting up to 50,000 hp.

The first propellers were carved from wood, or from multiple laminations (plies) of wood, or were simply bent pieces of aluminium or steel sheet riveted to a piece of tube. From 1930 the choice was usually between compressed wood laminates, bonded with resin adhesive at high pressure and temperature and with an erosion-resistant coating, and forged duralumin, an aluminium alloy. From about 1935 it became common for all except light aircraft to have variable-pitch propellers, the blades being pivoted in a hub which usually incorporated hydraulic (oil) pistons but sometimes a gearbox driven by an electric motor. The pilot selects fine pitch for take-off, so that the engine can turn at maximum rpm to give the highest possible power while the propeller blades give high thrust at low airspeed. As the aircraft climbs away the pilot coarsens the pitch, until in level cruise the blades are fully coarse matched to the high forward speed and lower engine rpm.

By the Second World War, propellers were often of the constant-speed type in which a c.s.u. (constant-speed unit) containing a centrifugal governor continually adjusts pitch to match the engine power and airspeed. If the engine were to fail, the pilot could quickly feather the propeller, turning the blades until at about mid-length they were edge-on to the airstream, so that rotation stopped. This gives least drag and also prevents possible further damage to the engine. Braking or reversing pro-

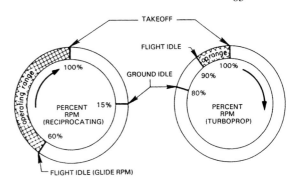

Aircraft reciprocating (piston) engines, like car engines, operate over a wide range of percentage speeds. In contrast, turboprops operate over a narrow band of speeds (many are constant-speed engines), power being varied by fuel flow (Pratt & Whitney).

pellers have a further pitch setting for use after landing. Once weight is on the wheels, a lock can be disengaged and the propeller put into reverse pitch. Opening the throttle then blows the slipstream forward, causing powerful braking which is doubly welcome on a slippery runway. Many propellers can be switched to a ß (beta) control mode in which the pilot can override the c.s.u. and adjust pitch directly, anywhere from forward to reverse. This provides almost instant response for landing and taxiing, and is especially appreciated when manoeuvring in confined spaces.

All of these features are found on turbine propellers, plus a few extra considerations. One obvious factor is that the power transmitted may

A typical operating pitch range for a turboprop propeller; to go into reverse the blade would rotate clockwise past the ground-idle to a large negative setting. Even at flight idle most turboprop propellers can exert a powerful braking effect (Pratt & Whitney).

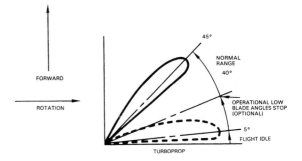

be much greater than in piston engines, necessitating blades of high efficiency, great strength, and large area. Provided there is room without hitting the ground or fuselage, modern propellers can have a diameter up to 22 ft (6.7 m). Power absorbed and thrust generated varies to some degree with 'solidity', the fraction of total disc area actually occupied by blades. Solidity of contraprops and coaxial propellers is obviously above average, and 40 years ago many turbine propellers were fitted with extremely broad rectangular blades with square tips which also had high solidity, up to 30 per cent.

Broad tips reduce the ratio of thickness to chord (breadth across the blade from leading to trailing edge). T/c ratio is one of the key factors in enabling wings to fly faster than sound. It may be difficult to make a wing thinner, so a designer may instead increase the chord, an extreme example being Concorde. When turboprops made possible more powerful propeller-driven aircraft, there still remained a limitation on speed caused by the fact that the tips of the propeller blades have an airspeed much greater than that of the aircraft itself. Shockwaves form on the blades, causing not only a deafening noise but also separation of flow and a loss of thrust. Blades with very broad tips alleviated this problem, and among other things enabled a Lockheed P–3 Orion to set a speed record at just over 800 km/h (500 mph). The even faster Tu-95 uses a combination of thin blades set in an exceptionally coarse pitch.

These broad but thin propeller blades may be manufactured by different methods. The first examples 50 years ago were either solid forged duralumin or hollow steel. The latter were made by pressing out the front and back of the blade from thin steel sheet and then brazing these together round the edge with a hollow steel spar down the centre. The interior was then filled with low-density foam to keep the thin skins from buckling inwards. Today's blades can be of composite construction using an adhesive binder to tie together strong fibres of glass, carbon, or a spider-web type material called Kevlar. Such blades are very light but not particularly hard-skinned, so a protective strip of abrasion-resistant material must be added along the leading edge.

Such propellers are used on most of today's local-service and commuter transports. These have to meet stringent noise regulations, so their propellers must not only give higher thrust than before, for a given diameter, but must also be quieter. The key to reducing the noise is to reduce the tip speed, which for a given diameter

means running at lower rpm. Instead of turning at over 2,000 rpm on take-off, today's propellers rotate at only 1,000–1,700. To transmit the high power they therefore need more blades, and five or six blades have become common.

Most early turboprops had the propeller gearbox centred in the inlet. A few early turbine propellers even had large hollow spinners so that air could be ducted past the compact hub to the inlet of the engine. Most of today's turboprops have an inlet offset above or below the propeller, but it is still important for the roots of the blades not to leave a turbulent wake. The broad blades of the Lockheed Electra of 1957 could be placed in reverse pitch, which could have blown the available air away from the engine inlet, so the inner portions of the blades were separated and fixed at a positive pitch to the hub. This feature is continued on the derived P–3 Orion.

A few aircraft have pusher turboprops. This normally requires a special version of the engine, with the reduction gear at the rear. There are important advantages in propulsive efficiency, in keeping the whirling propeller slipstream away from the tail, and especially in reducing noise in the cabin. On the other hand, the propeller blades have to withstand possible severe vibration caused by the turbulent wake coming off the wing, as well as the hot gas from the jet-pipe(s) playing on the highly stressed blade roots.

Since the basic task of a propeller blade is similar to that of a wing – to generate the largest possible force at 90° to the relative movement through the air – it might be thought that they have remained crude. Unlike wings, they have failed to sprout flaps, slats, spoilers and other movable auxiliary surfaces. The sole attempt to do better came in 1959, when George Rosen, of Hamilton Standard, proposed the variable-camber propeller. This comprised a single hub carrying two sets of blades, one immediately behind the other. Pitch of the front and rear blades of each pair could be controlled independently, so that each pair could operate either like a biplane or as a single blade with powerful camber. The idea was intended to give increased thrust (lift) for VTOL. It was tested in conventional flight on a B-17 in 1965, but dropped. So far the consensus is that propellers are not amenable to auxiliary devices.

As far as possible, the power (output) turbine of a turboprop extracts all the remaining available energy from the gas and transfers it to the propeller. Typically the propeller provides 90–93 per cent of the thrust, the remainder being the residual thrust from the jetpipe(s). If the jetpipe is directed straight backwards and sized for maximum thrust, the total horsepower from a turboprop – called 'equivalent horse-power' – is given by a rule-of-thumb approximation which adds to the measured shaft horsepower (shp) the jet thrust in lb multiplied by 0.3846. In SI units the ekW (equivalent kilowatts) is given by adding 68_{n2}, where F_n is thrust in kilonewtons. In many aircraft the residual jet thrust is ignored, as shown by the fact that the jetpipe is pointed sideways or downwards.

The point has been made that 50 years ago nobody could see much kinship between the turbojet and the turboprop, apart from the fact that both used a gas-turbine core. In 1936–46 Power Jets and Metropolitan-Vickers pioneered various kinds of ducted-fan jet engine, but, just as nobody could see the potential in Whittle's turbojet six years earlier, so did these eminently sensible engines fail to generate any more than passing interest. Ten years later Rolls-Royce at last introduced what the firm called 'bypass turbojets', but with such unambitious BPR that they behaved merely as slightly complicated turbojets. Only gradually did designers believe what had been staring them in the face ever since Whittle's calculations of 1936, and start making turbofans with BPR of 3 and more. GE led the way in 1965 with the TF39 (BPR 8), and still leads with the GE90 (BPR nearly 9).

As elaborated later, several factors limit the optimum BPR. One is obviously the engine's overall bulk and weight, which can be closely equated with cost. Another is the fan diameter, which does bear a relationship to installed drag, though as the air passes *through* the fan the drag of even a seemingly enormous engine pod can be surprisingly low. Around 1959 errors in calculating installed drag played a central role in restricting the BPR of British 'second-generation' turbofans to about 1.0. Another factor is the difficulty of matching optimum fan speed with that of the LP turbine driving it. This problem is largely responsible for the fact that in today's big turbofans the diameter of the LP turbine is increased to the greatest possible extent, causing a large bulge in the core cowling. The LP turbine of the GE90 has a mean gas-path

GE's rationale behind the relatively low fan pressure ratio (FPR) of the GE90: from the top, sfc and propulsive efficiency (compared with a typical competitor, shown as a square); next, fan thrust divided by drive power; next, cycle temperature; finally, core power needed for a given thrust (General Electric).

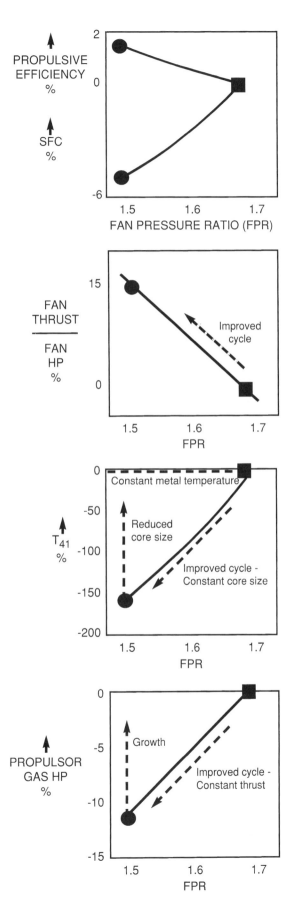

slope of 34°. Alternatively, in its turbofans derived from helicopter engines Lycoming has thought it preferable to stick with a reliable high-speed LP turbine and reduction gearbox.

Whereas the timid 'bypass jets' merely used slightly oversized multistage LP compressors, by 1960 it was realised that for peak propulsive efficiency it pays to use a much larger fan with a lower fan pressure ratio (FPR). Accordingly, today's turbofans use a single-stage fan more like a multiblade propeller than a gas-turbine compressor. It alone provides almost all of the installed thrust, the core's job being to provide the shaft power to drive it. FPR has in the past hovered in the range 1.6–3.0, but in the GE90 the chief basis of the design was the lowest practical FPR, and this was fixed at 1.5. For any given thrust, reducing FPR also reduces the required drive horsepower, which in turn means a smaller and/or cooler core, but at the expense of a bigger and heavier fan.

Nearly 60 years ago Whittle pointed out that a fan, then called an augmenter, could be mounted either at the front or back of the core.

At the front, it has to be driven from the LP turbine by a long shaft passing through the whole length of the core. An aft fan can be mounted around the perimeter of an independent LP turbine running in its own bearings in the hot gas immediately downstream of the core, the fan blades being 'double-deck' extensions of the turbine blades. Yet a third arrangement would be to use an aft fan in which the turbine blades form a ring surrounding the inboard fan blades.

Each arrangement has good and bad features. A front fan supercharges the core; if the FPR is 1.5 and the core (HP) compressor PR is 20, then the overall engine p.r. will be 30. The fan tends to protect the core against FOD (foreign-object damage) because any object sucked into the inlet will probably not seriously damage the large fan blades, which will centrifuge the object outwards to be shot down the bypass duct, instead of entering the more delicate HP compressor. A front fan should be durable, because it stays cold, and another advantage is that it can rotate together with low-speed LP compressor blades to supercharge the core even

Another set of GE curves designed to show how the GE90's combination of higher overall p.r. (OPR) and higher BPR give it a considerable edge in fuel economy compared with competitors. All engine makers produce brochures which attempt to show that their engine is the only one in sight (General Electric).

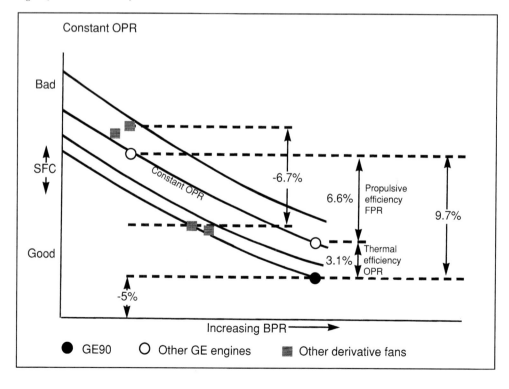

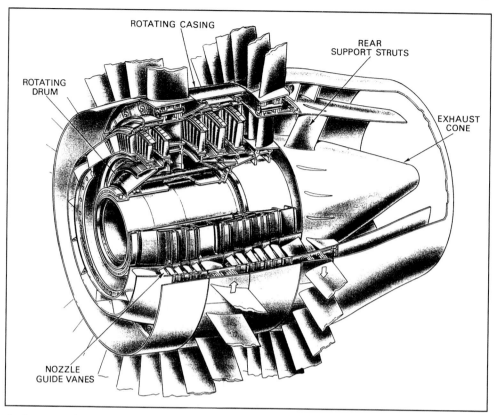

An integrated contrarotating aft fan, which would be mounted downstream of a suitable gas-generator core. Each set of fan blades (cut off here to save space) is driven by a six-stage turbine which forms the stators for its partner (Rolls-Royce).

further. On the other hand, it needs a long drive shaft which, because of the very high torque transmitted, has to be heavy and need large and firm bearings. Even if the rotational speed is modest, the large diameter of the fan means that in the cold airflow the tip Mach number will be about 1.5, so most of each blade will be supersonic. Another problem is that, if the fan is near the front of the nacelle pod, the engine may be troublesome to start in a crosswind.

Any aft fan has the problem of being in the hot part of the engine. If the fan blades are carried on the end of the turbine rotor blades then the latter will have to bear a greatly increased mechanical stress caused by the centrifugal force and aerodynamic reaction forces on the fan blades. There are likely to be fundamental speed-matching problems caused by the gas acting on the inner part of each double-deck blade to drive a fan carried on the end at a greater radius; ideally this calls for each fan blade to be carried on the periphery of a two- or three-stage

turbine. In contrast, the reverse arrangement, in which an aft fan is driven by a ring of outboard turbine blades – the so-called tip turbine – can easily achieve ideal speed matching, but only at the cost of a clumsy and aerodynamically inefficient system of ducts to channel the hot exhaust gas outwards and the fresh propulsive air inwards in an arrangement like intermeshing fingers. Whichever arrangement is chosen, a minor advantage is that ram compression in the inlet duct to the aft fan can heat the air to the point where transonic Mach problems are reduced or eliminated.

Over 30 years ago the only way to make a fan blade was to use solid metal, such as forged titanium. Once designers realised that the optimum BPR ought to be 3 or more, the sheer size of the blades posed problems. Everything possible was done to make them lighter, not only to reduce the direct weight of the fan but especially because, to pass certification tests, the entire front structure of the engine has to withstand prolonged running with one fan blade severed, imposing colossal out-of-balance forces. Despite this objective, the addition of a large front fan can more than double the total gyroscopic couple of the engine at take-off rpm.

As explained in Chapter 8, to reduce fan weight it was essential to make the blades with a high aspect ratio, with a short chord. Such long but narrow blades tend to be aerodynamically unstable and to suffer from potentially catastrophic vibration, called flutter. To prevent this it was necessary to fit each blade with a blunt projection on each side about two-thirds of the way from root to tip, called a part-span shroud, or mid-span support, or snubber or clapper. These projections all rubbed against those on each side, providing the required viscous damping to prevent build-up of flutter. Their obvious drawback was that, apart from adding mass in a place where every extra gramme adds about 50 grammes to the weight of the fuelled aircraft, it is highly undesirable to insert a blunt ring round the fan in a region where, at maximum rpm, the airflow is supersonic!

Just 30 years ago Rolls-Royce decided to try to develop wide-chord blades needing no mid-span snubbers, and to do so by switching from solid metal to either a hollow structure or a high-strength composite. They developed wide-chord blades with a sharp-edged – so-called lenticular – profile which promised to make possible a remarkable increase in fan aerodynamic efficiency, as well as a significant reduction in weight and in number of blades. Unfortunately, the method of construction, based on a proprietary carbon-fibre composite called Hyfil, proved unable to meet the vital bird-ingestion test in which unfrozen 4 lb (1.8 kg) chickens are fired at high speed into the engine while it is running at maximum rpm. For nearly 20 years the unimpressive 'American' style of solid-titanium blade with mid-span shrouds ruled the scene, before Rolls-Royce at last did manage to achieve a breakthrough with wide-chord snubberless hollow blades. Today GE claims to have regained the lead with enormous blades made of intermediate-modulus fibres bonded by a much tougher resin matrix.

Although the GE90 has the boldest BPR so far, it by no means represents the ultimate. Indeed, we have reached the stage where the gap between the jet engine and the turboprop has ceased to exist, in that engines of the next generation can be variously called propfans or UHB (ultra-high-bypass) jet engines. To some degree the name chosen depends on whether the propulsive blades are enclosed in a shroud or

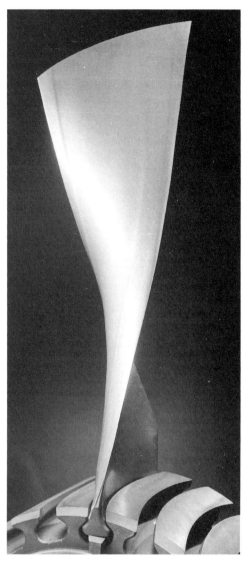

Rolls-Royce finally achieved a breakthrough with wide-chord, hollow, snubberless fan blades in the 535E4, certificated in 1983. No American rival has yet gone into service (Rolls-Royce).

left open as in a propeller.

Thus, from this point on, no distinction will be made as to whether we are discussing a 'fan' or a 'propeller'. The aerodynamics and blade construction have merged, and the only variables are: whether there is one set of blades or two rotating in opposite directions, whether the blades have variable pitch, whether they are at the front of the core or behind it, and whether a reduction gearbox is needed. Each choice is determined by required aircraft performance (notably the range, field length, and cruise Mach number), engine installed weight and drag, first cost, maintenance/spares costs, fuel burn, and noise. These in turn vary between one country and another, between one application and another and from one year to another.

For example, fuel economy is more important to a global long-hauler than to a commuter transport. Moreover, whereas in 1976 the price of crude oil made the installed specific fuel consumption a dominant factor, today it comes some way down the list. A night take-off from Orange County, California, has to be roughly a million times quieter than from the middle of Russia or Africa, for example. As in almost everything in aviation, it is easy to show that each choice is better than the other.

Considering the question of whether there should be a single set of blades or a counter-rotating pair, the answer is that a single-rotation propeller is simpler, lighter and cheaper, but runs out of efficiency at lower cruise Mach numbers. A Mach 0.6 aircraft has no need for counter-rotating propellers, provided the power can be put through the available diameter. At Mach 0.7 a traditional propeller is no longer acceptably efficient, and a gap in efficiency begins to open up in favour of counter-rotation. At Mach 0.8 the best single-rotation blades have fallen to a net efficiency of about 80 per cent, while counter-rotation is still close to 90 per cent.

Prolonged research by many engineering teams has confirmed that, for long-range aircraft cruising at about Mach 0.9, the lowest fuel burn is promised by a counter-rotation engine with a BPR of more than 25. How this should be arranged is anybody's guess. The first to fly, the GE/SNECMA UDF (Un-Ducted Fan), achieved the remarkable combination of counter-rotating variable-pitch blades with a BPR nudging 35 without the need for a gearbox. Alternative designs have used various kinds of front or rear gearbox, which for minimum engine size and weight can be integrated into the fan hub(s). Thanks to a Russian logical preference for numerical results rather than the whims of fashion, all propfan aircraft currently in active development are Russian or Ukrainian. Their engines include drive gearboxes and variable-pitch hubs carrying blades giving exceptional solidities of around 50 per cent, in both tractor and pusher forms. Despite a desperate shortage of money and almost everything else, the Russians and Ukrainians hope by 1997 to be testing aircraft with twin tractor propfans at the

Pratt & Whitney will at last get a modern fan blade into service on the PW4084 for the Boeing 777 in 1995. Its blade is seen here with that of the PW4168 (centre) and PW4000 (right) (Pratt & Whitney).

tail, twin pusher tail propfans, and four under-wing propfans, all with contra-rotating blades.

Purists may consider that the demarcation line between an HBPR turbofan and a propfan is not the omission of a surrounding cowl duct, but the incorporation of variable pitch. It will be recalled that all of the variable-pitch blades in traditional axial compressors were stators. The chief breakthrough in the UDF was the manufacture of thin wide-chord composite fan blades in a variable-pitch hub. When we consider the advantages in aircraft performance gained by the variable-pitch propeller it is surely remarkable that we have managed so long with fixed-

Classic curves published by NASA at the start of propfan funding in 1980. Asked whether today they would modify these curves, NASA said 'No'.

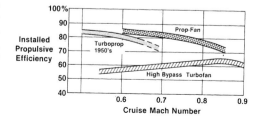

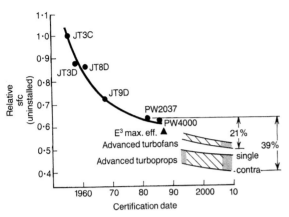

A different set of curves published in 1984, suggesting that by the end of the 1980s turbofans would begin to be replaced by propfans. Asked whether they would like to modify these curves, P&W said; 'We'd rather you didn't use it' (Pratt & Whitney).

There are still many ways in which an HBPR turbofan/propfan can be arranged. Initially several design teams inclined towards the conventional front-fan layout, merely increasing the size of the fan and adding variable pitch and a reduction gear and, in some cases, contra-rotation. So much effort has now been applied to alternative arrangements that they are no longer seen as posing high risk, though actual service experience has yet to begin.

These alternative schemes all promise higher propulsive efficiency. The big questions include: is a gearbox worth including; is a surrounding shroud worth including; and is it worth having two sets of contrarotating blades? Nobody denies that in all such engines there is a fundamental problem in matching the rotational speed of small turbines with giant fans. GE/SNECMA may go down in history as the only people to fly a large direct-drive variable-pitch propfan, and to do so with considerable success. Despite having open (unducted) blades it was acceptably quiet and achieved the expected performance, but at the price of an inelegant and generally undesirable mechanical arrangement of intermeshing multistage turbines and rack-and-pinion pitch change.

pitch fans in fixed-geometry ducts in the 747 and all other modern jetliners.

In fact, variable pitch is not urgently needed even to match such different conditions as sea-level take-off and high-altitude cruise, but it does have the great advantage of rendering a reverser unnecessary. Moreover, unlike today's turbofans, putting the blades in reverse pitch should give braking thrust almost as great as that available on take-off.

Just how much more efficient are geared fans depends on whose opinion is sought, but the margin appears to exceed 5 per cent. In 1986 Rolls-Royce published a series of studies showing how progressive refinement enabled a proposed engine with a 13,000 hp gearbox to be made lighter and shorter. At the same time the

*Three answers which were still being studied at press-time: **a**, a Rolls-Royce single-rotation geared propfan; **b**, a Pratt & Whitney geared contrarotating propfan; **c**, a Rolls-Royce aft contrafan with direct gas drive.*

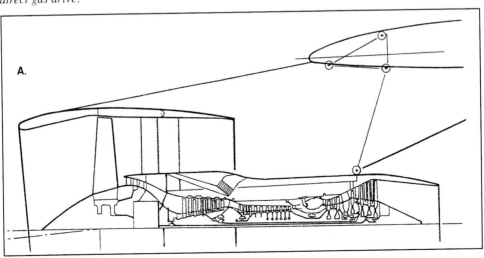

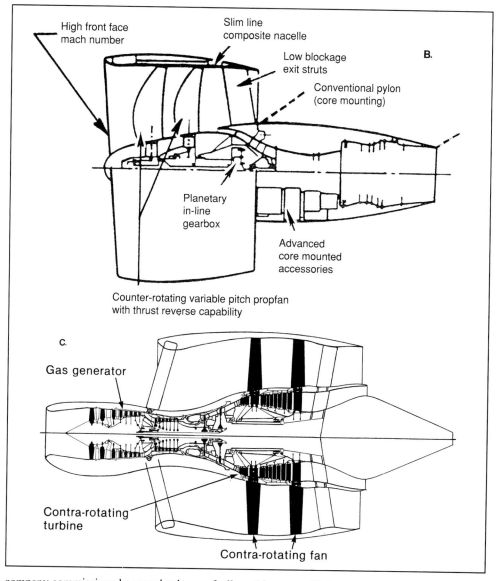

B.

High front face mach number

Slim line composite nacelle

Low blockage exit struts

Conventional pylon (core mounting)

Planetary in-line gearbox

Advanced core mounted accessories

Counter-rotating variable pitch propfan with thrust reverse capability

C.

Gas generator

Contra-rotating turbine

Contra-rotating fan

company commissioned a completely new facility for testing gearboxes up to 15,000 hp. This costly tool is a major expression of faith in the future of large fans and propellers, but if the market does the logical thing and accepts BPRs of 20 and above, then considerably higher powers will be needed. Initially such large and costly engines, offering a breakthrough in take-off thrust, fuel economy, and quietness, will appeal only to operators of the longest-ranged aircraft, and such aircraft are the biggest in the sky. At a guess, 60,000 hp is going to be needed, and this probably means collaborative engines.

There is no doubt whatsoever that engines of this type will become the norm for long-haul transport. In days gone by, new technologies were proven in military service (Lord Hives said he would never sell an engine to an airline unless it had long military experience behind it), but now the boot is on the other foot. The fuel-efficient PW2000 had millions of civil hours behind it before it reached the USAF in the C-17, and the RB.211 reached the RAF in the form of secondhand engines ex-British Airways! Thus the airlines must shoulder the burden of getting it right. To say 'Fuel is a bit cheaper' is nonsense.

7 Engine systems

In this chapter are grouped all the subsystems which feed, control and protect the engine. Some of them are visible in what is called the 'dressing' attached to the outside of the engine. The dressing can account for up to half of the installed weight of the engine, and something approaching half its cost. Today's engines are covered in a complex network of dressing, which must be carefully planned so that any faulty part can be replaced without disturbing other systems. First, however, we must return inside the engine to take a look at the internal air system (as distinct from the main flow of air that gives the engine its power).

Every gas turbine uses part of its airflow for auxiliary purposes. One of the most obvious, responsible for as much as one-tenth of the total mass flow, is to cool the discs and blades of the turbine. Another is to pressurize various internal cavities, to prevent the ingress of hot gas and,

Some idea of the problem of designing the dressing for a modern engine is afforded by the underside view of the Aviadvigatel D-30F6 (shown without afterburner), the 41,843 lb-thrust engine of the MiG-31M. A selection of items are: 1, hot bleed-air anti-icing manifold; 2, front mount; 3, regulator pump NR-3048M; 4, external gearbox; 5, fuel control RR-3048M; 6, pump RS-3048M for driving the giant afterburner nozzle; 7, drainage tank; 8, rear mount; 9, 10 & 11, auto fuel distributors ART-3048-2, 3048-1 and 3048-3; 12, igniter box for (14); 13, fuel/oil heat exchanger 9240; 14, gas-turbine starter TS-21S; 15, oil pump MNO-48; 16, centrifugal separator TsS-48; 17, oil tank; 18, unit RPPO-3048; 19, turbine PPO; 20, booster pump DTsN-76; 21, multipin cable connectors; 22, temperature transmitter TD-3048.

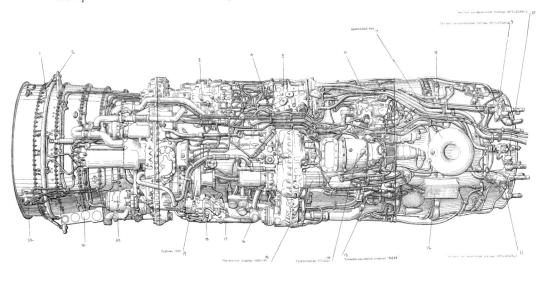

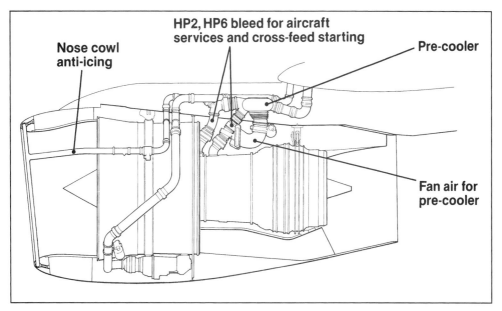

Bleed flows in modern engines are often greater than the entire airflow through Whittle's first turbojet! HP flows can approach 'red heat' and exceed 400 lb/in² pressure. Here are the main bleeds on a 535E4; one pipe feeds the starter under the fan case (Rolls-Royce).

especially, to try as far as possible to balance out the end loads on the shafts. We have already seen how bleed air can be blown on to a turbine casing to provide active clearance control. In most aircraft substantial airflows are bled away through lagged pipes to supply power to aircraft services, such as the ECS (environmental control system, incorporating cabin pressurization and heating or cooling) and airframe deicing.

Virtually all gas-turbine aircraft have to be certificated for flight in the most severe icing. Centrifugal compressors and the fan blades of modern turbofans generally experience only minor icing problems, but this is not the case with axial compressors with inlet guide vanes, especially in small engines. In any case, it is invariably essential to heat the inlet of the installed engine, because ice could form here in chunks large enough to disrupt the inlet airflow and, on breaking away, damage the fan blades or a fragile acoustic duct lining. Inlet deicing is conveniently done with hot bleed air, which is also often blown through the fan nosecone (spinner). In the case of smaller engines, especially turboprop and turboshaft engines, raw AC electric power is more common, because this is needed for the propeller anyway. The current is cycled (supplied intermittently) through the leading edge of the inlet and through the propeller spinner and blade leading edges. The

heater elements comprise patterns of metal foil or sprayed-on conductive material sandwiched between glasscloth or neoprene, bonded on to the inlet and finally given an anti-erosion coating. The on/off cycle time is selected to ensure that the ice shed during the active 'on' periods is never thick enough to damage the engine.

Jet aircraft usually fly high, and long-range aircraft soak at temperatures as low as –60°C. On a long flight the great mass of fuel can be cooled well below the temperature at which ice crystals can form, and even small crystals can soon dangerously block the fine-mesh filter as the fuel reaches the engine. In many engines the fuel is heated by making it cool the lubricating oil, as explained later, but even if this is the case the LP fuel supply may well also need a heater through which is passed compressor bleed air. All bleeds inevitably degrade engine performance, in that fuel is being burned for reasons other than propulsion of the aircraft. Accordingly, the hot bleed through the fuel heater is normally shut off; in British engines an automatic control is fitted, sensitive to filter pressure-drop (which rises rapidly as ice crystals build up), while many US engines rely on a cockpit warning light.

Since the earliest turbojets, designers have had to find ways to minimize the axial (endways) loads on the main shaft bearings. With

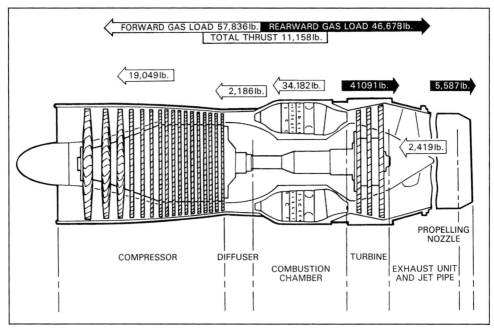

FORWARD GAS LOAD 57,836 lb. REARWARD GAS LOAD 46,678 lb.
TOTAL THRUST 11,158 lb.

19,049 lb.

2,186 lb.   34,182 lb.   41091 lb.   5,587 lb.

2,419 lb.

PROPELLING NOZZLE

COMPRESSOR   DIFFUSER   TURBINE
                        COMBUSTION   EXHAUST UNIT
                         CHAMBER   AND JET PIPE

Thrust distribution in a Rolls-Royce Avon, a single-shaft turbojet of the 1950s. It may seem strange to see that the main contribution to net thrust is made by the front of the combustion chamber, while the propelling nozzle actually pushes backwards!

three-shaft engines the problems might seem severe, but in fact they are usually made easier. However, the problem can be most simply understood by studying a single-shaft turbojet, which has an axial compressor with hundreds of tiny 'wings' all 'lifting' in a forwards direction and a turbine pulling in the opposite direction. The brochure figure for the engine's thrust of 11,158 lb masks the fact that the rearwards force trying to pull the turbine off the end of its shaft is 41,091 lb. The opposing force from the compressor blades is 19,049 lb, so the main drive shaft is always under severe tension. But the two forces are far from equal, leaving 22,042 lb (10 tonnes) to be reacted by the main thrust bearing, or location bearing, which is a large ball bearing fixed in the centre of the engine.

In practice, the designers can usually reduce this enormous and highly undesirable load very considerably. Some 40 years ago the author had to write descriptions of all major Western engines, and was astonished not only at the complexity of their internal air systems, but also by the fact that no two were alike. Even at that early date designers were feeding air bled from selected compressor stages to cool (sometimes to heat) and pressurize many parts of the engine.

Careful study of the drawings was needed to see cunningly arranged holes or slots, tiny gaps or impervious seals, and the way high pressures were often used to apply large end-loads on the sloping faces of conical 'discs' or on narrow peripheral rings.

Often the air carries with it fine droplets or vapour of lubricating oil. Bulk (liquid) oil is supplied under high pressure to all main shaft bearings and to the many gearwheels and auxiliary shaft drives. Seals are needed throughout the engine to control the escape of oil, and to maintain the air pressures in the various cavities. The most common are labyrinth seals, in which the rotating member carries a series of thin sharp-edged rings which almost rub against a light honeycomb structure or a soft abradable (rubbable) material or a surrounding rotating annulus of oil. A little air or oil may seep past the first ring, but the pressure in each successive cavity is less, and nothing comes out on the other side.

Various ring seals actually permit metal/metal contact, because when this occurs the surrounding ring moves away a microscopic distance. Thread seals 'screw' the air or oil in the direction opposite to the pressure difference. Hydraulic seals are used round major shaft bearings; centrifugal force keeps the surrounding film of oil in place, differences in air pressure merely causing slight differences in film thickness each side of the immersed ring. Carbon seals are actually pressed against the shaft by springs, the friction heat being taken away by the oil flow. Brush seals work effectively even with significant radial rubs; the thousands of fine wire bristles press lightly against a hard ceramic coating on the shaft.

Most engines have a self-contained oil system of the recirculating type, used for both lubrication and cooling. The oil is supplied from a tank which may be an integral part of the external accessory gearbox or even a cavity within the engine structure, but is usually attached externally. It flows through a filter to the main oil pump, which feeds it at a predetermined pressure to all the main shaft bearings, calibrated orifices controlling the flow to each bearing at all engine speeds and aircraft altitudes and attitudes. In the latest fighters the oil tank is rotated like a spin drier so that a pressure feed is maintained under prolonged zero-g.

Further feeds, often from an independent circuit, serve internal gears and external accessory drives. Hot oil is then returned via the scavenge piping through radiators cooled by air and/or fuel, while a deoiler (US word) or centrifugal breather (UK term) allows the system to be vented to atmosphere without losing any oil. The oil serves a crucial role, not only in lubricating bearings and all places where there is metal/metal contact, such as in gearwheels and shaft splines, but also in removing friction heat. In turboprops and geared turbofans the high power transmitted through the gear teeth requires large heat flows to be taken away and dissipated by the oil cooler.

To avoid metal/metal vibration round major shaft bearings the outer races are often surrounded by a thin layer of oil supplied under pressure. Such 'squeeze film' bearings minimize the dynamic loads transmitted from the main shafts to the bearing housing. The oil also has to carry away any foreign material which could damage the engine. Such material might conceivably enter with the inlet air, but normally comprises tiny particles of metal caused by any kind of wear or damage inside the engine. Abrasive particles would almost certainly cause progressive further damage, so even the smallest are flushed out by the oil and arrested at one of the filters or strainers in the oil system. These typically incorporate panels of fine wire mesh or resin impregnated with fibres, all supported against a background of perforated metal. The return line upstream of the scavenge filter nor-

Schematic diagram of the oil system of a typical modern turbofan (535E4).

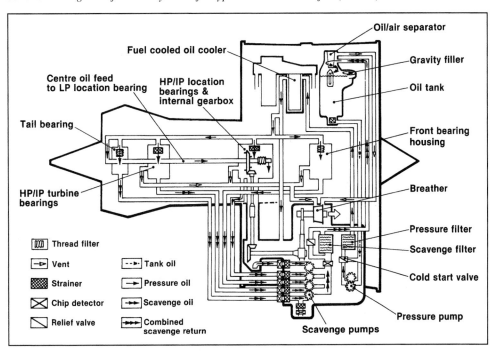

mally contains a magnetic chip-detector. This can quickly be unclipped from its self-sealing housing to see if any grains of metal are present, enabling an impending fault to be caught before the engine fails.

In most turboprops the oil is also needed for a torquemeter and to control the propeller. The torquemeter gives the pilot an indication of the power being developed, by measuring the torque (turning force) transmitted through the propeller gearbox. Some torquemeters use small pistons to react the turning load on the outer gear annulus, while others use the oil to react the axial end-load caused by helical teeth on the planetary pinions. High-pressure oil, in emergency at high flow-rates, is needed to operate the propeller pitch control mechanism. These extra duties sometimes result in turboprops using oil of higher than normal viscosity.

By far the most complicated subsystem in the engine, the fuel system has to supply clean, filtered fuel, free from foreign matter or ice crystals, to all of the main and afterburner nozzles at the exact pressures and flow rates needed for starting, acceleration and stable combustion at all conditions up to the maximum speed or altitude of the aircraft. The main controlling element can be hydromechanical or, in modern engines, a FADEC (full-authority digital electronic control). Hydromechanical controls can be of the pressure-control type, or of the proportional-flow type, or of the speed/acceleration type, or a pressure-ratio type, and there is no way that any can be described in full detail here. For example, the Hamilton Standard JFC-68 hydromechanical control, used on the Pratt & Whitney JT9D, contains 130 separate functioning items, not including parts associated with water injection!

This complexity reflects the fact that the system must feed the correct flow under all conditions of ambient air pressure, ram pressure, air temperature, engine speed, and pilot demand, and at all times limit turbine gas temperature, shaft speeds, and (probably) compressor HP delivery pressure. To do this requires a mass of speed governors, servo systems, sleeve and pilot valves, feedback and follow-up devices, pressure and temperature sensors, position adjusters, overrides, bypasses, rate controllers, metering

Airline captains fortunately do not have to understand the detailed working of the JFC-68 fuel control used on most JT9Ds. They may feel that switching to electronic FADECs has probably removed many sources of potential unreliability! (Hamilton Standard).

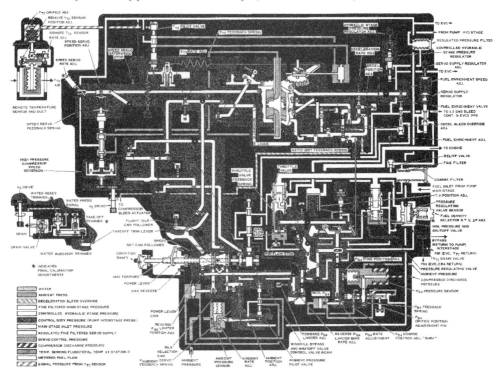

systems, and much more. Fortunately neither the pilots nor line mechanics need to know in detail how it all works.

Fuel is fed from the aircraft tankage by a booster pump in the bottom of each tank. To minimize unavailable fuel, the pump is generally at the lowest point in the tank, which emphasizes the overriding need to prevent the fuel from being contaminated by water, which is denser than the fuel and collects at this point. Fuel is delivered via an LP cock located outside any engine fire-risk zone, so that in emergency it can be unfailingly shut off.

If one represents the complex main flow controller as a mere box, it is possible to show some of the fuel-system elements and their interconnections in a greatly simplified form. The subject selected is a typical Rolls-Royce jet engine of the 1970s. In effect, this shows the main units interposed between the HP pump and the spray nozzles which adjust the fuel flow according to throttle position, ambient temperature and pressure, signals of engine speed (rpm), compressor delivery pressure and EGT (exhaust gas temperature), and the need for rapid acceleration or deceleration.

Of course, automatic control systems are needed for many engine functions in addition to governing the fuel flow. These include inlet and fuel-filter anti-icing, the entire starting cycle, water or W/M injection, continuous igniter operation, variable inlet vanes and stators, anti-surge bleeds, various valves controlling internal and external air flows (for example, heating compressor cavities and feeding air to shrink the turbine case), air supplies to aircraft services, signal outputs to aircraft monitoring and data systems, various emergency subsystems (EGT, overspeeds, oil pressure, and vibration among others) and, for afterburning engines, control of the nozzle. Turboprops can have subsystems for synchronization, to make all propellers turn at the same speed, and even synchrophasing, so that all propellers always have their blades parallel.

Water injection, just mentioned, was a common feature of both jet and turboprop engines, and is still found on some turboprops. Power is roughly proportional to the density of the airflow through the engine, and in most cases water injection was resorted to in order to maintain power during take-offs from very hot or high-altitude airfields, where air density is low. The technique is to inject a finely atomised spray of water or W/M (a mixture of water and methanol) into either the inlet to the compressor or into the combustion chamber. The water has to be very pure (so-called demineralized), because any impurities cause rapid buildup of hard deposits on vanes and rotor blades. It vaporizes rapidly, causing intense cooling. Methanol (methyl alcohol) is added as an anti-

A simplified fuel-system diagram for a two-shaft turbojet or turbofan. The broken lines indicate the sensing signals sent from the engine.

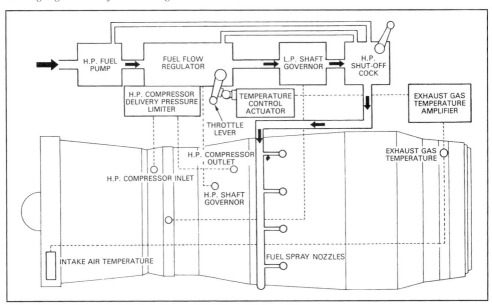

freeze, but it burns along with the fuel to give extra power.

Injected into the compressor inlet, the water both increases air density and also cools the turbine gas temperature, enabling extra fuel to be burned which further adds to the power. Injection into the combustion chamber gives better distribution and makes possible higher water flow-rates. It increases mass flow through the turbine, and reduces turbine temperature and pressure-drop, giving higher pressure in the jet-pipe. This increases thrust, which is augmented further by the cooler turbine making possible greater fuel flow.

With jet engines, water was sometimes used to boost thrust above the normal maximum take-off value, as well as enabling it to be maintained in the most adverse hot and/or high-altitude conditions. With turboprops, water injection is still often a means for not only restoring the normal take-off power, but even for boosting power to perhaps 110 or 115 per cent of that value up to quite high tropical temperatures. Water is falling into disuse, and most modern engines do not require it. The provision of a complete additional system of tanks (with fillers and drains, the latter to run-off any unused water, which would freeze at high altitude), pumps (often complicated air-turbine pumps with their own lubrication system), control valves, electrical interconnects with throttle

An external gearbox curved to fit round a fan case. All power comes in via a single spiral bevel gear on the radial driveshaft from the HP spool (Rolls-Royce).

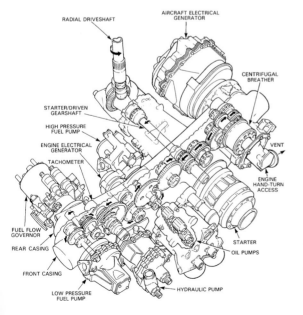

position and possibly the torquemeter, and various flow sensors and microswitches, meant a lot of bulk, weight, cost and extra maintenance.

The maximum thrust or horsepower of an engine with water or W/M injection was called the wet rating. At least with turbojets, 'wet' take-offs meant excruciating noise and usually a lot of black smoke.

Every engine has to power shaft-driven accessories. These must include fuel and oil pumps, and probably a tachometer (rpm indicator), but large engines can drive as many as 25 distinct devices. Most may be mounted on an external gearbox attached to the engine, some on a remote gearbox mounted on the airframe and driven by a shaft, and others, located anywhere in the aircraft, driven by high-speed turbines supplied with bleed air through lagged pipes.

One of the shaft-drive accessories is likely to be a starter. Starting systems must serve two completely separate functions: they must accelerate the engine from rest up to the speed at which there is adequate airflow through the combustion chamber(s) to support combustion, and then switch on the HP fuel and unfailingly ignite it. These two duties have to take place simultaneously, but it must also be possible to perform either independently. During maintenance checks of the engine and its accessory systems it must be possible to crank the engine without triggering fuel flow or ignition, and it must also be possible to operate the ignition system to relight the engine in flight without the need for externally-driven cranking.

Tiny gas turbines, up to about 100 hp, can be cranked up to self-sustaining speed by hand. Although the compressor and turbine may be small enough to slip into a raincoat pocket, it is hard work, because a step-up gear of about 15 ratio is needed in order to reach about 5,000 r.p.m. With most engines, diminishing assistance from the starter is necessary even after fuel ignition, until the engine reaches self-sustaining speed. Whatever the cranking method, the drive must incorporate some form of one-way clutch or throw-off jaws, so that, as soon as the engine begins to accelerate away, the starter drive is disconnected. In theory, a tiny single-shaft turboprop for an ultralight aircraft could be started by laboriously hand-turning the propeller, but the author has never heard of such an engine.

Virtually every engine used in manned aircraft is cranked up to speed by a mechanical starter, and there are many kinds. The most obvious is an electric motor, as used in cars, in

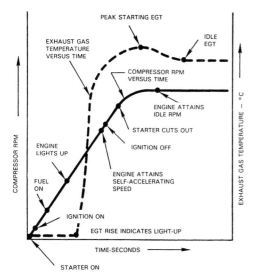

A typical starting cycle for an aircraft gas turbine. Light-up might be achieved in as little as three seconds in a fighter engine, or eight in an airliner engine (Pratt & Whitney).

many aircraft fed from the same 28-volt DC system as used for supplying the ignition and the starter-cycle timing and control system. Sometimes a single electrical machine can serve as a combined starter/generator, but usually an electric starter is a simple but powerful DC motor of traditional armature/commutator type. Some take full voltage throughout the timed starting cycle, in which case the drive torque starts at an extremely high value and progressively falls to a fraction of this at cutoff. Alternatively, the motor can be supplied through relays which allow the voltage to increase to a maximum at self-sustaining speed and then to fall away until the cutoff timeswitch is triggered.

The starter is geared to the main engine shaft, usually by being mounted on the accessory-drive gearbox. In a two-spool engine it drives the HP shaft. This runs up to speed and pumps enough air for combustion to start; once this has happened the airflow builds up extremely quickly, driving the LP spool as well. When a large turbofan starts, one can hear the HP spool accelerate to quite a high speed before the fan starts rotating; once combustion has started, the fan accelerates rapidly.

Electric starters for large engines are heavy, and in any case take a current that may be beyond the capability of the aircraft batteries. Front-line military aircraft have an overriding

need to be able to start in the shortest possible time, if necessary in the absence of any external power supplies. The answer is some kind of self-powered starter. Chronologically the first were cartridge starters, comprising a breech housing a formidable cartridge, usually filled with cordite, which is fired from the cockpit. The hot gas drives a small impulse-type turbine, which runs up to speed in less than a second, driving through a reduction gear and auto-disconnect mechanism. The cartridges are typically 'pint-sized', and because for various reasons the engine may fail to start on the first cartridge, some engines had triple-breech starters, reloaded with one, two, or even all three cartridges after the mission.

A variation, formerly common on many British warplanes, was the IPN (iso-propyl nitrate) starter, which replaces the cartridge by a combustion chamber. The electric control system first switches on an electrically-driven air scavenge pump which blasts air through the combustion chamber to sweep away residual fumes. Then it feeds the IPN fuel through a high-pressure switch to a spray in the chamber which is ignited by a high-energy discharge. The advantages are very high power for a rapid start, repeated as many times as necessary. The IPN tank will in any case be topped up after each mission. In the USA such units are called monofuel combustion starters, and some serve as an MEPU (monofuel emergency power unit) to drive hydraulic pumps or an electric generator following main-engine failure.

Another variation, seen on the latest US warplanes, is the JFS (jet-fuel starter). This is essentially a miniature gas turbine; indeed in some aircraft, such as the Harrier, the starter serves as an APU (auxiliary power unit) able to supply essential hydraulic or electric power on the ground, without running the main engine, or in emergency in flight following engine failure. Such units are self-contained (with their own starting, ignition, and lubrication systems), light, and capable of starting the main engines as often as necessary. For emergency in-flight power they must unfailingly light up at high altitude.

An even simpler and lighter arrangement, by far the most common on commercial transports, is to dispense with a self-contained gas source and drive the starter turbine with compressed air. Such pneumatic starters can be powered from an on-board APU, but normally rely, at least for starting the first engine, on a hose from a ground power truck. This feeds through a pressure-reducing valve which is opened electrically when an engine start is selected, and closed at a preset starter speed above engine self-sustaining speed. Once the first engine is running, the ground supply can be disconnected, and the remaining engines started by cross-bleed from the running engine. At least for large engines, air starting is cheaper than any other method.

Many small engines used in helicopters, trainers and cruise missiles are started hydraulically. Oil pressure from a ground supply is used to drive an engine-mounted pump or pump/starter, driving through a reduction gear and clutch. On completion of the start cycle the electrical control system operates valves to make the starter thereafter behave as an hydraulic pump. The simplest of all starting methods is air impingement. High-pressure air, from a ground supply or a running engine, is simply directed through a non-return valve and nozzle on to the engine turbine blades. This again is found only on small engines.

Ignition has to be sure-fire, even at high altitude where the intense cold greatly reduces the volatility of the fuel sprayed into the combustion chamber(s). Virtually every aircraft gas turbine uses a dual ignition system serving two igniters (igniter plugs) mounted far apart in carefully selected locations in the combustor(s). Usually each system is served by a hermetically-sealed igniter box, in the USA often called an ignition exciter. This is supplied with low-voltage DC from the aircraft, which charges a powerful capacitor operated by a trembler mechanism or a transistor chopper circuit. Alternatively, the box can be fed with AC (alternating current), passed through a transformer and rectifier.

The output is always a powerful pulse of electricity, typically with an energy of from 4 to 30 joules. This is fed through shielded electrically-screened HT (high-tension) harnesses to the igniter plugs. These are sometimes basically enlarged versions of a car spark plug, though because of the greater distance across which the spark must jump (the air gap) the potential difference has to reach 25,000 to 28,000 V. The alternative surface-discharge igniter feeds the high voltage down the centre to a tungsten tip from which electricity leaks across the surface of a semiconducting pellet, usually silicon carbide. This provides a pre-ionized path for the main discharge, which in consequence can flash over at only about 2,000 V.

In either case, the result is a large and intensely hot spark of such power that the surrounding atomised fuel is unfailingly lit. During the start-ing cycle each igniter flashes about 1.5–2 times per second. The flame swiftly propagates through the boundary layer from burner to burner – in some engines from can to can – until the whole combustion system is burning stably.

Unlike a piston engine, the ignition system is normally required only during the starting cycle. However, there are circumstances which preferably call for the igniters to fire continuously, notably flight through heavy rain, snow or icing. Many pilots would welcome continuous operation during take-off, landing, or overshoot (in the USA called a go-around), and with some types of engine, with tricky compressors, even in severe turbulence. To avoid drastically shortening the life of the ignition boxes and plugs, continuous operation is usually at a reduced power, such as 3–7 J. In any case, the entire system of power supplies, exciter boxes, and cables is invariably air-cooled.

Most of the engine-powered accessories provide power for the aircraft's own systems, notably electric, hydraulic and pneumatic (compressed-air) power, deicing, and the ECS (environmental control system). The last-named provides input power and air to pressurize the cabin and continuously supply it with air at a comfortable temperature and humidity. Both this and the aircraft pneumatic systems use compressor bleed air, though in the ECS this is normally used to drive turbomachines which feed fresh air to the cabin, because engine bleed may contain minute traces of oil. In any case, most turboshaft and turboprop engines cannot supply sufficient bleed air, and so they instead drive mechanical compressors.

We need not get too involved in the aircraft systems, except to note that in order to supply radars, computers, and other electronic items with electric power at an absolutely unvarying frequency, such as 400 Hz (cycles per second), the main generator (alternator) is often driven via an infinitely variable hydraulic gearbox. Typically this comprises a variable-stroke pump driving a corresponding hydraulic motor. To reduce bulk and weight, all of the newest aircraft use an IDG (integrated-drive generator), in which the drive and generator are a single unit. The objective is to drive the generator at constant speed no matter how the speed of the engine spool varies, from take-off to flight idle.

We have already seen that in a two-spool engine the main accessory gearbox is almost always driven off the HP shaft. The latter invariably incorporates a high-capacity bevel gear immediately in front of the HP compressor and as close as possible to the HP main location

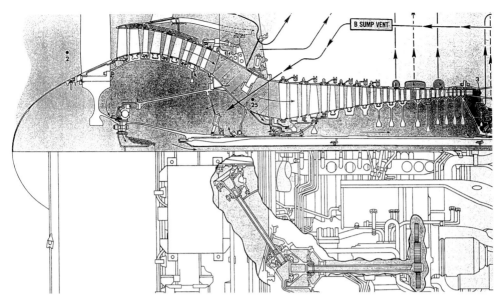

GE's large engines mount the main external gearbox on the core. This portion of a drawing of the CF6-80C2 shows the way the drive is taken off the front of the HP spool (General Electric).

bearing. This is so that there shall be virtually no permissible axial movement between the bevel gear and the mating gear on the radial shaft (USA term, 'tower shaft') which takes the drive outwards through a faired strut to the external gearbox.

The external gearbox is usually mounted under the engine, though in many fighter engines it is on top to permit easy access through doors in the top of the fuselage. Helicopter engines are usually mounted above the fuselage, in which case the gearbox is on top or on the outer side. In the case of high-BPR turbofans the designer has a clear choice. On almost all GE and Pratt & Whitney engines the external gearbox is tucked under the core, while on the CFM56, V2500, and Rolls-Royce engines it is mounted on the fan case. In a few engines both the HP and LP shafts carry bevel drives to separate external gearboxes, either to split the required torque into two (which can reduce overall weight) or to fit a large number of accessories in different places round the engine.

The external gearbox is self-contained, with its own oil system which is circulated by pumps and then cooled and returned via a centrifugal breather. The whole box is carefully sealed to prevent both oil loss and seepage into particular accessories. Both sides of the box may contain drive pads, one of which may be for hand-turning gear so that the engine can be manually cranked over during maintenance. Another can drive a remote auxiliary gearbox mounted on the airframe. Many fighters have two engines whose HP spools are linked by cross-shafts (sometimes also geared to an APU) so that either engine can drive both gearboxes.

One of the items mounted on the gearbox may be a miniature electric generator which senses the speed of the main engine shaft(s). Such a tachometer generator feeds 3-phase current, whose frequency is proportional to rotational speed, to a tiny synchronous motor in a cockpit instrument. This motor drives a drag cup or rotary magnet which pulls a spring-loaded pointer round a dial calibrated in rpm or in percentage engine speed. In an alternative system, the actual engine shaft(s) carry a phonic wheel, a ring of small square 'teeth' projecting evenly round the periphery. As each 'tooth' passes a variable-reluctance probe mounted almost touching the projections, it generates a small current which is amplified and fed to the indicator. The speed-probe current is suitable for driving modern colour displays which can simultaneously indicate the speeds and percentage speeds of every spool in four engines, each with three shafts.

These displays can also show EGT (exhaust gas temperature), oil temperature and pressure, fuel flow, vibration, and various other parameters. EGT is sometimes called JPT (jetpipe temperature) or TGT (turbine gas temperature), and

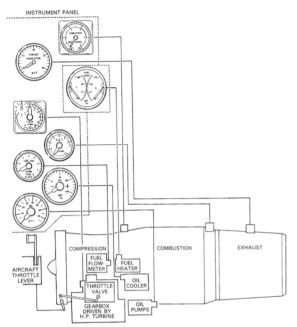

A schematic arrangement of instrumentation. The thrust indicator actually measures either jetpipe pressure or an engine p.r. (EPR), sensing pressures at inlet and exhaust (Rolls-Royce).

it can be measured either immediately in front of or behind the turbine section, or even at the inlet guide vanes of an intermediate stage. The sensing element is a thermocouple, a heat-resis-tant tube containing two wires of different materials, such as one of nickel-aluminium and one of nickel-chromium. They are joined at the end of the protective tube, which projects into a suitable point in the hot gas. Usually there will be from two to ten thermocouples arranged round the turbine section, plus others to sense the inlet air temperature. Each pair of wires is joined again inside a cool box, and it is the difference in temperature between the hot and cold junctions which generates a small voltage which drives the cockpit indicator.

Both the fuel and lubricating-oil systems incorporate various temperature and pressure sensors, pressure-sensitive switches and, in the case of fuel, flowmeters which indicate the actual flow. All transmit to the cockpit in the form of electrical signals. Vibration transmitters are transducers of the magnetostrictive or piezo-electric crystal type, which send out signals to a sensitive milliammeter with a dial calibrated in terms of vibration amplitude, the unacceptable region being red. Often signal filters are inserted so that the pilot can check on vibration in particular bands of frequency corresponding to N_1 (LP speed), N_2 (HP speed), or some other rpm value.

Thermocouples can give warning of excessive temperatures, and avoid costly damage to the turbine and other hot parts of the engine. Warnings can also be given by thermally triggered switches in strategic locations such as overboard air vents. In addition, the entire installed powerplant must be protected against fire, as outlined in Chapter 9.

8 Construction

Apart from the turbine itself, early gas-turbine engines were made by traditional methods using mainly aluminium-alloy castings and sheet steel, with the shafts machined from steel forgings. The turbine, like the valves of a piston engine, posed a major challenge from the outset, because when most metals and alloys are heated beyond a bright-red temperature their load-carrying properties fall almost to zero. Thus the materials and manufacturing processes vary enormously from the cold inlet to the hot exit(s).

It hardly needs stating that the designer wishes to make the engine as light and compact as possible, whilst preserving structural strength and rigidity, and freedom from oxidation or corrosion, over a lifetime which in modern engines may be 40 years. He can choose to make most parts from any of a wide range of materials, by different methods. Cooler parts may be made from low-density metals such as Al (aluminium), Mg (magnesium) or Mg-Li (magnesium-lithium), or from a composite based on fibres of carbon, glass or Kevlar. Hot parts require first Ti (titanium) alloys and then stainless steels and 'superalloys' of nickel and cobalt, culminating in the HP turbine rotor, which probably demands either exceedingly sophisticated and costly alloys or almost equally special ceramics.

In the HP turbine rotor conditions are so challenging that even the material density or cost become secondary considerations. In most of the engine these factors are very important indeed. The variation in density of available materials is considerable: Mg-Li has a specific gravity (density in g/cm^3) of about 1.3, followed by Al-Li at 1.48, Mg 1.74, CFRP (carbon-fibre reinforced plastics) 1.6–2.0, GFRP (glass-fibre RPs) and Kevlar 1.95–2.4, Al and alloys 2.7–2.8, important ceramics 3.1–3.3, Ti and alloys around 4.5, and steels and 'superalloys' 7.8–8.0. It is logical to open this chapter by discussing these materials and the various processes by which they are converted into engine parts.

Although today there are exciting alternatives to metal, metals are at present used for more than 90 per cent of most engines. Which metal is chosen for a particular part depends primarily upon the maximum temperature reached by that part. Cool parts, such as the inlet, fan case, LP compressor casing, and the casings of a fan, accessory, or propeller gearbox, can be made from aluminium (US spelling aluminum) or an aluminium alloy. In an increasing number of engines they are even made of composite material, though until recently such materials have not easily achieved the necessary structural rigidity. With the development of high-strength

Today's engines are generally hotter inside than those of 30 years ago, so they contain more nickel and titanium. From now on, metals will begin to be replaced by composites. These curves come from a 1994 paper by Sir Ralph Robins, Chairman of Rolls-Royce.

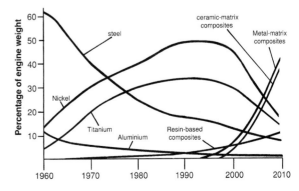

composites this is now changing.

In some applications secondary considerations supervene. For example, most fan cases have to be made extremely strong to meet the certification requirement of fan-blade containment. This is a very severe design case, and if 40-in (1-metre) blades were to come adrift and be uncontained the results would be awesome. Several aircraft have crashed because a fan break-up either caused a fire or destroyed the flight-control hydraulics. To prevent this, modern engines have a fan case in which the part surrounding the fan is of high-strength steel, or an alloy such as Armco, or immensely strong Kevlar composite (described later).

In many early engines various aluminium-bronze alloys were used for compressor blading to avoid corrosion in naval aircraft. Bronze is itself an alloy (of copper and tin, and sometimes other elements), the technology being based on ship propellers and similar marine parts. Oxidation and corrosion resistance is a basic requirement for everything in the gas path, and materials which meet the demand also survive in a marine atmosphere, though engines for naval aircraft do usually have some parts made of special materials.

To meet the demands of the mid-section of an engine, such as the HP part of a compressor and its delivery section, a common family of materials 40 years ago were the RR (Rolls-Royce) alloys originally developed for pistons. These are aluminium alloys with 1–2.2 per cent each of copper, magnesium, nickel, and iron, plus a

Specific strength (ultimate tensile strength divided by density) plotted against temperature for special nickel alloys and various future composites. Even though some of the plots are for strength along one axis only, the potential for improvement is huge.

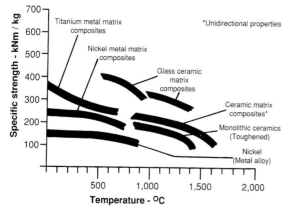

little silicon and titanium. Whereas plain Al is of little use even at 200°C, these alloys retain almost full strength at 250°, and they were used for most of the airframe of Concorde. Today compressors can operate at much higher pressure ratios, and demand far greater high-temperature capability. Titanium alloys play a leading role in many modern mid-sections, while the HP spool of the GE90 uses such costly alloys as Inconel 718, R88DT and Udimet 720. Only a few years ago such materials would have been considered necessary only for HP turbines!

The hot end of an engine, comprising the combustor, turbine(s) and jetpipe, was formerly manufactured mainly from high-nickel alloys, because these were the only materials able to retain adequate strength at the prevailing temperatures of 700° to 850°C. Turbine discs were an exception, made from forgings in special stainless steels. Stainless steels are iron with significant proportions of nickel and chromium, plus several other elements such as carbon, silicon and manganese. In modern turbine discs much of the iron has been replaced by exotic metals such as molybdenum, cobalt, and niobium, all making a contribution to strength and oxidation resistance at red-hot temperatures. Today the materials and techniques used to manufacture the hottest parts of engines are so advanced that, together with fan blades, they are discussed separately at the end of this chapter.

The processes used are many and varied, some for the broad forming of a metal part to shape, some for finishing, others for joining parts together, and yet others for changing the condition of an external surface. Other, totally different, methods are used to produce parts in composite materials.

One of the familiar traditional techniques is presswork, used to form parts to shape from metal sheet. Especially with the high-strength metals used in gas turbines, pressing may have to be done with the material at high temperature, yet even then it is very difficult to achieve the desired dimensional tolerance. The metal tends to warp, spring back, twist, or in some other way deform so that lengthy (and therefore expensive) hand working may be needed to correct the shape, and even then there may be locked-in stress. Better isothermal (constant-temperature) presswork is now being achieved using heated dies, of metal or ceramic, which repeatedly produce accurate stress-free shapes. Good results are also being achieved by hot blow-forming, in which the sheet-metal blank is forced into the die under the pressure of argon fed under microprocessor control to maintain the correct strain-rate. Argon, one of the inert gases, does not react with metals even in their molten state, and so plays an important role in the manufacture of aero engines.

Another related process is superplastic forming (SPF). About 30 years ago it was discovered that some metals with suitably fine grain structure can be subjected to tremendous ductile (tensile) deformation without tearing. With careful control of temperature and strain rate, SPF parts can be made in aluminium, titanium alloy and particular 'superplastic' steels by deep drawing. The presses have to be specially made, but the forming pressures are quite modest. The finished part can have very small bend radii and suffer such large changes in shape that the metal literally flows. For example, a billet can be squeezed into a thin-walled part with integral stiffeners. SPF is often combined with diffusion bonding (DB, described later) to produce complex components which in effect are a single piece of metal, instead of being made by joining perhaps a dozen separate parts.

Another ancient craft is spinning sheet metal, to make bowls and vases. The modern equivalent is flow-turning, in which a workpiece, initially usually a flat disc (a blank), is forced by computer-controlled rollers to bend to shape around a central rotating die called a mandrel. The result is almost any desired conical or even cylindrical shape, exactly to size with no joint. Previously, such a part had to be made by wrapping and welding sheet, followed by drawing and sizing to correct the shape.

Another familiar technique is machining, in which hard tools cut away material from the workpiece. There are various kinds of machining. In turning, the part is rotated on a lathe while being cut by a tool which slowly moves into or along the work. In milling, it is the work which is slowly moved past a rotating cutter. Jig boring is a kind of high-precision vertical milling. Broaching involves pulling or pushing a cutter past the workpiece to machine a linear slot such as a fir-tree root or a spline along a shaft; the broach is a linear cutter with many teeth, each of which approaches a little closer to the finished profile. All machining is today likely to be numerically controlled (NC). The machine tool is controlled by a computer, into which is fed a tape appropriate to the particular part. This greatly saves time, and makes possible the rapid machining of complex shapes which previously might have had to be forged, cast, or assembled by joining many parts together. It also virtually eliminates human error, so 'scrap' has almost become a thing of the past.

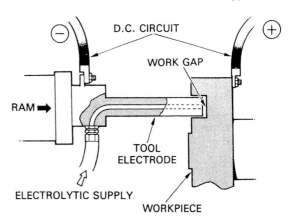

ECM is very widely used to shape materials that are almost impossible to cut in the normal sense.

There are many other techniques which can be employed to shape a part. In grinding, the cutting is performed by millions of exceedingly hard particles projecting microscopically from the surface of a wheel or drum. In electrolytic grinding, the wheel is electrically conductive and, with the workpiece, is immersed in a bath of electrolyte (conductive liquid, usually a solution of salts). The rotating wheel does not quite touch the workpiece, but removes small particles by electrochemical reaction. The rotation of the wheel sweeps away the by-products, which would inhibit the reaction.

Electro-chemical machining (ECM), is not machining at all. The tool is the cathode, the negative pole, in a DC electric circuit, and the workpiece the anode. The whole assembly may be immersed in an electrolyte bath; in any case it is vital to pump electrolyte under pressure through the tool, whose shape is an exact mirror-image of the finished part. Under computer control, an hydraulic ram feeds the tool into the work at a rate corresponding to the metal-removal rate. To save time, the current is usually enormous: 10,000 amperes removes 1 cu in per minute. ECM is used even on heavy forgings, and can produce parts to exact finished size with a mirror-like finish devoid of the slightest burr or scratch. It is widely used to shape the special alloys found in gas turbines, which are almost impossible to machine by mechanical shearing action.

An alternative for super-hard alloys is electro-discharge machining (EDM). Here, the tool is again a mirror-image of the finished workpiece, and the whole assembly is immersed in a dielectric liquid, usually paraffin (kerosene). A large electric power supply is converted by a condenser into a very rapid succession of giant pulses applied to the tool cathode. Each pulse breaks down the resistance of the dielectric in the small gap between the tool and workpiece, and results in a powerful spark. As each spark hits the workpiece, the metal is locally vaporized with explosive violence. The explosion is quenched by the surrounding liquid, which sweeps the residue away. Again the result is quite rapid shaping of complex profiles, sometimes of a dozen identical parts simultaneously, in material too hard to be readily cut.

Today, combustor liners and HP turbine rotor blades need to incorporate large numbers of holes through which cooling air can pass. These holes have to be small in diameter and drilled with great precision. In the turbine blades they have to be very long in relation to their diameter, and in the case of combustor liners they have to be drilled through the thin material at an extremely acute angle so that the surface area of the hole is very large in relation to the diameter, to provide the maximum cooling with the smallest possible airflow. Traditional drilling methods would be useless, not least because of the materials involved. A photograph in Chapter 3 shows the multi-hole combustor liner of the GE90, drilled by laser.

In stem drilling, holes are cut rapidly by immersing the work in a dielectric bath and feeding into the workpiece a tool cathode resembling a hypodermic needle made of titanium, through which is pumped a 20 per cent solution of nitric acid. The tool is insulated except at the very tip, which penetrates faster than in EDM. It can drill holes down to about 0.6 mm (0.024 in) diameter with a depth 200 times the tube diameter. A related process is capillary drilling. Here the tool is glass, with the conductive core a wire of platinum; the pumped electrolyte is again 20 per cent nitric. Capillary drilling is restricted to a hole depth of about 40 times the diameter, but can make holes as fine as 0.22 mm (0.009 in).

Chemical etching, or chem-milling, involves no tool or mechanical action at all. Instead, the workpiece, usually made from sheet metal, is 'passivated' (masked) over various areas and then immersed in a bath of acidic or caustic solution. Over a period, the areas exposed to the liquid are etched away chemically to reduce the wall thickness. The parts protected by masking remain the original thickness. By slow immersion or slow progressive removal, the thickness can be tapered in any desired manner. A typical result is a part which is thin and light except

round the edges or at various stiffening webs, all made rapidly without stress and with perfect surface finish. Many parts which have been machined conventionally are then briefly chem-milled merely to perfect their surface.

Among the basic forming methods, casting and forging are supreme. In heavy engineering, forging means heating a rough billet or slab of metal almost to melting point and then squeezing it to shape either in a giant press or by using blows from a steam-powered hammer. In gas-turbine manufacture almost all forging is die-forging, in which the workpiece – in some cases almost white-hot, in others at room temperature – is squeezed in a press between upper and lower dies. Such parts as main drive shafts, compressor casings or half-casings, combustor rings, rotor discs, blades, and gearwheels can be forged very close indeed to the finished shape and dimensions. Die forging is an economical way of producing parts which, in the case of blades, incorporate thin aerofoils with twist and camber that would be difficult to make by other methods, apart from ECM. In isothermal hot forging the dies are in a furnace, held at a constant temperature. Such precision forging demands exact control of forging temperature and absolute cleanliness of the dies. Forged parts are also known as wrought parts.

A particular type of forging, much used to make blading, is upsetting. Bar stock is fed into a machine which, at high speed, electrically brings the working end of the bar to forging temperature and then, by hydraulically controlling the feed of the stock and the withdrawal speed of an anvil forced against the end, leaves the end of the bar with a particular irregular profile. This profile distributes the metal correctly to make the root, tip, and any shrouds or snubbers in subsequent forging.

A process somewhat akin to forging is extrusion. As the name indicates, this is forming a linear part of constant cross-section by squeezing it like toothpaste through a die of the correct shape. It has been used for many engine parts, including rings used to stiffen casings, which of course require bending to circular shape and then joining the ends. Most metals are not difficult to extrude, but steels were a challenge until the French Ugine-Séjournet firm discovered 40 years ago that molten glass could be used as the lubricant.

In casting, the metal is melted and run as a liquid into a mould. Thus, the problem of shaping a refractory (heat-resistant) alloy is side-stepped. Very hot parts, such as the flaps of an afterburner primary nozzale, used to be welded from sheet, but today are more cheaply cast in one piece. Casting is also used for making such parts as aluminium gearbox casings. In traditional casting the mould is produced in sand by a pattern which is a replica of the part to be made. In die casting a permanent mould is used. A particular form of casting much used to make circular parts is centrifugal casting. Here the die is usually water-cooled metal, for faster solidification, and it is rotated at high speed on a vertical axis to give a finished part of high density devoid of flaws.

For turbine blades the most important method is now investment or 'lost-wax' casting. Used by the Chinese around 2000 BC, it begins by making a multi-piece steel die containing a highly polished internal cavity having the exact inverse shape of the finished part. Molten wax is carefully injected to fill the die completely, and allowed to set to produce a replica of the finished part. Several – typically from 2 to 20 – of these identical patterns are then assembled on a 'wax gating tree' in Christmas-tree style. This is then dipped in slurry, a liquid ceramic, which quickly dries. The tree is dipped several more times until the ceramic coat is about 6 mm (0.25 in) thick. The wax is then melted and run out, care being taken to ensure that every ceramic shell mould is completely free from wax by firing it at over 1,000°C. The red-hot mould is then filled with the blade alloy, which has been electric-induction melted and brought to an exact temperature. After cooling, the ceramic shell is removed and the blades are cut away from the cast gating tree, chemically cleaned, and carefully inspected by numerous methods. Investment casting is used for many engine hot parts, but it is especially important for turbine blading, and how modern blades are made is outlined later.

Engine parts are joined by several methods. Wherever a joint may have to be dismantled there is little alternative to drilling holes and fitting bolts and nuts, the latter being of one of the patented 'stop' types that cannot work loose in service. Accessories – even some large items such as gearboxes and oil tanks – and bleed connections, fuel injectors, cover plates, and many other items are retained by similar stop nuts held (like all others, with the correct torque) on studs welded to the engine casings. Many engines contain numerous countersunk screws. These always have cross-type heads, such as the Parker-Kalon series. Locking wire, used by the mile in the Second World War, is now seldom seen.

Where parts will not need to be separated, the almost universal answer is some form of welding or bonding; rivets are rare. As in so many processes, human welders have tended to be replaced by computer control. Once the software has been perfected, this should mean absolutely repeatable joints without the shrinkage or distortion that was previously hard to avoid. Among the more traditional forms, used to join sheet metal, are spot and seam resistance welding. In spot welding, twin electrodes, usually of copper, are brought lightly together at each joint and a large DC current passed to cause intense local heating; almost at once the desired temperature is reached, the current is switched off and the electrode pressure increased to make the joint. In seam welding, the parts are moved past copper wheels which by the same process make a continuous joint.

Traditional manual arc or gas welding is almost extinct in engine production, but considerable use is made of tungsten inert-gas (TIG) welding. The electrode torch, which forms a DC cathode, is made of consumable thoriated tungsten and fitted with a 'gas lens' which shields the weld with inert argon, even though in many cases the entire operation takes place in an atmosphere of high-purity argon. To avoid the workpiece cracking, the arc is started without torch contact and the finishing current is tapered off in a programmed manner.

Electron-beam welding (EBW) faintly resembles an electron microscope, in that a beam of electrons is focussed magnetically on the target in an evacuated (vacuum) chamber. It differs in the much higher power of the beam, the weld being made by the heat generated by the electron impacts. Under computer control, perfect welds can be made between dissimilar materials with virtually no distortion or shrinkage. EBW is the preferred method for such tasks as joining the turbine nozzle vanes to their rings, or the discs, rings, and spacers of a compressor rotor,

EBW compared with a traditional gas/arc weld.

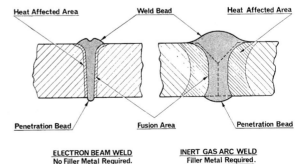

or attaching the HP turbine stub shaft (made of stable bearing steel) to the expandable material mating with the turbine disc.

Plasma-arc welding differs from traditional arc welding in that the arc carries with it ionized gas aimed at the weld through a small nozzle. The gas travels at several thousand mph at a temperature of 12,000–18,000°C to produce a weld that is much finer, more accurate, faster and distortion-free, and needs less filler material. Plasma welding does not need a vacuum chamber (though the weld is protected from oxidation by inert gas) and can be done manually, so it is handy for irregular joints, especially where only a few welds must be made and automated operation is not justified.

Diffusion bonding (DB) is a seemingly simple process in which metal parts are heated in a vacuum furnace to the diffusion point, just below the melting temperature. At this point the metal molecules are mobile, so that when the parts are brought together under high pressure the molecules migrate in both directions across what had been the interface. In effect, the two parts become a single piece of metal, even though the two components might be of dissimilar metals such as a high-nickel alloy joined to a cobalt-based alloy. As there is no local heating, there is no stress or distortion. DB is often combined with superplastic forming (SPF), using balloon-like inflation by argon between hot dies to produce startlingly complex engine and nacelle parts which would otherwise have to be made by joining numerous smaller components. An alloy particularly amenable to SPF/DB is 6Al-4V (6 per cent aluminium, 4 per cent vanadium) titanium.

Of course, the ubiquitous laser is also capable of making superior welds, and laser-beam welding (LBW) is rapidly gaining in popularity. One reason is that it offers all the advantages of EBW without the time-consuming need to 'pump down' a vacuum chamber, so it saves time in series production. Typically, a high-power (eg, 2 kW) continuous CO_2 (carbon-dioxide) laser is used, under open-shop conditions.

Inertia bonding is an economical way of joining circular parts such as shafts and discs. One of the parts is spun at high speed whilst attached to a heavy flywheel. At an exact predetermined speed it is suddenly forced under high pressure against the mating surface, which is fixed. The instantaneous friction heat and pressure bonds the parts without melting, to give a joint with outstanding physical properties, though finish-machining will be needed to remove a ring of metal squeezed outwards at the joint.

Many of the principal parts of an engine require heat treatment. This can involve placing the part in a furnace and bringing it at a controlled rate to a programmed temperature, possibly holding this for a specified time, and then cooling it at a prescribed rate. A common form is annealing, in which the temperature reached and subsequent slow cooling are designed to reduce internal stress caused by welding, machining, or other operations, or to reduce hardness and thus improve machinability, facilitate cold-working, or produce a desired microstructure or physical property. Solution heat treatment, for example using a bath of molten salts, holds the part at a particular high temperature long enough to allow one or more of the constituents to enter into solid solution, followed by rapid cooling by some form of quenching so that the constituents are held in a supersaturated unstable state. Many heat treatments have as their objective overall hardening of the part.

In carburizing, the part is heated for a timed period in a furnace with a source of carbon atoms. These migrate into the surface of the workpiece and make its outer layer significantly harder and more wear-resistant. Another form of surface-hardening is to apply a very hard solid material, such as tungsten carbide, chromium carbide or Nichrome, to a particular region subject to wear, such as the abutment faces of mid-span or tip shrouds. The material can be melted on from a welding rod using a manual TIG torch. Another method is plasma spraying, in which the facing material is a fine powder which is mixed with argon and blown in a controlled manner on to the work by an electric arc. As an alternative, a defocussed CO_2 laser can be used to blow the powder on to the work, building up the coating in a series of overlapping beads.

Other parts are protected by plating, in which a dissimilar metal is deposited on the surface in the form of ions (electrically charged atoms) transported by electric current in a bath of liquid electrolyte. Corrosion resistance is improved by cadmium or nickel-cadmium plating, splines and sleeves are protected against galling and fretting by a coating of silver, and seal faces are given enhanced abrasion-resistance by chromium. Turbine blades can have ceramic thermal-barrier coatings, as described later. Some parts have a skin of nickel or gold, plated in layers to provide a filler for subsequent brazing.

Brazing is akin to soldering, but the jointing metal is run in as a melt into a joint space between two parts heated in a furnace. The technique is widely used to bond metal honeycomb cores to a metal skin on each side. A particular application is manufacture of Transply, patented by Rolls-Royce for making combustion-chamber liners. Two sheets of refractory metal are chem-milled on one side with a network of grooves like city streets, with small holes drilled at various junctions. These etched faces are then brazed together to form a sandwich. Cooling air enters through the holes in the outer sheet, but then has to pass along the passageways in order to reach a hole from which it can escape through the inner sheet. Thus much more cooling is obtained from less air.

Other kinds of sandwich are increasingly being used to save weight in the largest engine-related structures such as inlets, pod cowls, reversers, and access doors. To make large sheet structures adequately rigid requires heavy gauges of material and heavy reinforcing members, especially round the periphery. Ill-fitting cowl panels with projecting edges and gaps can double the drag of the complete installed engine. Sandwich panels, either of composites or made from thin metal sheet stabilized by a core of aluminium or stainless-steel honey-

comb, provide adequate strength and rigidity for a fraction of the weight, and also suppress noise.

Today the most exciting developments are probably the progressive introduction of non-metallic composites and ceramics. In fact, some engineers believe that early in the next century it will be economic to build almost non-metallic engines!

Composites are materials made up from two or more dissimilar components. In practice, almost all are fibre-reinforced. They can be regarded as either strong plastics reinforced by adding even stronger fibres, or as masses of fibres stuck together by adhesive. Compared with metals, most are softer, and so are less able to withstand erosion (for example when taking off in a sandstorm), and they also cannot at present be used for the hottest parts of engines. Within these limitations, they can enable parts to be much lighter for equal rigidity, cheaper to make from fewer parts, structurally more efficient, and also better able to damp out noise.

Rolls-Royce pioneered such construction with the RB.162 lift jet of 1962. To reduce weight to a minimum, this very simple turbojet, used at take-off and landing, had almost all the cooler parts made from resin-bonded glassfibre

Early combustion chambers and flame tubes were welded from thin sheet, with cooling holes or 'wiggle strip' joints (left). Today's liners may be machined rings, with many thousands of laser-drilled holes, or incorporate Rolls-Royce's Transply construction (lower right).

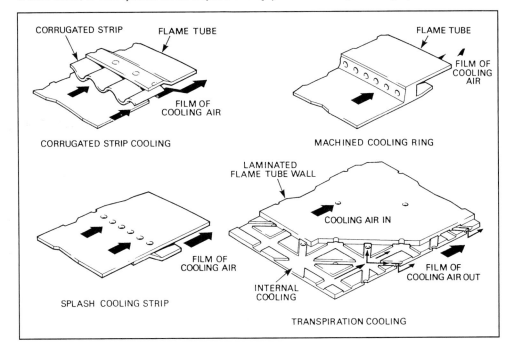

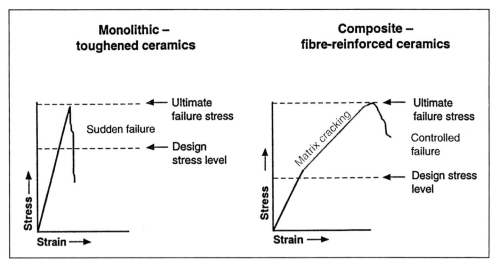

Whereas monolithic (homogenous) ceramics can be incredibly strong, and made tougher by fine particles, they suddenly snap when overloaded (left). Composites display metal-like linear elastic behaviour to the point where microcracks appear in the matrix, failing thereafter in a controlled manner (right).

reinforced plastics (GFRP). Subsequently, though GFRP remains important, generally stronger and much more rigid composites have been introduced, notably CFRP (carbon fibre), graphite fibre composites, and Kevlar, in which the filaments resemble spider-web material.

All use fine filaments ranging from one to three thousandths of an inch in diameter, which is generally finer than human hair. All far surpass the strongest metals in specific tensile strength (strength divided by density), but they cannot be used alone. They have to be bonded together by an adhesive, of which the most common are various epoxy resins. The fibres are first wetted with the resin and then made into the part by various methods. One technique is first to form strips or sheets called prepregs. Pieces cut from these are then assembled in a stack and bonded together. Alternatively, a single filament, or rather a cable of some hundreds of filaments or a flat tape made from thousands, is then wound all over a mandrel – a former having the shape of the finished part – until the required thickness is built up.

Filament winding is always used for such items as high-pressure containers and rocket-engine nozzles, and is increasingly used for drum-type structures such as bypass ducts. For example, the bypass duct of the GE F404, formerly a distinctive ribbed component made by chem-milling titanium, is now made lighter and more cheaply from a CFRP composite.

One of the many advantages of such composites is their record specific strength and stiffness, which, compared with metal, can reduce a part's weight by 20–50 per cent. Another is that the axis of the fibres can be tailored to lie parallel to the direction of the applied load. In prepregs all the fibres lie parallel, but when laying up the plies to form a finished part the various layers can have the fibres running in different directions. Another advantage is that manufacturing cost can be significantly lower than with metals. The chief limitation on the use of composites has been their inability to operate at very high temperatures, though in recent years several new adhesives, notably polyimides, have extended the upper temperature limit from some 140°C to at least 290°C.

In the 1950s, before CFRP was invented, engine firms were experimenting with composites based on aluminium. The reinforcement comprised either long filaments or small 'whiskers' of Bo (boron), SiC (silicon carbide) or Al_2O_3 (alumina). The author has on his desk a test piece of Al/SiC produced at Derby in 1959. It looks much like a bar of aluminium, except at the end broken in tensile testing, where about a million fine filaments protrude slightly. The Al matrix limits operating temperature to about 400°C, while glasses (which interface extremely well with SiC) can often go higher. Using titanium as the matrix extends the upper limit to about 800°C, with 1,000° in prospect, usually using alumina fibres. For higher temperatures recourse must be made to ceramics.

Germany pioneered ceramic turbine blades during the Second World War. Over 30 years ago the Ford Motor Co predicted that they would sell millions of cars with simple gas-turbine engines with hot blading of SiN (silicon nitride) 'as cheap as sand on the beach'. It has yet to happen, but today many ceramic-bladed turbines are on test around the world, and for the more distant future engineers look to ceramic-based composites.

Modern engineering ceramics have little in common with china and pottery, but comprise a variety of oxides, nitrides, carbides, and borides. So far, almost all test running has taken place either with coatings on metal or with monolithic parts (ie, single chunks). By far the most experienced ceramics are SiC and SiN, which at over 1,000°C are stronger than any of today's superalloys, besides having a density of only about 3.2 compared with 7.9. They would be magnificent were it not for their rigid crystalline structure. This makes them far more sensitive than bulk metal to flaws such as crystal dislocations. They thus have poor impact strength, and if for any reason they are overloaded they cannot begin to yield plastically, but fracture.

Engineers are using powerful electron microscopes, especially of the scanning type, to find out more about the crystalline structure of these potential wonder materials. Thus they hope to make them less brittle, and more damage-tolerant. In parallel, they are working on fibre-reinforced ceramics. In addition to carbon fibres, many carbides, nitrides, and oxides are being investigated as reinforcing fibres or whiskers in ceramic matrices, one mix being SiC/SiC, the fibres being the same material as the bond.

The consensus of opinion is that in 15 years' time new military engines, at least, will have composite compressors and ceramic turbines, and that hardly any metal will be found anywhere. Of course, other engineers are trying to make useful parts out of single-crystal metal, as has already been done with turbine blades. That would revolutionize every engineering structure.

There is one more technique to be outlined: sintering, or powder metallurgy. Since 1970 many of the most refractory metals have had such high melting temperatures, and have remained so strong at white heat, that they have been almost impossible to form into parts such as HP turbine rotor discs. Today, powder metal-

lurgy is a mature shop process; GE recently announced that well over 10,000 of their engines in service had powder-metal rotor parts. The ultra-clean powder is first extruded into dense fine-grained billets, which are then isothermally forged to produce the desired disc shape. The objective is to forge an almost finished disc, needing little subsequent work other than broaching the blade-retention slots.

The promising alloy Mg-Li (magnesium-lithium, the lightest of all engineering metals) is a special case. The elements are melted in an atom ratio 60/40 and the melt is then pulverised by a high-energy stream of argon to produce particles from 0.02–0.25 mm diameter. These are then compacted under heat and pressure to the exact finished form of the part. With a density of 1.3, the resulting parts appear almost to float in air, yet the material retains strength to about 200°C. Mg-Li can hardly fail to be important in future engines.

In small engines, a relatively high proportion of the bulk and mass is taken up by inlets, gearboxes, and accessories, and even by such ancillaries as inlet filters and jetpipe muffs for infra-red countermeasures (IRCM). In contrast, in large engines more than 90 per cent of the mass is composed of parts that are either exactly circular or arranged radially so that their tips form a perfect circle. Accordingly, rings, discs, and circular casings account for a high proportion of each engine.

Such parts had played only minor roles in aircraft piston engines, and for 20 years from 1937 aircraft gas turbines suffered from inexperience in design techniques, reliance on a very small range of special alloys whose fabrication was possible only by costly and labour-intensive

Today we are firmly on the road towards the non-metallic engine. This forecast for a next-generation fighter engine is by Prof Peter Hancock of Cranfield. CMC = ceramic matrix composite; MMC = metal; PMC = polymer.

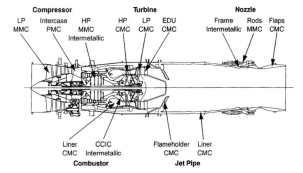

Few people, even engineers, would suspect that the discs of the GE90 HP turbine were made out of powder. The material is René R88DT, which for a given stress and creep life gives a margin of 17–53°C over R95 and over 110°C over DA718 (General Electric).

methods, and the inability of production test and inspection methods to guarantee the absence of flaws in manufacture.

Even such seemingly undemanding parts as the rotors (impellers) of centrifugal compressors posed problems. Despite long experience with piston-engine superchargers, the first turbojets required impellers much larger than any previously made, and running at speeds which pushed the technology to the absolute limit. The only answer was painstaking machining from a forging in a closely controlled aluminium alloy. Great care was taken to achieve optimum properties in all radial directions, and after machining to the finished form the surface was either chemically etched or physically polished to remove any traces of tool marks or scratches, which by locally concentrating stress would promote the start of a crack which would spread by fatigue until the part failed catastrophically. Some designers calculated that boring a hole down the centre would result in hoop stress at the edge of the hole close to the material's safe limit, so they left the impeller solid and drove it by a flange bolted on the end (even these small bolt holes were deeply worrying). This made it

difficult to inspect, and many impellers exploded at take-off rpm because of a forging flaw lurking in the centre of the mass.

As noted earlier, today designers have the option of making a centrifugal impeller from titanium alloys. Although of greater density than aluminium, the considerably higher material properties enable the thicknesses – what old-time engineers call scantlings – to be reduced, so that the part may well come out lighter. This reduced bulk, coupled with vastly improved non-destructive testing (NDT) methods, can today guarantee the absence of manufacturing flaws. This, combined with the increased strength of the material, enables modern impellers to have radial vanes that are curved instead of straight, giving higher efficiency, and to run at much higher speed to achieve pressure ratios around 10.

Quite apart from the desperate problems of their aerodynamics, early axial compressors were often inelegant mechanically. It was not uncommon to find discs of steel with a large central hole shrunk on to a heavy drum and carrying blades of steel or, for naval aircraft, aluminium-bronze riveted round their periphery. Some engines, such as the Armstrong Siddeley Mamba and Double Mamba, even had the blades riveted between pairs of discs. To make the rotors of some engines, hollow forgings had to be produced with a length up to almost a metre and an outside diameter of 0.5 m (20 in). Such a forging might weigh a ton, and more than 90 per cent of it would be laboriously machined away during manufacture.

Today's axial-compressor spools are invariably assembled from discs and spacers. Each disc (US spelling, disk) is very thin but has a strong, thick inner periphery. This may be attached to the drive shaft, but in two-shaft engines the disc hole is comfortably larger than the LP shaft passing through it. Some rotors are EBW'd into a single spool, while most of the large ones have two or even three bolted joints. The precision, which is demanded by active clearance control between the casing and the tips of the blades, would have amazed designers of even the late 1960s. It is maintained by careful temperature control, using internal airflows for both local heating and cooling.

Compressor casings are machined from centrifugal castings or forgings. Almost always they are split longitudinally with bolted joints into upper/lower or left/right halves, and often also into front/rear or even front/centre/rear, the material varying from aluminium at the inlet to titanium or high-nickel alloy at the delivery. It is worth interjecting here that joining dissimilar metals, anywhere in the aircraft, must be done with care if chemical or electrochemical attack is to be avoided. Some form of passivating or barrier material must be added at the interface. Variable stators can be held in plain bearings in raised turret-like projections on the outer side of the casing, while fixed stators are slotted into dovetail grooves and retained by set screws or other locking devices. Usually the stators are EBW'd into groups, and the casing has to be formed from inner and outer shells. This increases rigidity, and provides collection chambers for bleed air tapped from various stages. Internal bleeds are often led through inward radial pipes.

The basic structure of the engine, in which the main shaft bearings are carried, is usually made in various materials by flow-turning or other methods in a number of bolted sections. Each main bearing occupies a chamber (US term, a sump) with controlled air pressure and temperature, and a metered oil supply. The entire carcass of modern fighter engines has to be loaded in giant 'torture rigs' to ensure that no blade rubbing will occur even in $9g$ manoeuvres.

Combustion chambers are made from refractory alloys. Few engines use simple wrapped and welded sheet. Centrifugal casting, wrapped/welded extrusions, flow turning and forging are all found, invariably with final machining or grinding to shape and with welding used to join the various rings and sometimes also the HP turbine nozzle ring. Lasers or electrolytic methods are increasingly used to drill the complex patterns of holes that provide air film cooling, and some liners have ceramic or other refractory coatings, applied for example by plasma spraying.

Turbine rotor discs were traditionally made in special steels, typically with 13 per cent each of Ni/Cr/Mo, 10 per cent Co (cobalt) and smaller amounts of several other metals. The blank would be forged in a large press and then machined, and the fir-tree slots made by broaching. By the 1960s high-nickel alloys were being used, and today increased rotational speeds have been made possible by making discs by expensive powder metallurgy in exotic defect-tolerant alloys, the extruded blanks then being isothermally hot-forged almost to the finished shape. Usually the disc is forged integral with a stub shaft which may be of a different alloy.

Turbine nozzles (US term, vanes) have to withstand the highest temperatures anywhere in the engine, but remain stationary. Thus their manufacture is dominated by heat-resistance, and they are generally cast in small groups in high-nickel alloy. Inner and outer endwalls are shrouded to form continuous rings, through which cooling air is passed to emerge from numerous holes. These holes are closely spaced all over the leading-edge region to form a protective air film; radial rows face downstream further aft, and the rest of the air is discharged from the sides of the trailing edge. The vanes or groups are located in ways that permit thermal expansion, especially in a radial direction. Some HP vanes have ceramic coatings, and they are obvious candidates for all-ceramic construction eventually.

HP turbine rotor blades are one of the greatest challenges ever faced by engineers. Today they must withstand gas temperatures higher than the melting point of the metal of which they are made, with occasional violent thermal shocks (sudden changes in temperature) whilst under colossal tensile and bending stresses, and go on doing this for, in some cases, over 25,000 operating hours.

In the Second World War, when 100 hours was an acceptable life, British designers carved blades laboriously from austenitic-steel or high-nickel bar stock, while the Germans used tubes or tapered-thickness sheet wrapped and welded or deep-drawn to the aerofoil profile so that cooling air could pass down the centre. By the mid-1950s improved alloys had enabled entry gas temperature to rise from about 750° to 950°C, but further progress was limited by the creep strength of the material. The obvious answer was to try to incorporate internal passages for cooling air, which – discounting the sometimes large differences in pressure across the gas flow – would be automatically pumped from root to tip by centrifugal force.

Rolls-Royce pioneered cooled blades, and was testing crude wrapped/welded blades in 1952. Next, fantastic cooling was achieved by adding a small proportion of finely atomised water to the cooling airflow, but for various reasons this came to nothing. By 1954 blades were being forged in Nimonic alloy with various bulges running the length of both faces. Holes were then drilled the length of the blade in these bulges so that, when the blade was forged to the correct aerofoil profile, the previously circular holes formed narrow passages for cooling air.

These cooling passages were difficult to make precisely repeatable, and their narrow width limited the speed of the air. To do better, a block of Nimonic was drilled with numerous accurately positioned holes, each of which was then filled with a steel pin. The block was heated in a furnace and extruded with glass lubricant through a die of sausage or banana profile, the holes and pins becoming flattened ovals. The blade was then machined to the exact profile, and the steel pins were leached out by hydrofluoric acid. RR persisted with such blades into the 1970s with early RB.211s, but finally recognised that wrought blades were becoming increasingly difficult to produce in alloys specially designed to retain their shape at high temperatures.

In 1966 RR had bought Bristol Siddeley Engines, and since 1950 Bristol had been successfully making blades by investment casting. Since then, HP blading has progressed along three routes: new alloys, new methods of casting, and ever-better internal airflow. Until the late 1970s the alloys were based on nickel, containing such elements as aluminium, titanium, and hafnium. Some of the alloying metals combined readily with atmospheric oxygen when in the molten state, so by 1960 the blade alloys were melted and cast under vacuum conditions. Today's alloys are often based on nickel and cobalt, and the latest incorporate small amounts of yttrium for enhanced oxidation resistance and rhenium for extra strength.

Apart from eliminating the atmosphere, casting techniques have been revolutionized by controlling the way the metal solidifies from the melt. Ordinary bulk metals cool from the liquid phase to the solid by separating out billions of discrete crystals, which are randomly orientated (as in a sack of apples the stalks point in random directions). Each crystal grows into its neighbour, so that there are no gaps, but the strength of the joints is a small fraction of the strength within each crystal. Thus, ordinary equi-axed cast objects are far weaker than their theoretical strength. The first advance, made from 1972, was directionally solidified (DS) casting. The cooling blades were carefully drawn down out of the casting furnace into a water-cooled lower chamber in such a way that the crystals tended to form columnar grains extending along the entire length of the casting. Thus, grain boundaries across the blade, perpendicular to the principal tensile stress in service, were virtually eliminated. This at a stroke enabled high-temperature creep resistance and blade life to be greatly enhanced, giving the designer a choice of higher gas temperature, increased rpm or extended blade life, or any combination of these.

The ultimate in casting appeared to be reached around 1980, when the first single-crystal (SC) blades were produced. The exact processes are commercial secrets, but clearly involve modifying the alloy to eliminate elements which tend to strengthen grain boundaries, and also to introduce cunning modifications into the DS process so that the very first crystal to separate out from the melt can be persuaded to grow throughout the melt. Of course, every detail is computer-controlled. SC blades simply look and feel like very nice

Three as-cast HP turbine rotor blades, before any machining; from left to right: traditional multi-axed, directionally solidified, and single-crystal.

 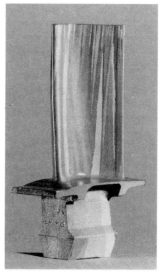 

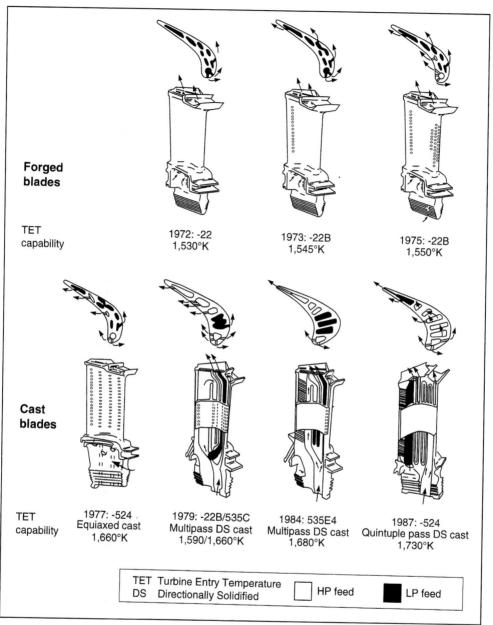

Forged blades

TET capability

| 1972: -22 1,530°K | 1973: -22B 1,545°K | 1975: -22B 1,550°K |

Cast blades

TET capability

| 1977: -524 Equiaxed cast 1,660°K | 1979: -22B/535C Multipass DS cast 1,590/1,660°K | 1984: 535E4 Multipass DS cast 1,680°K | 1987: -524 Quintuple pass DS cast 1,730°K |

| TET | Turbine Entry Temperature | | HP feed | LP feed |
| DS | Directionally Solidified | | | |

Rolls-Royce has come a long way since it put the first forged aircooled turbine blades into production in 1956. Here are the blades used in civil RB.211 turbofans (Rolls-Royce).

bits of metal, and apart from their featureless surfaces give few clues that they are such a technological miracle. It might be added that with SC technology it should be possible to build a cross-Channel bridge, or an aircraft with a wing of 747 thickness and a span of 1,000 ft, should such structures be needed.

The third route, improving the internal cooling airflow, has demanded a massive research effort in the creation of ceramic cores. In this case the ceramics do resemble traditional pottery, but the materials had never previously been made with engineering precision in forms which began complicated and got steadily worse. An accompanying drawing shows in simplified form how the cooling of Rolls-Royce HP blades has been improved over the period 1960–85. At one time air impingement on to inner walls was important in some engines, but today the main objective is to envelop the blade so completely in a film of cooling air that it remains unaware of the white-hot gas rushing past it.

Cooler outer casings can be made of aluminium alloy, and may eventually be ultralight Mg-Li. In 1963 Pratt & Whitney began making bypass ducts for military engines with deep chem-milled rectangular pockets to leave a thin wall stiffened by an integral grid. Today this technique is seen on the latest fighter engines such as the P&W F119 and Eurojet 200, though the stiffening grids are now coarser and triangular. Rolls-Royce has pioneered the use of CFRP for the outer casings of fan ducts, starting with the Tay; the F404 has now followed suit.

The casings over the hot parts of the engine are usually made with the highest attainable dimensional accuracy in material selected as much for its low coefficient of thermal expansion as for its heat-resistance. They can be forged, rolled, flow-turned or centrifugally cast, and finish-machined with integral stiffening. Where active clearance control is fitted, the cooling air was originally fed through plain surrounding pipes, in some cases of square section, but today the manifolds are tending to be integral with the casing structure.

The core jetpipe has to be titanium or a nickel alloy. For higher propulsive efficiency some engines mix the core and bypass flows, and the resulting integrated nozzle is a bonded honeycomb to combine adequate rigidity with light weight. Noise-suppressing construction is outlined in the next chapter.

Afterburner jetpipes are usually double-skinned, with cooling air induced between the inner and outer shells by the ejector action of the jet. Variable afterburner nozzles are a major challenge, demanding the production of large parts in the most thermally resistant materials possible, with the highest dimensional accuracy, especially around the mounting and actuation bearings, and strong enough to resist severe gas loads at white heat. Powder metallurgy and ceramic-matrix composites are among the answers for future engines.

There remains the question of fan blades. In the earliest 'bypass turbojets' the question hardly arose, but American designers soon showed the way to go, and in the late 1950s the BPR began climbing. In the JT3D Pratt & Whitney used fan blades machined from titanium forgings, and this quickly became the universal method until, in 1968, Lockheed picked the RB.211 for its future L-1011 TriStar. A key factor in the choice was a dramatically superior fan made of CFRP. As related in Chapter 16, this failed to be certifiable, and Rolls-Royce had to fall back on the traditional type of blade of solid titanium.

What killed the CFRP blade was inadequate resistance to foreign-object damage (FOD) ranging from hailstones to birdstrikes. This was a major blow, because CFRP promised to free the designer from the constraints of solid metal. Perhaps the most severe single requirement in the certification of an airline turbofan is fan-blade off (FBO). The test engine has to be run at maximum power and an explosive charge fired at the root of one of the fan blades, severing it from the disc. The fan case has to contain the severed blade and all subsequent debris. In a poor design this could include most of the engine structure, because the severed blade could break off those next to it, the out-of-balance force then wrecking the engine.

The objective is to sever the blade without stopping the engine from running and giving useful power. Containment demands a surrounding case of great strength. Even if the blade weighs only 7 kg (15 lb), its impact on the inside of the case has been likened to 'a car travelling at 40 mph'. The blade could hit anywhere, and making the entire case contain such kinetic energy – possibly not from one blade but from a succession – puts up the weight alarmingly, because this is the biggest part of the engine. There is thus tremendous pressure to make the blade as light as possible, and if it is solid titanium the only possible way to do this is to reduce the chord.

A further restriction on chord is imposed by the strength of the hub, which like the blades is machined from a forging in titanium alloy. At take-off rpm even a 6.8 kg blade exerts a centrifugal pull of 60 tonnes. At the back of the root the load is increased by the bending load caused by the thrust that is propelling the aircraft. Any attempt to increase blade chord imposes such a load on the disc, simultaneously increasing the size of the blade-retention slots, that very soon there is no way of getting enough metal there to retain the blades, and the disc cannot be designed.

The answer thus had to be to make the blade long and narrow. Such blades are aerodynamically unstable and prone to flutter, which would very quickly cause them to fail in fatigue. To prevent this, they had to be stabilized by adding a snubber or clapper at about two-thirds of the way from root to tip, which complicated the forging, added weight and formed a blunt barrier to the transonic airflow. For a given thrust, this demanded an increase in fan diameter, making the whole engine heavier, and fan weight was further increased by the large number of blades needed, typically 30 to 50.

The CFRP fan developed at Derby from 1966 tantalisingly showed the great advantages of switching to a low-density composite. The pro-

The first wide-chord snubberless fan blades to go into service, on the 535E4 in 1984, were made by diffusion bonding thin skins on each side of a honeycomb core (Rolls-Royce).

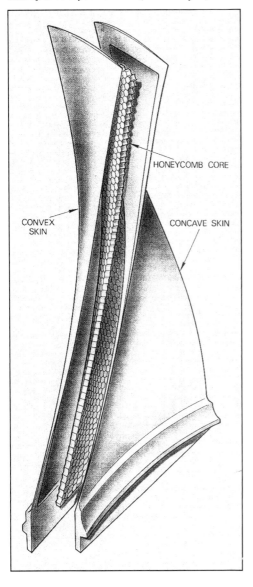

HONEYCOMB CORE

CONVEX SKIN

CONCAVE SKIN

totypes of the RB.211 had only 25 blades, each of wide chord but very thin – so-called lenticular form, with sharp edges – and inherently stable and thus needing no snubbers. Each was made by laying up numerous plies of prepreg, each with the axis of the fibres carefully chosen, and the final pack bonded under heat and pressure to cure the resin adhesive. The result appeared superb, aerodynamically much more efficient, and half the weight of the narrow titanium snubbered blade.

Eliminating snubbers increased the efficiency of the blade as a lifting surface, eliminated the extra blockage and profile drag of the snubber, avoided the shockwave interaction in the accelerated flow trying to get past the snubber, and by removing unwanted mass enabled the blade inboard of the snubber to be thinner. Among other benefits was improved core performance, wider surge margin and, as explained in the next chapter, significantly less noise.

With such goals so nearly within their grasp, Rolls-Royce kept trying. There was nothing wrong with the high-modulus carbon fibres, but the resin bond was relatively weak. The best way to go seemed to be to try to make hollow metal blades, somewhat resembling the hollow steel propeller blades of 40 years previously. After years of research at Derby, the first wide-chord snubberless blades entered airline service on the 535E4 engine in October 1984.

The blades were all-titanium sandwiches. Inner and outer skins were made from flat plate by forging, hot isothermal forming, and then chem-milling to reduce thickness except round the edges, so that when the skins were brought together the interior would be a hollow cavity. This was filled by a honeycomb made by crimp-rolling and resistance-welding Ti strip. Each blade was then assembled by activated diffusion bonding, a patented liquid-phase process. Similar but larger blades were soon also introduced to the 211-524G/H, and smaller ones on the IAE V2500. In many millions of hours not one of these blades has failed in service.

In 1980 RR began studying alternative configurations for the super-power engine that became the Trent. The final choice was again all-Ti, but with a core in the form of a single large sheet formed into what RR call a Warren girder (in a bridge or a biplane Warren struts slope diagonally, and in the fan blade the skins are joined by sloping webs). The skins are made by SPF and joined to each other and to the core by DB. The effort involved is demonstrated by the fact that the first blades were tested in 1984 and the Trent entered service on the A330 in January 1995.

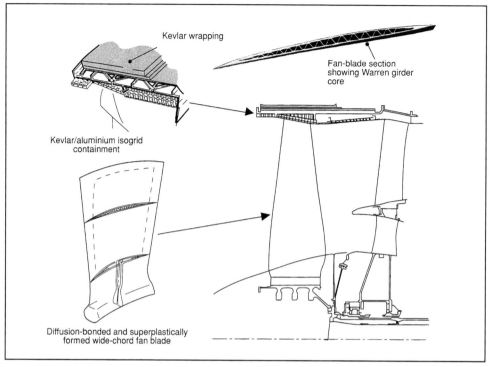

The Rolls-Royce Trent has very large fan blades made by similar SPF/DB methods but with the honeycomb core replaced by an integral Warren-girder structure. The resulting blade is some 15 per cent lighter than the former wide-chord type, and about 380 lb per engine is saved by the Kevlar/isogrid containment system (Rolls-Royce).

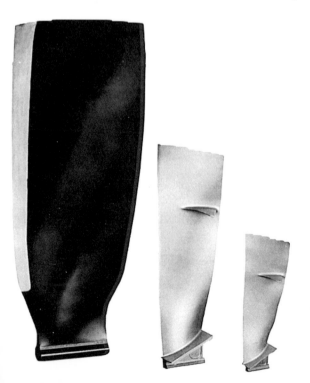

The only rival to such blades is that developed by GE for the enormous GE90. Experience with various test engines funded by NASA and the USAF gave GE confidence to switch to non-metal construction. Intermediate-modulus carbon fibres were selected, but a key development was the availability of improved resin systems incorporating toughening agents offering about three times the shear-strain capability and delamination-suppression of previous, more brittle epoxy adhesives. The huge blade is given a polyurethane coating for protection against erosion, UV light, and aircraft fluids, and the leading edge is protected by a bonded Ti guard. The attachment dovetail needs only a single outer layer of dry bearing material which holds the blade in the hub. It is expected that the repeatability of production GE90 blades will be perfected by robotic fibre placement, using the Cincinnati Milacron process previously used on airframes and large rocket casings.

A fan blade for the enormous GE90, compared with blades for a CF6-80 and a CFM56 (General Electric).

9 Installation

The earliest turbojets not only allowed aircraft designers to escape from the previous limitation on maximum flight speed, but also gave them unprecedented choice in the engine installation. Indeed, it is difficult to think of any kind of engine location that has not already been flown.

This is not the place to pontificate on different aircraft configurations, but it is desirable to comment on what the engine requires and how the aircraft designer tries to provide it. For a start, we can begin with intakes and inlet ducts and follow through to the nozzle.

For subsonic aircraft, the ideal inlet is the shortest possible, because that delivers ram air with the smallest loss in energy caused by the duct. It also should meet the condition of delivering at a uniform pressure distribution all over the face of the fan or compressor. One of the few problems with such an inlet is that on a gusty day with the aircraft parked crosswind the flow into the engine is inevitably distorted, and this can make starting difficult.

Some of the first jet fighters were in effect conversions of piston-engined aircraft, with the engine(s) mounted low in the forward fuselage. This had the advantage of a simple inlet and short duct, followed by another desirable attribute, a short jetpipe. It also tended to result in a downsloping top to the fuselage ahead of the cockpit, giving excellent forward view. On the other hand, the low thrust-line could result in large changes in longitudinal trim between flight idle and full power.

At first glance some later fighters, such as the F-4 Phantom II, might appear to have a similar installation but with the engines behind the wing. In fact the engines in these aircraft were installed at an angle, fed from inlets passing over the wing. The inclined thrust-line virtually eliminated trim changes caused by the engines, and in any case the nozzles were not very far below the aircraft longitudinal axis, though this was disguised by the way the tail was mounted entirely above the engines.

Most twin jets with the engines inside the fuselage have the nozzles side-by-side at the extreme tail. A few aircraft, notably the Lightning, had the nozzles superimposed, though in fact the engines in that aircraft were widely separated, one being in the top of the rear fuselage and the other under the wing and joined to its afterburner by a long pipe. In single-engine flight this swapped yaw asymmetry for trim change in pitch. In the Tu-105, later the

Left-hand engine installation of an F-4 Phantom, showing the variable inlet with boundary-layer bleed and large secondary flow bypassing the inclined engine.

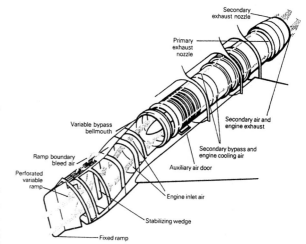

Tu-22, the engines were mounted externally above the rear fuselage, one each side of the fin spars. This was in some ways an efficient solution, especially as the entire front part of the inlets could translate forward, leaving an all-round gap to admit extra air on take-off. The penalties were slightly higher drag than internal mounting, and the need to overcome nose-down pitch at high power. The same is true of a dorsally mounted engine amidships. This was first seen in the hastily contrived He 162; the Fouga Gémaux of 1951 had two fuselages and either two small jets or one bigger one. An even stranger arrangement was seen in Boeing's first thoughts on the XB-47, with four engines inside the humped top of the fuselage, two blasting past each side of the fin.

In the Ju 287 V1 of 1944, two engines were hung under the wings and two more low on the sides of the forward fuselage. Seven years later Martin put two engines low down outside the forward fuselage of the XB-51, and a third inside the tail end, fed by an S-duct. In September 1955 the prototype Caravelle introduced an arrangement which has since become common: an engine hung externally on each side of the rear fuselage. This led in 1958 to pairs of engines; all such aircraft so far built have the engines of each pair side-by-side, though superimposed rear-fuselage engines have several times been projected. Later came three rear engines, and then one rear engine and two under the wings. In most of the modern tri-jets the centreline engine has been internal, but in the DC-10 and MD-11 it is above the rear fuselage in a minimum-length nacelle. This can fractionally increase propulsive efficiency, but complicates engine removal.

Some early designers found it hard to shake off piston-engine practice, and in the He 280, Me 262, and the third Meteor the twin turbojets were just hung well out under the wings. In all other Meteors a short centrifugal engine was bolted between the wing spars and encased in a nacelle centred on the wing. Engines with double-sided centrifugal impellers have to be boxed inside a plenum chamber, so the Meteor had no inlet duct as such, but merely let the air flow past the front spar into the engine. The nacelle was not significantly larger in diameter than the engine, but it was soon realised that its aerodynamics were poor, and in the Meteor IV the nacelle was not only given a bigger inlet and nozzle, matched to the more powerful engine, but was also made much longer at the front and rear. Later it was realised that the inlet was still not quite large enough.

In passing, in 1943 Michael Daunt learned the hard way that it is possible to get sucked into a Meteor inlet. That fine test pilot lived a further 50 years, but with other aircraft, such as the A-7 Corsair, some victims were less fortunate. Several former Soviet ground crew are alive today because of the inlet screens on their powerful fighters. Ingestion of foreign objects, from people and birds to stones and spanners, is often a cause for concern among designers. In the case of 'gravel-airstrip' 737s the answer was a powerful air jet, fed by compressor bleed, to blast down from a pipe ahead of each inlet.

In 1942 the Bell Airacomet was powered by engines tucked under the wing roots, to combine accessibility with absence of asymmetric problems. This became a common layout, with the occasional oddball in which the engines were above the wing roots. For larger aircraft, British designers concluded that the best installation was actually inside the wing roots, the inlet ducts and jetpipes passing through 'banjo frames' in the spars (sometimes structurally a $\cap$ closed by a 'dogbone' across the bottom) and the engines being between the spars for easy removal or, as in the Victor, behind them. This attractive low-drag installation was fine for slim engines and wings of sufficiently large root profile, but it went out of fashion and would be nonsensical with high-BPR turbofans.

A few smaller aircraft, such as the McDonnell XFD (FH-1) Phantom, tried the same idea with small fighter wings, because they had particularly slim axial engines. In 1943 de Havilland flew the Vampire, with wing-root inlets ducted to an engine in the fuselage. When Australia and France built Vampires with Nene engines with double-sided compressors they at first added ungainly 'elephant-ear' inlets above the fuselage to feed the rear face of the impeller. Later Nene-engined aircraft managed with single inlets either in the wing roots or in the sides of the fuselage. Some aircraft, such as the Sud-Ouest Espadon family, even had a single ventral inlet ducted up to the plenum chamber. The Sea Hawk had a unique bifurcated (divided) jetpipe leading to a half-size nozzle at the end of each wing root. This enabled an additional tank to be put in the rear fuselage.

A few four-jet bombers used superimposed engines. In the Su-10 all four were above the level of the wing, while in the Short SA.4 Sperrin two were above and two below. But the team that got it right was Boeing. In 1946 they wisely threw out the idea of four engines in the top of the fuselage and eventually created the Model 450, with a graceful wing of high aspect

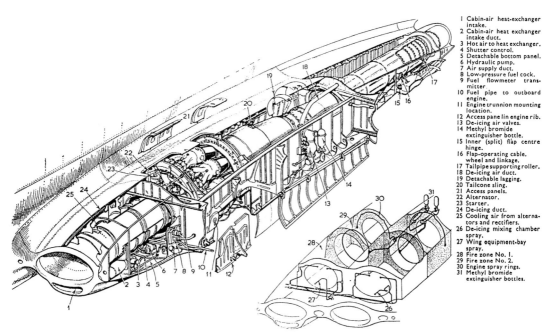

1 Cabin-air heat-exchanger intake.
2 Cabin-air heat exchanger intake duct.
3 Hot air to heat exchanger.
4 Shutter control.
5 Detachable bottom panel.
6 Hydraulic pump.
7 Air supply duct.
8 Low-pressure fuel cock.
9 Fuel flowmeter transmitter.
10 Fuel pipe to outboard engine.
11 Engine trunnion mounting location.
12 Access panel in engine rib.
13 De-icing air valves.
14 Methyl bromide extinguisher bottle.
15 Inner (split) flap centre hinge.
16 Flap-operating cable, wheel and linkage.
17 Tailpipe supporting roller.
18 De-icing air duct.
19 Detachable lagging.
20 Tailcone sling.
21 Access panels.
22 Alternator.
23 Starter.
24 De-icing duct.
25 Cooling air from alternators and rectifiers.
26 De-icing mixing chamber spray.
27 Wing equipment-bay spray.
28 Fire zone No. 1.
29 Fire zone No. 2.
30 Engine spray rings.
31 Methyl bromide extinguisher bottles.

In 1947 the D.H.106 Comet, the first jet airliner, pioneered the concept of turbojets (in this case D.H.Ghost 50s) buried inside the roots of a long-chord wing. Now unfashionable, such an arrangement has much to commend it.

ratio, swept at 35°, carrying pairs of engines in single pods inboard and single engines well outboard. The inboard pods were unusual in being hung on thin pylon struts well below and ahead of the wing. At first this radical idea caused raised eyebrows, and British designers hooked on buried engines came up with an impressive catalogue of reasons why the underslung pod, totally outside the airframe, was an unattractive solution. History – or perhaps mere fashion – has proved them wrong.

One of the most dramatically 'futuristic' aircraft was the Northrop XB-49. A pure flying wing, it had eight turbojets buried internally, with a 'letter box' inlet along the leading edge and only the ends of the jetpipes projecting at the trailing edge. An earlier version had been flown with piston engines, and the giant pusher propellers had conferred the essential yaw (directional) stability, so the XB-49 had to have fins added to replace the effect of the propellers. Apart from this it was a breathtakingly clean aircraft, so it was no small surprise when Northrop later replaced the eight turbojets by six more powerful engines, four inside the wing and two hanging underneath in external pods!

Obviously, the external pod can be hung on any wing with adequate strength; the aerofoil profile is immaterial. Inlet and jetpipe efficiency can be perfect, and installed weight can sometimes be lower than with a buried installation. Access can be excellent all round, reversers are easier to fit, and there is no problem in switching to a different type of engine. The pod and strut improve airflow over the wing, and an extremely important advantage is that the massive pod hung well out on the wing greatly reduces the wing bending moment and is even more advantageous in damping out flutter. Today Airbus is marketing the A330 with two big engines and the A340 with four small ones, and the mass distribution across the wing of the A340 enables it to fly 50 tons heavier than the 330 with the same wing.

Fifty years ago designers realised that jet engines opened up the prospect of aircraft able to fly faster than the speed of sound waves through the atmosphere. This prospect made most of them go back to school and read what theoretical aerodynamicists had been discovering since the beginning of the century. We have already seen how, at exactly the speed of sound, the basic rules for airflow are reversed, so that flow is slowed down and increased in pressure in a convergent duct and speeded up and reduced in pressure in a divergent one. This and other changes profoundly affect the design of engine installations for supersonic aircraft. For example, the lips of the air inlet can no longer be rounded, but must be sharp-edged, and the shape and profile of both the inlet and nozzle must be variable in flight.

The first supersonic aircraft, the never-completed Miles M.52, had the cockpit housed in a conical centrebody. A centrebody is simply a body in the centre of the engine air inlet, and in supersonic aircraft it serves to create oblique shockwaves which convert some of the relative air velocity into pressure in the inlet. Even sharp-edged, an ordinary 'pitot' inlet in the nose of a fuselage or engine pod inevitably falls off in pressure-recovery as Mach number rises far beyond unity, because of the strong 'normal' (90° to the airflow) shockwave across the whole inlet through which the air must pass, almost instantaneously being slowed to below the speed of sound. At Mach 1.6 the loss is 10 per cent, but by Mach 2.5 it has risen to a prohibitive 50 per cent. The answer is to make the air first pass through an oblique shock, which makes the downstream normal shock much weaker.

The ideal is to design the inlet so that the air is slowed through a succession of weak oblique shocks, and this keeps up the pressure recovery in the inlet to Mach numbers as high as 3.5 or even 4.0. This can be done by making the centrebody of Oswatitsch form, in which the angle of the cone gets progressively steeper as the lip of the inlet is approached, so that numerous oblique shocks are focussed on the lip. Alternatively, the inlet can be two-dimensional, with a centrebody of sharp wedge shape focussing the shocks on a straight lip. This lip can be nearly vertical, as in the F-106, F-4 Phantom, Tu-22M2, Su-15 and -24, B-1, and Tu-160, or horizontal, as in the F-14 and 15, MiG-25 and 29, Su-27, Tu-22M3, Tornado, and Concorde.

This is only part of the battle. The inlet has to operate subsonically, even if only at take-off and landing. Here the traditional rules apply, and the inlet should be as unobstructed as possible. Almost certainly, additional air will have to be sucked in through doors in the wall of the duct. Depending on the installation, these auxiliary doors can be anywhere around the duct, and they are often arranged to hinge open from either the front or the rear to serve as both an inlet or, in supersonic flight, to dump excess air overboard.

In supersonic flight the inlet and duct have to

A simple (but sharp-edged) inlet is adequate for Mach numbers up to about 1.5, though many aircraft, including the F-16 and F/A-18, use them at up to Mach 2. At Mach 2 such an inlet cannot have a pressure recovery greater than about 72 per cent, as the graph shows. Fast aircraft therefore use multi-shock inlets.

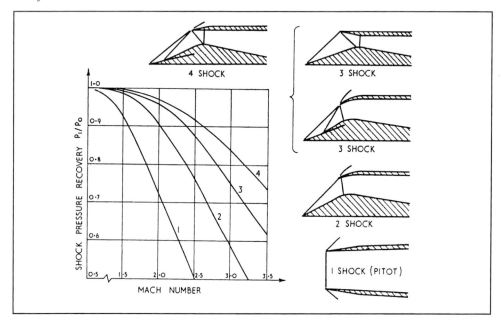

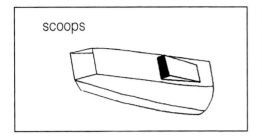

scoops

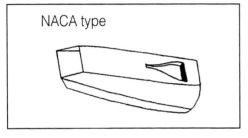

NACA type

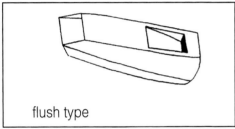

flush type

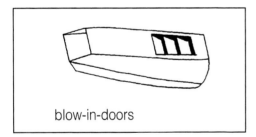

blow-in-doors

Contrasting auxiliary inlets downstream of a vertical wedge inlet (Aérospatiale).

change their shape so that the air is slowed down and increased in pressure in a convergent front portion. This ends at the narrowest part, the throat, where the normal 90° shock reduces velocity below that of sound. The rest of the slowing down and increase in pressure, with subsonic rules applying, requires a divergent section. Aircraft with a circular inlet thus have a centrebody which starts as a cone and then tapers again to a point or a sharp 'trailing edge'; it often continues as a tube faired into the hub of the engine compressor. With a rectangular inlet

one wall is made in the form of hinged panels which for supersonic flight can be pushed into the inlet to reduce the throat area to a minimum. Even with this constriction, a significant proportion of the airflow will have to be sucked out and dumped overboard, sometimes through perforations in the centrebody or through perforations or slits in the variable wall or the fixed opposite side of the duct.

With any aircraft, subsonic or supersonic, as the air flows across its surface it becomes pulled along with the aircraft by its own viscosity. The thin layer actually in contact with the skin behaves as if it were glued to the aircraft. Successive layers are less and less pulled along with the aircraft, until at a distance from the skin of perhaps 50 mm (2 in) the relative velocity is almost the speed of the aircraft. The sluggish, turbulent layer surrounding the aircraft is called the boundary layer. In airships it could be so thick that gunners on top of Zeppelins in 1916 could kneel upright and feel just a 70 mph gale around their heads. In jet aircraft the boundary layer is just what the engine does not need, so, whenever an inlet is a long way downstream of the nose, this layer is diverted away, for example by placing a splitter panel about 150 mm (6 in) away from the fuselage wall and putting the inlet on its outer side.

Some of the most difficult inlets to design are those of agile fighters. In the F-15, rectangular inlets are used with the variable throat panels and spill doors in the top, plus the unique feature of making the entire front section pivoted so that it can hinge down for take-off and low-speed flight. The MiG-29 and Su-27 have a similar geometry, but without needing the pivoted front section because they are located under the wing to obtain better recovery at extreme angle of attack (AOA). Also called α (alpha), this is the angle at which the wing meets the oncoming air). Uniquely, the Su-27 has demonstrated AOA up to about 120° in the 'Cobra' manoeuvre without compressor stall. These aircraft have inlet screens to prevent foreign-object damage whenever the engines are running on the ground. The Su-27 has a fine-mesh screen hinged up from the floor of the duct, while most MiG-29s have a perforated door which hinges down from the upper wall, air then being admitted by large louvres in the top of the duct.

In contrast, the F-16 manages with a simple fixed inlet standing well below the fuselage, reflecting the USAF's belief in uncontaminated paved runways. Such an inlet is superficially simple, but needs careful design. In the Eurofighter a chin inlet with sharp lips and a hinged lower lip is used to feed twin engines. This is far more difficult, needing an enormous amount of computer modelling, partly to ensure that neither engine can affect the other, for example in a surge. Separate inlets are easier.

Complex variable inlets were essential for such aircraft as the Lockheed SR-71 and MiG-25, which spent a significant proportion of each mission at Mach numbers exceeding 2.5. Even the complicated SR-71 system was on many occasions unable to prevent what was euphemistically called an 'inlet unstart'. The position of the huge inlet spike on this aircraft was so critical that it took years to make the control system sufficiently precise and fast-acting. An unstart could be triggered by any mismatch between inlet geometry and engine demand, or by an external disturbance, especially one moving the normal shock just outside the inlet lip. The result was a loud bang and such violent yaw that 'the airplane tried to swap ends. You got tired of the canopy thumping your head, and your eyeballs hitting their limit switches.' Yet almost all supersonic military aircraft hardly ever exceed Mach 1, so the F-16, F-18, Viggen, Gripen, Jaguar, Rafale and many other types have simple plain inlets.

In Chapter 5 it was noted that in some of the fastest aircraft the outer nozzle is part of the airframe, not the engine (though it is certainly part of the propulsion system). In the Viggen, though the main air inlets are simple, the engine was installed upstream of an ejector nozzle mainly because this left three slots round the periphery through which a thrust reverser could expel the jet. Normally these slots are open, to entrain extra air, but should a Viggen wish to go supersonic they can be closed. In the Tornado the engine itself incorporates a thrust reverser which projects behind the fuselage. In almost every supersonic military aircraft the variable nozzle projects behind the airframe, and in most cases the engine is withdrawn directly aft through the rear fuselage. Not many aircraft still require removal of the rear fuselage and tail.

Engines for jet-lift V/STOL are discussed later. The USAF so clung to the belief it would always have 10,000 ft paved runways, in some way immune to missile attack, that it has shown little interest in such engines. In recent years it has admitted that vectored thrust can be useful, not so much to reduce field length as to enhance combat manoeuvrability. In the F-15S/MTD (STOL/Maneuvering Technology Demonstrator) the engines were fitted with 2D (two-dimensional) nozzles able to vector up/down or go into full reverse. Two-dimen-

sional means the nozzle is square or rectangular, with a constant profile across its width. Pratt & Whitney developed this nozzle with inner and outer sets of flaps and petals, all in refractory alloys, able to vector in flight and even to go into reverse for violent deceleration. It assisted development of the rather different P&W nozzle for the F-22 Lightning II, in which the flaps vector the jet only over the range ±20°. The policy is 'We demand the ultimate in combat agility, even if we have nowhere to land after the mission'.

Another long-runway aircraft is the same maker's F-117A. The design of this pioneer 'stealth' warplane was determined primarily by the need for minimum radar cross-section (RCS), and this inevitably compromised propulsion efficiency. The inlet probably took longer to get right than any other part of the aircraft. The initial service standard was covered by a 25 x 15 mm (1 x 0.7 in) mesh, to admit air without allowing radar to 'see' the front of the engine. The grid was made with triangular 'wires' with the sharp edge at the front, heated to avoid icing. The long jetpipe leads to a box, triangular in plan, which changes the section from circular

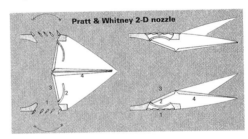

The 2D nozzle developed for the F-15S/MTD comprised: **1**, reverser vanes; **2**, arc valve; **3**, convergent nozzle/blocker; **4**, divergent/ vectoring nozzle.

Four different configurations of the nozzle of the F-15S/MTD.

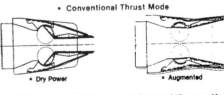

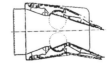

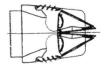

to a flat duct 1,650 x 100 mm (65 x 4 in), with 11 guide vanes turning the gas outwards and also stopping the top and bottom of this unusual nozzle from blowing apart. The nozzle is surrounded by thermal tiles, similar to those which cover the Shuttle Orbiter.

Whereas the F-117A reflects hostile radar signals from numerous flat facets, all set at carefully selected angles, the next-generation B-2 exhibits no facets but has a computer-designed shape of pristine smoothness. Well back above the wing are the inlets to the pairs of engines, each with a fixed-geometry zig-zag lower splitter plate to divert the boundary layer, and an upper lip whose angles accord precisely with those of the leading edge of the wing and the auxiliary inlets above the ducts which are open on take-off. The ducts curve down to the engines, the nozzles of which are recessed into the upper surface well forward of the trailing edge. As in the F-117A, it is difficult for defenders to 'see' hot nozzle structures. As it is considered that the B-2 could attack some targets from high altitude, visible contrails are reported to be eliminated by injecting chlorofluorosulphonic acid into the jetpipes. So far as is known, IR-homing missiles cannot lock on to the hot gas of the jet, though the flattened jets of the F-117 and B-2 must be deliberate.

British buried installations clearly would have been good from the 'stealth' viewpoint, whereas the B-52, especially in its final (B-52H) version, has a radar signature like the proverbial barn door. Buried installations become difficult to achieve with the fuel-efficient BPRs of the latest big-fan engines. The GE90 fan has a diameter of 3,124 mm (123 in), and the company's 747 testbed emphasizes the bulk of this monster engine. Such diameters restrict the range of possible installations, and even the small CFM56-3, with fan diameter reduced from 1,735 mm (68.3 in) to only 1,524 mm (60.0 in), could not fit under the wings of the 737 without flattening the underside of the inlet and nacelle!

Large engines with a BPR of 15 and upwards are increasingly going to be barred from today's favoured underwing location, except in the case of high-wing aircraft such as the Ukrainian An-70. In 1971 the prototype VFW 614 introduced turbofans mounted on high pylons above the wing, well back with the inlets at almost 50 per cent chord. This allowed the aircraft to sit very low on the ground, but failed to set a fashion. Now, however, we may see propfans installed in a similar way, though with the propellers ahead of the leading edge.

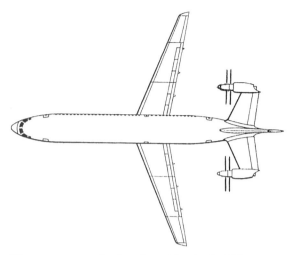

The arrangement finally adopted for the An-180 (ANTK O.K. Antonov).

Already propfans are being installed in a previously rare location on the tips of the horizontal tail. This is the arrangement finally adopted for the Ukrainian An-180, after prolonged study of different layouts. In this case the engines, contra-rotation D-27s, are installed as tractors on a fixed tailplane. In the rival Yak-46 it is intended that the pusher version of the same engine should be used, attached to the fuselage. A pusher engine is theoretically rather more efficient, and if a gearbox is used it also moves the main noise sources further behind the cabin, but the jetpipe has to discharge past the roots of the propeller blades. Traditional turboprops are also getting into the act, examples being the Avanti and CBA-123 Vector.

Other turboprops are tractor engines, and the main choice in designing the installation has been the inlet. Turboprops are usually small engines, and thus turn at high rpm, but almost all their power has to be transmitted through a gearbox with a speed-reduction ratio of anything from 10 to 30. Thus, the gearbox is relatively large and massive, and it also has to react the very large thrust, torque, vibration, and gyroscopic loads from the propeller. Almost always the gearbox is placed next to the propeller, but it has to be driven by the turbine, or by a separate LP power turbine, which except in pusher installations is at the other end of the engine. There are various possibilities.

De Havilland (later Bristol Siddeley, then Rolls-Royce) developed the Gnome P.1200 from a helicopter engine with no provision for a shaft down the centre. Accordingly, the free power turbine, rotating at 20,000 rpm, drove a primary gearbox at the back which turned a shaft at 6,000 rpm. This shaft took the drive across the top of the engine to a secondary gearbox above and ahead of the inlet to drive the propeller at about 1,250 rpm. Most early turboprops – the Rolls-Royce Dart and Tyne, General Electric XT31, Pratt & Whitney T34, Turbomeca Astazou, and Kuznetsov NK-12 – were able to adopt a more obvious layout. The reduction gear was placed in the centre of an annular air inlet and driven from the front of the single spool or by a separate LP shaft down the centre of the engine. With this arrangement it is easy to use a ducted spinner, as seen in several piston engines. Pratt & Whitney considered such a spinner 'best as far as airflow and aerodynamic characteristics are concerned', but after trying it rejected it in favour of a conventional spinner with an annular inlet just behind it.

Another turboprop pioneer, Allison, adopted what seemed astonishing at the time, because it had never been necessary with a piston engine, but which has become a common answer. The reduction gear was moved right away from the power section and carried on the casing of the long drive shaft and braced by two tubular struts to the compressor casing. This allowed the inlet – not part of the engine, but of the nacelle cowl – to be offset and connected to the engine by an S-duct. In the C-130 Hercules and E-2 Hawkeye the inlet is below the spinner, and in the Electra and P-3 Orion it is above. Underscoop inlets are also seen in almost all other turboprop aircraft

One of the latest turboprop nacelles is that of the Saab 2000. The engine is the Allison AE 2100 with high gearbox for a chin inlet, the propeller by Dowty, prime nacelle contractors Westland and Hispano-Suiza, and three suppliers are indicated.

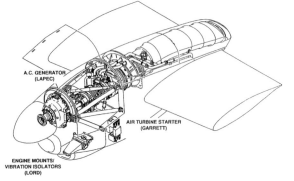

today, though in the PT6 engine – by far the most popular of all turboprops – the duct goes to an inward-radial inlet completely surrounding the engine behind the gearbox. As PT6 aircraft often work from rough strips, this inlet has a mesh screen.

Many turboprops, including the PT6, are also available in turboshaft form for helicopters. In most helicopters the engine incorporates a simple gearbox turning an output shaft at about 6,000 rpm. This is a speed high enough for the drive torque to be modest, so the shaft and drive casing can be quite light. But the main rotor gearbox of the helicopter reduces this speed to about 250–300 rpm, so that massive construction is needed downstream. In modern helicopters the engine installation is usually above the fuselage, with the jetpipe(s) turned out sideways. Where the helicopter will encounter sand, dust and other particles it is important to fit a particle separator over the inlet. Military and naval engines are now designed with such a separator forming an integral part of the installation. In some the air is turned through large angles before entering the compressor, so that all solids, even minute dust particles, are flung outwards and extracted. In the GE T700 the accessory gearbox drives a blower which acts as a vacuum cleaner, sucking out the extracted material. In the outstanding T800 engine the air merely makes a sharp inwards turn to reach the compressor, 'over 97 per cent' of the particles going straight on to be collected in a surrounding box. Many Russian helicopters have simple hemispherical deflectors ahead of the inlets, while Sea King/Commando helicopters can be fitted with a plain flat plate which does little to protect the engine except in cruising flight.

Thanks to terrorism, even civil helicopters are often seen with infra-red countermeasures (IRCM) intended to defeat heat-seeking missiles. The simplest answer is to add a plain box round each jetpipe so that the missile cannot 'see' any hot metal. For something closer to real protection, various kinds of 'suppressor' are in use, most having the form of a large box which entrains fresh air which mixes with the hot gas before being ejected upwards or downwards. For additional protection a pulsed IRCM beacon can be added, usually above the rear fuselage. The intense IR emission is pulsed, for example by rapid-acting window shutters like some lighthouses, and this confuses the missile and causes it to break lock.

Helicopters have to be able to fly slowly or even hover for long periods. This precludes the usual use of ram air for cooling, and demands

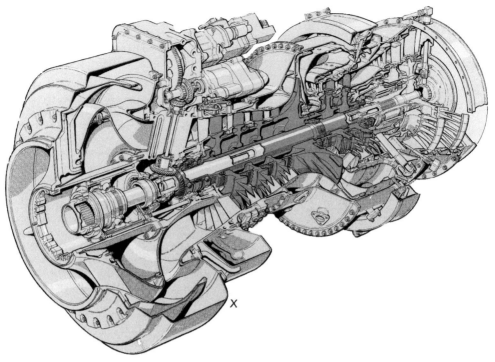

The largest part of the Rolls-Royce Turbomeca RTM 322 helicopter engine is the particle separator. Air entering at the left passes at high speed round a sharp bend, where centrifugal force flings solid matter outwards into surrounding collector X.

special provisions to prevent overheating of the engine compartment and engine auxiliaries. Some use can be made of the rotor downwash, but the vital oil cooler needs forced draught. In a helicopter the lubricating oil has to carry away the waste heat not only from the engine but also from the entire high-power gear train from the engine and main-rotor gearboxes, and from the various bearings and bevel gears in the drive train to the tail rotor. This demands a large oil cooler with an engine-driven fan to pump air through at high velocity, no matter what the external aerodynamics might be. In some helicopters this fan absorbs 10 per cent of the power.

Every installation of a modern aero engine has to be protected as far as possible against icing and fire. Any accretion of snow should be dispersed by heat and airflows as soon as the engine is started, but freezing fog and rain at below 5°C can be serious, even with the aircraft parked. Previous chapters have described how hot (usually compressor bleed) air is the simplest answer for jet-engine inlets, and electrothermal heaters for turboprop and shaft

engines. All draw their energy from burning fuel, which is thus no longer available to help propel the aircraft.

Fire protection is something else. Even up to recent times the history of aviation has been punctuated by catastrophic engine-attributable fires, to the extent that many passengers have been killed before the aircraft could even take off. The nub of the problem is that some parts of the engine are extremely hot and are fed with combustible fluid. Combustion of this fuel is designed to be contained, but even without enemy action it has in the past been not uncommon for engines to burst into flame by themselves. A glance through accident reports reveals a host of reasons; 'failure of No 1 thrust bearing . . . failure of the HP shutoff valve following LP disc disintegration . . . mechanical failure of the combustor liner . . . fracture of No 2 hydraulic line downstream of the pump . . . buildup of burning fuel in a bay where no detector was thought necessary . . .' and, after all this, 'selection of the incorrect extinguisher handle' (on many occasions).

It goes without saying that the engine is, as

far as possible, installed inside a box made of fire-resistant material – such as steel or titanium – with proper ventilation and with flame or smoke detectors at every location where the passing airflow could escape. The surrounding structure is divided by fireproof bulkheads into zones, each with its own carefully designed cooling/ventilation and fire/overheat/smoke detection and extinguishing systems. A few aircraft still in use were designed on the premise that in flight there would be ample cooling airflow round the engine to sustain the strength of an all-aluminium surrounding structure, but this is foolish. In any case, we have learned the hard way that fire on the ground can be just as lethal as in the air.

There have been an extraordinary number of cases where the passengers not only were aware of an engine fire but knew which engine was affected, while the flight crew were either in blissful ignorance or shut down the wrong engine. Despite this, there are so many protection systems that the mind boggles at these tales of disaster. As early as 1934 military aircraft were being fitted not only with engine-bay fire

extinguishers but also with actuation systems triggered automatically by unnatural (eg, forced-landing) deceleration, inversion of the aircraft on the ground, or structural deformation caused by impact. For 50 years we have also had various detectors triggered by abnormal temperature, or by an unnatural rise (or rate of rise) of temperature, or by sudden concentration of smoke, or by sudden emission of electromagnetic radiation covering not only the wavelengths of visible light (flames) but also from ultraviolet (UV) to IR. Detectors can be anywhere in each zone, but the most useful location is across the aperture through which cooling or ventilation air escapes.

In the author's experience, at least half of all fire warnings are false alarms, but every single one has to be taken seriously. Radiation sensors are the least likely to give false indications, and the main problem here has been in setting the radiation wavelength limits. Most are designed to be triggered only by the combination of frequencies (including UV but not far-IR) emitted by burning jet fuels. Only military and larger commercial aircraft have such sensors, but near-

Fire detectors are usually positioned wherever cooling or ventilation air escapes. There can be many such places, as this drawing of secondary flows through a typical PW4000 nacelle indicates (Pratt & Whitney).

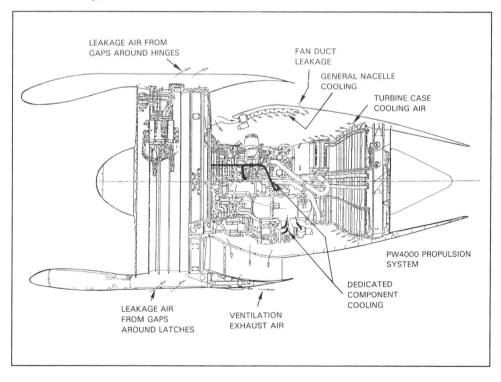

LEAKAGE AIR FROM
GAPS AROUND HINGES

FAN DUCT
LEAKAGE

GENERAL NACELLE
COOLING

TURBINE CASE
COOLING AIR

PW4000 PROPULSION
SYSTEM

DEDICATED
COMPONENT
COOLING

LEAKAGE AIR
FROM GAPS
AROUND LATCHES

VENTILATION
EXHAUST AIR

ly all have some form of continuous-element detector. The commonest are thermocouples, in which warnings are triggered by a rise in temperature-difference between the cold and hot junctions joining two different metals. Thermistors are triggered by the change in resistance of a semiconductor through which a current flows. Capacitance detectors have a voltage applied to a conductor surrounded by dielectric (insulating) material in a tube; heating the dielectric changes the capacitance. Gas-filled detectors are tubes filled with a material which absorbs a large volume of gas; heating causes a rapid rise in pressure. In each case the detector should be wrapped round each zone, the final electrical output triggering strident visual and aural warnings in the cockpit. It is then usually left to the pilot or flight crew to decide whether to operate an extinguisher. False warnings may be caused by electrical chafing, short circuits, moisture, loss of gas or various other problems.

Fire extinguishers are usually high-pressure bottles filled with a compound based on Freon, located outside a fire-risk zone. Each circuit usually has two bottles, discharged individually by a cockpit twist-lever which fires a cartridge-actuated valve. The liquid extinguishant is piped to one or more spray nozzles or perforated pipes so arranged that – in the absence of severe damage, such as caused by an uncontained fan or core breakup – every part of the fire receives a predetermined concentration for a certain minimum time, such as 1.5 seconds. It is the captain's responsibility to decide if and when to fire the second bottle.

The normal exhaust gas temperature (EGT) sensors, and thermocouples or thermal switches in the cooling-air outlets, may be equipped to warn of temperature rise above a safe limit. Turbine overheat is not an emergency, but if allowed to continue can cause costly damage. Smoke detectors are usually not engine-related.

On the other hand, visible smoke in the jet is today regarded as an environmental problem. In the 1950s, especially during take-offs with water injection, turbojets emitted varying intensities of smoke, as well as noise sufficient to cause physical pain and at least temporary deafness. Gradually these became environmental issues. The high-BPR turbofan virtually eliminated the noise problem, as related later, but smoke required prolonged efforts, especially on 'the favorite engine of the world's airlines', the Pratt & Whitney JT8D. This was purely for aesthetic reasons, because eliminating smoke does not noticeably increase thrust or reduce fuel burn.

Jet smoke is caused by sub-micron particles of carbon. These are harmless, and could be burned to the monoxide or dioxide, but remain visible because they are expelled through the turbine and jetpipe before combustion can take place. Serious efforts to reduce smoke began about 1965, and naturally were concerned to premix the fuel and air, increase aeration throughout the length of the chamber, and generally give all the carbon in the fuel time to combine with the ample available oxygen. The problem was complicated by the discovery of various other chemicals in the jet, such as oxides of nitrogen, which unlike carbon are harmful to humans. The objective here is to reduce the residence time, so that the N is prevented from combining with the O. Suffice to say that the improved JT8D chamber went into service in July 1968, and in modern engines combustion virtually eliminates visible smoke. We saw how designers are dramatically reducing the harmful emissions.

Noise, however, is an ongoing problem with many civil aircraft built, say, before 1980. Military aircraft have never had to meet any noise standards, and it would be a foolish government that would introduce such legislation because slim military engines for combat aircraft are inherently noisy, and silencing measures would degrade their performance. In a war there is little to be gained by coming a good second!

In contrast, all commercial transport aircraft have to meet noise legislation, the international limits first being decreed in US Federal Aviation Regulation Part 36. The International Civil Aviation Organisation also issued Annex 16 to its Treaty to define noise certification standards, which is binding on member countries (virtually all those of aviation importance), but predictably it is the US document which has come to be the one that matters. FAR.36 has been issued in increasingly severe Parts, the first (issued in 1969) merely arresting any increase in noise. Subsequent Parts have applied to each generation of new-build aircraft with increasing severity, whilst progressively being retroactively applied to older aircraft. As a result thousands of older jet transports have been either expensively modified or retired early, which is just what Boeing likes.

The legislation is complex, different parts applying with varying force to different aircraft, with limits determined by aircraft gross weight, number of engines, age and other factors. Moreover, the noise nuisance is divided into Landing (measured under the glidepath 1 n.m.

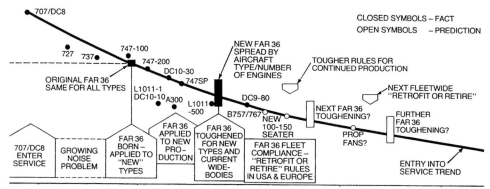

Noise legislation was prompted by the unacceptable thunder of early jetliners, especially on take-off. Noise is measured in dB (decibels), on a logarithmic scale; 6 dB up or down means doubled or halved sound pressure level (SPL).

from the downwind end of the runway), Sideline (measured 0.25 or, for 4-engined air-craft, 0.35 n.m. from the runway centreline), and Take-off (measured 3.5 n.m. from the upwind end of the runway). The rules are designed solely to reduce disturbance at air-ports, and are made even more complicated by the possibility of climbing steeply on take-off and cutting back power before passing over urban areas, choosing particular climbout paths to avoid residents, and adopting steeper landing approach glidepaths.

Most of the noise from turbojet aircraft comes from the jet itself, as described earlier. These were the aircraft that either had to be modified to meet noise legislation or else junked. The best answer was to fit a modern engine, but to save money upfront (even if it meant continuing to burn more fuel) many operators merely revamped the nacelles. This typically involved extending the inlet and jetpipe surround, and lining the interior with materials specially designed to absorb sound waves, converting the energy of the waves into heat.

Almost all of these acoustically absorbent materials are based on the concept of the Helmholz resonator of 1849. This is an enclosed volume communicating with the exterior through a small aperture; for example, a box of air or gas with one small hole. Sound waves impinging on the hole cause the air inside the volume to be compressed. When the system is tuned to resonate at the applied frequency, the conversion of sound to heat is considerable. The trouble, of course, is that the noise is a jumble of countless time-variant frequencies, and the resonator cannot be effective against all of them.

The basic method is to line the inlet and jet-pipe with a sandwich comprising an outer skin, a honeycomb filling, and an acoustically porous inner wall. Only one hole is needed for each honeycomb cell, but in most cases life is made easier by using an inner skin with uniform fine perforations all over. Alternatively, an attempt at broad-band absorption can be made by using an inner sheet of porous metal whose randomly varying pores admit the energy to fixed cavities whose resonance is trimmed by being lined with a different absorbent material.

The inlet can be lined with acoustic sandwich made of carbon or glass fibre, with an alumini-um honeycomb or polyurethane-foam absorbent and an inner skin of close-weave wire cloth. The jetpipe has to be a sandwich made of stainless steel, titanium or high-nickel alloy, the outer skin being larger in diameter but bearing the structur-al load, and the perforated inner skin replacing the original pipe. Everything in the engine instal-lation has to be non-combustible, and long-term resistant to all the aircraft fluids that may be encountered. The perforations in acoustic linings cause a little extra drag, slightly increasing fuel consumption or reducing thrust, and of course all addition of noise-suppressing features adds to cost, maintenance cost, and weight.

With turbofans the problem is enormously eased by the reduced jet velocity, and the higher the BPR the better. Engines of modest BPR are clearly candidates for a mixer nozzle surround-ed by a full-length bypass duct, because this both improves propulsive efficiency and reduces jet noise. With BPRs greater than 3 or 4, the designer has to weigh these advantages against the weight of the extra bypass duct and streamlined cowl, which may well be of more than 3 m (10 ft) external diameter. Full-length cowls are also a challenge to the designer of the

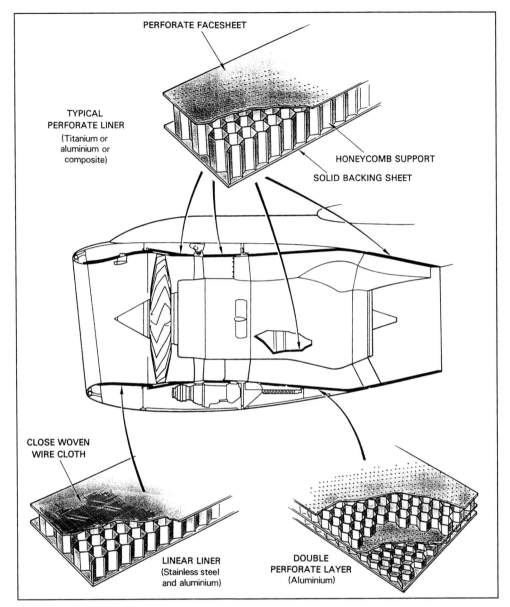

PERFORATE FACESHEET

TYPICAL
PERFORATE LINER
(Titanium or
aluminium or
composite)

HONEYCOMB SUPPORT

SOLID BACKING SHEET

CLOSE WOVEN
WIRE CLOTH

LINEAR LINER
(Stainless steel
and aluminium)

DOUBLE
PERFORATE LAYER
(Aluminium)

Typical noise-absorbing materials, and an indication of where they might be used in an airline turbofan nacelle (Rolls-Royce).

translating type reverser, which is one reason for the introduction of the pivoting-door type.

With high-BPR engines the dominant noise sources are in the turbomachinery, notably the fan and turbine. This noise is made up of two types of component. One is the general noise of the high-velocity air and gas flow through all parts of the engine, including the noise of combustion hundreds of times more intense than such sources as cookers or blowlamps, plus all the auxiliary pressurizing and cooling flows, together with the noise of gears and auxiliaries such as pumps and generators. There are so many sources that this can be generally called broadband noise. The other components are those caused by the fan, and, less importantly, the compressors and turbines.

The loudest man-made noise sources are

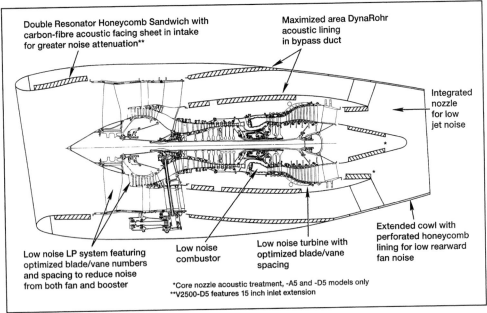

Double Resonator Honeycomb Sandwich with carbon-fibre acoustic facing sheet in intake for greater noise attenuation**

Maximized area DynaRohr acoustic lining in bypass duct

Integrated nozzle for low jet noise

Extended cowl with perforated honeycomb lining for low rearward fan noise

Low noise turbine with optimized blade/vane spacing

Low noise combustor

Low noise LP system featuring optimized blade/vane numbers and spacing to reduce noise from both fan and booster

*Core nozzle acoustic treatment, -A5 and -D5 models only
**V2500-D5 features 15 inch inlet extension

Modern airline engines are inherently quiet, and are then made quieter. These are features of the V2500 nacelle (International Aero Engines).

sirens, and these operate by chopping through jets of air at audio frequencies. This is exactly what happens inside a gas turbine, showing the challenge the noise-suppressing designer has to face. Early engines with inlet guide vanes made the fan or first-stage compressor chop through the wakes from the vanes, and removing them from modern engines provided a bonus in eliminating this loud discrete-tone noise. There are still bound to be blade-passing tones emitted by the fluctuating aerodynamic pressures caused by the fan wake on the outlet guide vanes and fan-case support struts. As in every part of the engine, powerful computers enable the designer to find the best possible combination of number of fan blades, number of outlet guide vanes and struts, fan speed, axial spacing between the fan, vanes and struts, and the geometry of the duct, whilst maintaining the highest propulsive efficiency and minimum weight.

Usually a blade/vane numerical match that generates little noise does involve some loss in engine performance. Likewise, noise is always reduced by increasing the gap between the fan and the guide vanes and struts downstream, and between all the other blade rows in the core, whereas a primary objective is at the same time to make the engine as short and compact as possible. Opposing objectives abound in engine design. For example, noise emitted to the front

of the engine through the inlet can be almost eliminated by providing a constriction in the inlet which accelerates the airflow to the speed of sound. No sound can get out through a choked inlet, except a tiny amount through the boundary layer on the walls. But no extra air can get in either, so on the approach our quiet aircraft had better not overshoot (go around) unless there is an absolutely foolproof way of immediately retracting the constriction. Yet another dichotomy is seen in the combination-tone noise – loosely called 'buzz-saw' noise – projected forward through the inlet from the supersonic fan tips. It would be easy to run the fan slower and eliminate the shockwaves, but to restore the thrust the engine would have to be bigger, heavier, and burn more fuel.

Bearing in mind the multitude of noise sources and the colossal power of modern engines, the insignificant noise problem from today's commercial jets is almost unbelievable. By far the greater part of this achievement is due to increasing the BPR, which by increasing the fan airflow reduces the required fan pressure ratio and jet velocity. In addition, by taking more energy from the core gas flow to drive the fan, it reduces the core jet velocity. As the noise from a jet moving at about 1,000 ft/s is proportional to the 8th power of jet velocity, these effects are very powerful.

PART II
THE HISTORICAL STORY

10 The pioneers

'Scientific investigation into the possibilities of jet propulsion has given no indication that this method can be a serious competitor to the airscrew/engine combination'.
BRITISH UNDER-SECRETARY OF STATE FOR AIR, 1934

Sir Frank Whittle once wrote to the author; 'The invention was nothing. The real achievement was making the thing work.' For centuries the world has been full of people who were unable to do anything with what may have been a good idea. Quite a lot of them were inventors of jet propulsion for aircraft.

Histories of jet propulsion invariably start with the Aeolipile invented by a Greek mathematician named Hero (or Heron), who lived in Alexandria about 2,000 years ago (AD 60 ±200 years). It comprised a boiler delivering steam to a freely rotating drum or sphere spun round by two tangentially mounted jets. One account says it was used (presumably with pulleys and cords) to open temple doors, but we have no proof that it was ever built.

In any case, waterwheels and windmills (the first gas turbines) ante-date the Christian era. Other primitive gas turbines were used to rotate painted shades over lamps and turn spits over fires. In 1629 Giovanni Branca in Italy illustrated his scheme for a steam jet from a boiler squirting on the vanes of a turbine geared down to drive a stamping mill. Around 1700 Gravesande illustrated Newton's 'action/reaction' law of motion with a hypothetical carriage propelled by a steam jet controlled by the driver. In 1791 John Barber patented a Watt-type beam engine driven by a primitive gas turbine via sets of reduction gears. The first self-powered turbine known to have operated was a 5 ft 'steam wheel' made by W. Avery of Syracuse, New York, in 1831. In the 19th century there were dozens of patents for steam turbines, and also for propellers and helicopter-type lifting rotors driven by tip reaction jets.

Already we can see that there are two totally distinct fields of endeavour: jet propulsion and the gas turbine. As far as practical hardware is concerned, the gas turbine came first. In 1888 Sanford Moss, a 16-year-old mechanic in San Francisco, began studying burning fuel in compressed air, and his 1900 doctoral thesis was on the subject of the gas turbine. He joined General Electric, and by 1907 had an operating gas turbine, but its performance was so poor that the idea was abandoned. Prompted by Frenchman Auguste Rateau, who developed an exhaust-driven turbosupercharger in 1916 and had it flying in 1918, Moss produced an improved turbosupercharger in 1921, starting work which led to over 160,000 turbos for aircraft piston engines in the Second World War.

The other pioneers of the gas turbine were C. Lemale and R. Armengaud, who worked for the Société Anonyme des Turbomoteurs (SAT) in Paris. They designed a crude engine in 1900, and by 1906 were achieving an efficiency of 3 per cent, going into liquidation in 1909. The obvious problems were the poor efficiency of compressors and turbines, and the pathetic limits on pressure and gas temperature; but where heat was otherwise going to waste, gas turbines began to make sense. From 1905 until after 1940, Brown-Boveri in Zurich (who had made the multi-stage centrifugal compressor for SAT) not only made but commercially sold gas turbines for pressurizing blast furnaces, running on the waste gas, and later electric-generating plant and, in 1941, a 2,200 hp locomotive with electric transmission. Later such locomotives were run by British Railways and other operators.

By 1920 the stationary gas turbine was an established type of prime mover, and many engineers wondered if it could be adapted for aircraft. In 1919 Britain's newly formed Air Ministry asked the director of its South Kensington Laboratory, Dr W. J. Stern, to report on the prospects. Stern, who played an equivocal role until after the Second World War, studied a hypothetical engine of 1,000 shp. He simply took accepted industrial practice, with peak gas temperature 500°C, and overlooked the possibility of making an engine lighter and more compact. For example, he allowed 1,250 lb for fuel pumps and drive gears! The engine was thus totally uncompetitive.

He wrote; 'The internal-combustion cycle cannot be utilised in a turbine power plant without large losses . . . In its present state of development, the IC turbine is unsuitable for aircraft on account of weight and fuel consumption . . . the weight of a 1,000 hp set comes out to something of the order of 10 lb per hp, the fuel consumption being 1.5 lb of oil per bhp hour.' Nobody with any technical knowledge questioned these nonsensical conclusions, which accordingly became carved in stone. The result was that this official report was used to justify instant dismissal of any subsequent proposal to develop British gas turbines for aircraft propulsion.

Turning to the alternative concept, jet propulsion, the author is astonished that, throughout the 19th century, would-be aviators ignored the rocket (Chapter 12). Instead, they preferred heavy and complex steam, electric, or gas-powered machines.

So far as the author knows, the first self-powered jet engine to be proposed was patented by René Lorin in 1908. He simply reasoned that, if

Two of Marconnet's patented ideas, the first using a Roots blower and the second tandem centrifugal blowers; in each case, C is the combustion chamber.

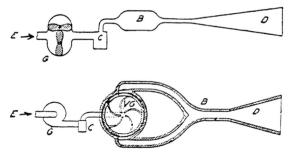

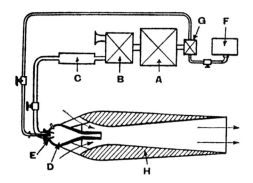

Morize's 1917 scheme is described in the text.

a conventional Otto-type piston engine was not required to do any useful work, the energy in its exhaust could provide useful thrust. He envisaged a multi-cylinder engine lying flat inside a wing, each exhaust valve admitting the hot gas direct to a divergent nozzle. Of course, unless the velocity could be supersonic, he really needed a convergent duct; moreover, the intermittent firing and trivial mass flow made the idea a nonstarter.

In 1909 another Frenchman, Marconnet, had proposed several schemes for pumping propellants into a combustion chamber with a divergent nozzle. In 1913 compatriot Hayot devised an S-shaped arrangement of inlets, diffusers, auxiliary (ejector) inlets and combustion. In the same year Lorin patented different forms of ramjet.

Romanian Henri Coanda built a biplane with a Clerget inline piston engine which, instead of turning a propeller, drove a centrifugal compressor blowing air to the rear. The thrust was said to be 220 kg (485 lb), a figure the author disbelieves. On 10 December 1910 the aircraft thus powered inadvertently became airborne, crashed and burned. Often called 'a turbine aeroplane', this was of no more significance than the Campini aircraft mentioned later, and Coanda wisely decided to switch to a propeller.

Several inventors sought to augment a jet by entraining fresh air. In 1917 O. Morize patented several variations on a basically simple jet engine in which piston engine A drove an air compressor B (and pump G drawing fuel from tank F) which via an equalizing chamber C (to smooth out delivery pulsations) fed nozzle E in the combustion chamber D to which G also pumped fuel. Fresh air was entrained to augment the jet discharged through ejector tube H.

Another 1917 proposal was patented by H. S. Harris of Esher, England. He made engine A

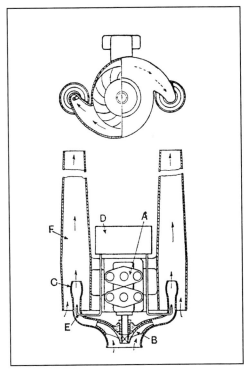

Harris (1917) used a piston-engined centrifugal to blow air down twin combustion pipes.

drive a compressor B to blow air through combustion chambers C fed from tank D to fuel

injectors E to create propulsive jets in divergent/convergent ducts F. This would have worked, had it been tried, but not as well as propellers. He apparently made no use of the piston-engine exhaust.

To another Frenchman, Maxime Guillaume, goes the credit for the original idea of making a gas-turbine jet engine able to start from rest; in other words, a turbojet. His Paris patent dated 3 May 1921 shows an axial compressor A driven by an axial turbine B. Crank 1 turns gearwheel n' and high-speed gear k to turn shaft a carried in bearings j. As the spool speeds up, air is compressed by rotor blades b with intermediate stators d. Combustion chamber f, with refractory lining g, is fed with fuel from tank p through needle-valve q, ignition being provided by plug o fed from magneto n. Guide vanes h' direct the gas on to turbine rotors c and stators e, the jet escaping on the right at i.

Guillaume said that his propulsion by reaction would not only be 'plus souple' (more versatile), but would permit 'très grandes vitesses se rapprochant de celles de l'obus' (very high

The historic drawing of an axial turbojet in the patent by Maxime Guillaume, submitted on 3 May 1921 and awarded on 13 January 1922. This was the first known description of a turbojet, though Guillaume lacked both aerodynamic and mechanical-engineering knowledge, and could not have constructed a successful engine.

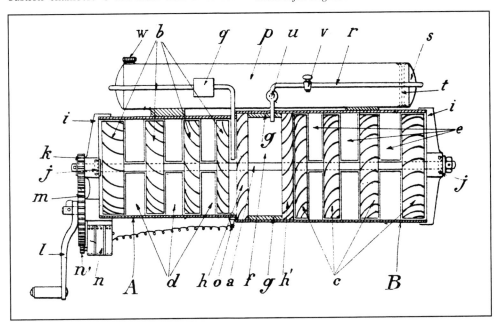

speeds approaching those of artillery shells). He even suggested taking off vertically with the engine axis vertical and then, depending on how the slipstream made the elevators effective, gradually rotating the thrust axis horizontal. Unfortunately he had nothing but the idea. He had no aerodynamic, stressing or mechanical engineering knowledge – he drew his turbine blades back-to-front – and even if he had found a sponsor could not have built a running engine.

Another visionary was Dr A. A. Griffith. Like Guillaume, he was not burdened by much technical knowledge, but he did have the advantage of being a highly qualified aerodynamicist and mathematician. Working at the Royal Aircraft Establishment (RAE) at Farnborough, England – which until its recent closure had been the source of more aeronautical progress than any other site on Earth – Griffith spent the early 1920s evolving a new analytical treatment for axial turbines. Traditionally, the gaps between the blades were treated as flow passages; Griffith instead considered the blades as aerofoils. He published this in 1926, and received permission to conduct experiments on a cascade of aerofoils in a wind tunnel, and on a simple rotating test rig.

This rig consisted of a single-stage axial compressor driven by a single-stage axial turbine. Air was sucked through by an exhauster pump, so that the air passed successively through the turbine stator, turbine rotor, compressor rotor, and compressor stator. It was tiny, just 4 in diameter and with blades 0.5 in long, but the overall efficiency was said to be 91 per cent. There was no way of separating this into component efficiencies, but supposing the turbine efficiency to have been 95 per cent, the compressor figure would have been 96 per cent, which are fantastic values. Indeed, bearing in mind the proportional effect on such tiny rotors of leakage round the blade tips, the author finds these figures incredible.

This remarkable result encouraged Griffith to propose, in November 1929, the development of what we today would call a turboprop. He proposed an exceedingly complex engine based on the contrarotating contraflow principle, as shown later in this chapter. He calculated that such an engine could be made smaller, lighter and more fuel-efficient than piston engines of equal power. This might well have been possible, because the proposed engine was in some respects more advanced than anything flown today, more than 65 years later. What Griffith failed to comprehend – or rather, what he dismissed as of no interest – was that such an engine was beyond the industrial state of the art.

When faced with a thorny decision, the British have a reputation for reaching a compromise. Here was an eminent worker in an official establishment proposing an engine so advanced that few in authority could understand it, though some probably noticed it was of the species Dr Stern had decreed could not rival established piston engines. The special panel of the Aeronautical Research Committee, chaired by H. T. (later Sir Henry) Tizard, did not reject the whole scheme, but were afraid to sanction the cost of building the engine, so the decision was made to build and test the proposed axial compressor and also carry out combustion research.

Almost at once, someone must have regretted permitting any money to be spent at all, so Dr Griffith was packed off to the South Kensington laboratory, where there were no manufacturing facilities. Within days of his arrival there he was asked to pronounce on another scheme for a new kind of engine which had come out of the blue from a totally unexpected source. The inventor was RAF No 364365, Fg Off Frank Whittle.

Just 22, and barely 5ft tall, this officer from a humble background had completed a tour on Siskin fighters with No 111 Sqn, and was such an exceptional pilot that he had been picked to do the crazy-flying act at the RAF Pageant. He was now at the Central Flying School at Wittering on an instructor's course. While there he had worked on his scheme for using a piston engine to drive a compressor to blow air past fuel jets and out of a jetpipe. He correctly reasoned that such a power plant could in theory release aircraft from the speed limitation imposed by propellers. Unlike other inventors, Whittle possessed all three vital capabilities of mathematics, aerodynamics, and mechanical design. Despite this, his scheme always came out inefficient and much too heavy. Unlike Campini, who unknown to him was working on the same idea, Whittle suddenly had a flash of inspiration. In October 1929 he realised it should be possible to increase the blower pressure ratio and replace the piston engine by a turbine in the jet of gas.

This was the answer he had been seeking. He quickly produced a report, with calculations and drawings, and took it to the CFS Commandant. Gp Capt 'Jack' Baldwin was impressed, and within days Whittle was asked to report to Adastral House in London. Here W. L. Tweedie, in the Directorate of Engine Development, said, in effect; 'We know from Dr Stern's report that gas turbines are not worth bothering with, but we had better take you to see the expert, Dr Griffith.' The outcome was almost beyond belief. Griffith was himself eager to get on with his own complex gas turbine scheme, and he should have seen that the young officer had invented a far more practical engine which could actually be built and run. Instead, his comments were entirely negative. He dismissed Whittle's tip speeds, temperatures, and component efficiencies as foolishly optimistic. His attitude was one of scorn, as if this simple engine was beneath contempt.

Whittle said; 'I was well aware of the many prior failures in the gas-turbine field early in the century, but was convinced that the causes of these – low compressor and turbine efficiency, and lack of suitable materials – could be overcome in due course, especially in the case of aircraft, where the very low high-altitude air temperatures substantially offset the adverse factors.' He was puzzled at his dismissal.

Later, on 8 December 1929, Tweedie wrote; 'Dear Mr Whittle, there has now been an opportunity to examine in some detail . . . no reason has been found for any alteration of the preliminary . . . criticism advanced against your various assumptions . . . It must be remembered that a tremendous amount of work is being done . . . and you may rest assured the criticisms of your scheme were made with the full knowledge of the results achieved by actual experiment.' This was palpably untrue, because at that time no work whatsoever was being done in Britain on jet propulsion, and nothing was intended. The wording was merely intended to make Whittle think the officials knew far more than he did, so that with luck he would not bother them again.

Many years later, the other pioneer of the turbojet, Hans Pabst von Ohain, said; 'If the British experts had had the vision to back Whittle, World War 2 would probably never have happened. Hitler would have doubted the Luftwaffe's ability to win.'

Part of the trouble was that the 'bible' Stern Report was backed up by what we can today recognise as the pathetic performance of piston-engine superchargers, which were the only compressors available to the Air Ministry. This was a time when novel ideas were viewed with even more suspicion than usual, and when money for any kind of 'defence' application was almost non-existent. Whittle's assumptions did push the available technology to the limits, but nobody in the Air Ministry cared to comment that two or three years of research on modest budgets would transform the achieved intensity of constant-pressure combustion, the aerodynamics of compressors and turbines and, most importantly, the metallurgy of turbine blades. A key factor was that the universal body of expert opinion found it hard to believe in an 'out of the blue' scheme proposed by a junior officer – the least likely people to invent anything – of such youth and small stature. Had Whittle been a silver-haired six-footer he might have been treated less patronisingly.

Back at Wittering, Whittle rechecked his calculations, and, encouraged by Fg Off W. E. P. 'Pat' Johnson, he applied for a patent. It was filed as No 347,206, *Improvements in Aircraft Propulsion*, on 16 January 1930. Today we would call his engine a single-shaft turbojet. The parts shown in his patent drawing were numbered: 1, inlet casing; 2, inlet guide vanes; 3, compressor rotor; 4, two axial rotor stages; 5, two axial stator stages; 6, he meant to show the centrifugal inlet guide vanes; 7, centrifugal impeller; 8, outlet diffuser; 9, 90° elbow and outlet vanes; 10, multiple tubular combustion chambers; 11, fuel burner; 12, turbine entry guide vanes; 13, turbine rotor; 14, two stages of axial turbine rotor blades; 15, two stages of turbine stators; 16, drive shaft; and 17, multiple Laval-type convergent (choked)/divergent nozzles.

Whittle said that his invention was 'a method of propulsion in one direction by the reaction caused by expelling fluid in the opposite direction . . . particularly adapted for aircraft', the heat cycle 'consisting of one or more stages of compression, one or more stages of expansion

The drawing by Fg Off Frank Whittle, RAF, in his Patent of 16 January 1930. The key is in the text.

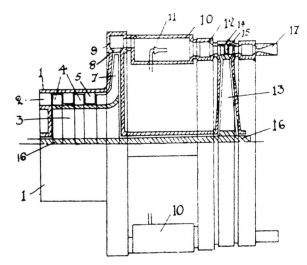

and a heat addition between the end of compression and the beginning of expansion, part of the work done in expansion being employed to do the work of compression and the remainder to provide the fluid reaction'. He believed 'that an embodiment of this invention will provide a large thrust in proportion to its weight, that it will perform at greater altitudes than are at present obtainable, that it makes possible higher speeds than have up to the present been obtained, that it will operate with any fuel now in use, and that it will have a reasonably low fuel consumption. Further, that simplicity and convenient external form is achieved . . . It can be demonstrated that the efficiency as a propulsive engine will not be reduced by reduction of the density of the atmosphere, and owing to the low temperature of the upper atmosphere may actually be enhanced . . . Controlling means may include fuel control, gas flow control or mechanical control of the blower and/or its mover. The final emission of gas may perhaps be directionally controlled for manoeuvring purposes.'

So in 1929 Whittle had opened the way not only to jet fighters and SSTs, but also to the Harrier and the vectoring agility of the F-22! He tried to interest various companies, but the response was totally negative. Accordingly, he got on with his next job of testing seaplanes and flying boats; he also made 46 test launches from ship catapults, as well as an open-sea ditching trial (he could not swim). In his spare time he wrote a paper, published by the Royal Aeronautical Society, on superchargers, in which he suggested using a fuselage-mounted engine to do nothing but supercharge the wing engines of a large aircraft (an idea put to use in Germany and the USSR but not in Whittle's own country). He also wrote a detailed *Case for the Gas Turbine*. In 1932 he began the Engineering Course at RAF Henlow with the unprecedented aggregate of 98 per cent marks in all subjects, whilst keeping his hand in with 'crazy flying' at air displays. In 1934–36 he was at Peterhouse College, Cambridge, graduating with First-Class Honours in the Mechanical Sciences Tripos despite taking two years instead of the normal three.

In January 1935 his patent for the turbojet

From the outset, Whittle was bothered by the high jet velocity of a turbojet, which with the slow (under 400 mph) aircraft envisaged in the 1930s meant poor propulsive efficiency. This 1936 patent was one of his attempts to 'gear down' the turbojet, in this case by using a central gas-turbine (compressors 58, combustion 59, turbines 60) to drive large fans 50 producing relatively slow jets 52A.

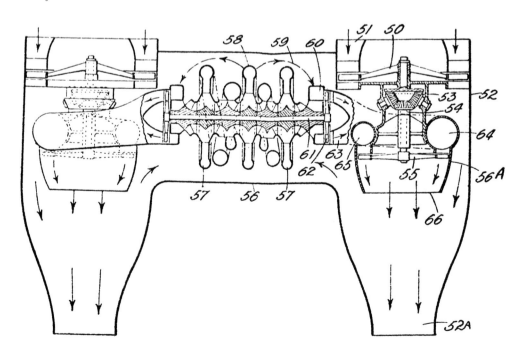

came up for renewal. The Air Ministry refused to pay the required fee of £5, and Whittle decided he could not afford to. Barely three months later R. D. Williams, who had often flown with Whittle but left the RAF owing to ill health, enquired about his 'aeroplane sans propeller'. Whittle replied; 'I have allowed the patent to lapse. Nobody would touch it on account of the enormous cost of the experimental work . . . though I still have every faith in the invention.' After many false starts a firm of investment bankers, O. T. Falk & Partners, provided funds to launch, in March 1936, a tiny company called Power Jets Ltd. Whittle was allowed one postgraduate year at Cambridge, with permission to act as honorary chief engineer of the company, provided he never devoted more than six hours to that work in any one week!

Williams's letter triggered several fresh gas-turbine patents, one of which (No 456,980 of 18 May 1935) was for the outline design of a turbojet. Whereas the original patent had proposed the principle, this new patent was for a bench-test engine. It was called the WU, from Whittle Unit. It differed from the 1929 scheme in that the compressor was a single rotor but of the double-sided centrifugal type, it was driven by a single-stage axial turbine with a water-cooled disc, there was a single large combustion chamber and a single central jetpipe and nozzle. At the front was a wheelcase with gears driving the fuel and water pumps and other auxiliaries.

With the assistance of A. A. Hall of Clare College (later Sir Arnold Hall, Director of the RAE and Chairman of Hawker Siddeley), Whittle designed the compressor rotor. It was to be machined from a forging in High Duty Alloys RR.56 high-strength aluminium alloy with a diameter of 19 in (482.6 mm), with 30 vanes on each side. It was to be driven by a turbine of 16.4 in diameter at 17,750 rpm. At the design tip speed of 1,500 ft/s it was to have a mass flow of 26 lb/s at a pressure ratio of 4.4. For comparison, Stern had assumed a compressor measuring 3 x 3 x 6 ft, weighing 2,000 lb and handling 500 lb/min, adding 'the dimensions of the rotary compressor assumed above are certainly optimistic, and probably a much bigger machine will be required to get an approach to 70 per cent efficiency'. Whittle's compressor, handling more than three times the airflow, weighed one-tenth as much and had roughly one-eighth the bulk.

The main area of uncertainty was the combustion chamber. On 19 February 1936 Whittle went to the British Industries Fair, where, without mentioning his engine, he outlined his

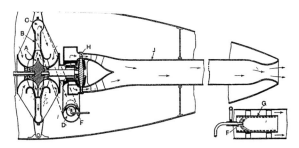

Drawing from Whittle's patent of 18 May 1935 for an actual engine: A, double-sided compressor; B, diffuser; C, air delivery scroll; D, combustion chamber; E, omitted; F, burner; G, flame tube; H, turbine; J, jetpipe.

requirement to burn about 200 gal (240 US gal) per hour in a volume of about 6 ft^3 to numerous firms specializing in burning liquid fuels. Most just laughed, because (he was told) the heat-release per unit volume was about 24 times greater than had ever been achieved. But Laidlaw, Drew & Co of Edinburgh were prepared at least to experiment, and in due course received a small research contract. With their help the WU was given a single huge chamber curled round the rotating part.

Whittle also talked to such other firms as Firth-Vickers (Stayblade turbine rotor blades and disc forged integral with shaft), Alfred Herbert (compressor rotor machining), and Hoffman (bearings), but the contract for the detailed design and construction of the WU was awarded in June 1936 to the British Thomson-Houston Co (BTH) of Rugby. The contract was on the basis of cost plus an agreed percentage. But it was clear that Falk's £2,000, plus a further £725 added in July 1936 – of which £200 came from an old lady who ran a corner shop near Whittle's parents in Coventry – would not be enough. It was virtually impossible to attract other investors, especially as the project was a government secret. People who were approached and given a hazy outline said; 'Surely, it's up to the Air Ministry to fund it'. But a direct request to Air Ministry for a research contract in October 1936 brought flat rejection, Deputy Director of Scientific Research D. R. Pye saying; 'It's hardly likely you will succeed where so many better-equipped people have failed.'

Pye's attitude was that there might be some justification for pure research to obtain some basis of quantified data and background knowledge. He failed to see that Power Jets had a mere £2,000, and that allowing this to be swallowed up in test rigs would leave the company with nothing but paper reports. Whittle could see that the only possible way to proceed was to take the gigantic gamble of running a complete engine.

On the proverbial shoestring, the WU was designed at BTH Rugby, and built there mainly by BTH chargehand Bentley and his assistants Bailey and Berry. It was finished in March 1937, and mounted on a truck connected to the wall by a spring balance. The starter was a two-cylinder Sprite, but the jerky torque distorted the drive shaft. The only alternative was a 20 kW electric motor, so heavy that the truck wheels had to be removed, which in turn made it impossible to use the spring balance to measure thrust. The jetpipe pointed out of an open window. The WU was controlled from a remote panel, and fed with diesel oil from a pump which for initial tests was driven electrically. First, the WU was motored round by blowing compressed air on the turbine. At 9,300 rpm the impeller fouled its casing. There was no money for new parts, so the damage was rectified by hand. (Had this rub occurred during actual running it would have been more serious.)

After repairs, the engine was again ready on 12 April 1937. It was one of the great moments in man's history; it was also memorable for what happened. Whittle signalled 'fuel pump on' and then 'engage starter coupling', followed by 'start'. In his own words;

The starter motor began to turn over. When the speed reached about 1,000 rpm I opened the control valve which admitted fuel to a pilot burner in the combustion chamber, and rapidly turned the handle of the hand magneto to ignite the finely atomised spray of fuel which this burner emitted. An observer peering through a quartz window in the chamber gave me the thumbs-up to show the pilot flame was alight. I signalled for increased speed, and as the tachometer indicated 2,000 rpm I opened the main fuel valve.

For a second or two the speed increased slowly. Then, with a rising shriek like an air-raid siren, the speed began to rise rapidly, and large patches of red heat became visible on the combustion-chamber casing. The engine was obviously out of control. Realising what this meant, the BTH personnel watching escaped at high speed; a few took refuge in large steam-turbine casings. I screwed down the control valve immediately, but this had no effect and speed continued to rise. Fortunately the acceleration ceased at about 8,000 rpm, and slowly the revs dropped again.

This first run of a totally new type of engine made some of those present wonder just what they were letting loose. How could an engine

The WU as originally built. First started on 12 April 1937, this was the first turbojet in the world to run.

accelerate with the fuel shut off? In any case, the colossal power being unleashed seemed hardly containable within such a thin shell of sheet metal, especially with large areas glowing red. Next morning they adjusted the fuel control and rechecked everything, and tried again. The result was even more frightening; the engine shrieked away out of control from a mere 1,500 rpm, without Whittle opening the main fuel valve. Sheets of flame belched from the jetpipe, and the glowing casing ignited clouds of fuel vapour from leaking joints, so that flames were dancing in mid-air. Again, at about 8,000 rpm, the acceleration stopped abruptly.

The explanation was simple. Every time the fuel pump was tested, a small amount of fuel was injected into the combustion chamber, and after a day of testing a lake of fuel had collected in the bottom of the chamber. Once this had caught fire, nothing could slow the engine down. For the third start the chamber was drained beforehand. Amazingly, the engine ran away again, this time because a spring in the fuel burner had been weakened by the heat. On the fourth attempt the engine ran under control to 7,600 rpm, but was then shut down when one of the glowing patches set fire to the ignition cables. Further tests showed that the main problem was combustion. Increasing the fuel flow at 8,500 rpm merely caused the extra fuel to burn downstream of the turbine, and there was no money for a replacement turbine.

Indeed, there was little money for anything. While the RAF backed Whittle in every way they could – for example, by not requiring him to take the usual examination for promotion to Squadron Leader – the Air Ministry contributed nothing to Power Jets until May 1938, and Whittle had to watch every penny. He nearly cracked under the strain, which in fact was to get worse for seven years, not because of the problems in developing the engine, but from the suspicion and enmity with which he was regarded by officials and manufacturers, and by the outrageous behaviour of the company picked by the Air Ministry to produce his engine.

In contrast, a young German, Hans Pabst von Ohain, had seen his turbojet pass from the germ of an idea to running hardware in just two years, 1935–37, and then to the start of flight testing in two further years, 1937–39. He was relieved of the argument, stress, and financial worries, and merely had to tell a big factory what to make.

Von Ohain studied physics at the University of Göttingen, where in November 1935 he patented a method for putting a sound track on film and a new form of aircraft engine, a turbo-

jet. The latter featured a single rotor with vanes forming a centrifugal compressor on the front and an inward-radial turbine on the back in a casing having the size and shape of a bass drum. There was to be some form of combustion chamber to add heat between the compressor and turbine.

Von Ohain experimented with a small model, called S 1 (*Strahltriebwerk* 1), made for him by Göttingen engineer Max Hahn, but realised he needed proper backing. He wrote to aircraft manufacturer Ernst Heinkel. In contrast to British firms, Heinkel was already looking for new methods of high-speed propulsion. In April 1936 von Ohain and Hahn joined his company. With Heinkel engineer Wilhelm Gundermann, they led a team which was allowed to select the best men from each workshop, so that within six months 17 engineers were assigned to the HeS 2. Heinkel did not inform the RLM (air ministry).

The S2 was not an engine but a demonstration rig, even the rotor being made from sheet metal, using an exhaust-valve steel for the turbine side. Hahn was the combustion expert, but this was regarded as so difficult that it was sidestepped by piping in gaseous hydrogen from the shop. This rig was first run in September 1937, and before long was giving a thrust of 250 kg (551 lb).

Von Ohain assured the author he had not seen Whittle's closely similar patent, published five

Max Hahn with the first proof-of-concept model at Göttingen.

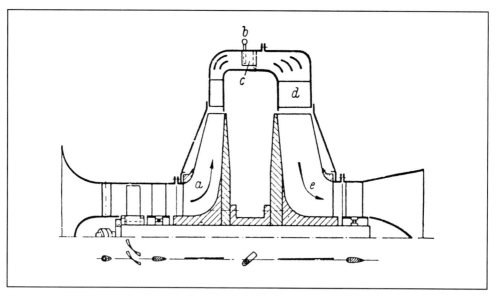

Flow through the hydrogen-burning S2: a, axial followed by centrifugal compressors; b, gas fuel pipe; c, burner baffle; d, guide vanes; e, inward-radial turbine.

years earlier in German, French, and Swedish magazines, but Gundermann later wrote; 'Naturally we kept track of other patent applications . . . and became familiar with the very similar work of Frank Whittle in England and Milo AB in Sweden.'

Heinkel immediately sanctioned the development of a flight engine, the HeS 3, plus two examples of the world's first jet aircraft, the He 178. The HeS 3 had an axial stage upstream of the centrifugal, and a folded annular combustion chamber which formed a bulging ring round the inlet. Air from the compressor was divided by a splitter, about half being deflected forwards into the combustion chamber and the rest to the rear to dilute and cool the combustion gas. The chamber was fed with kerosene from 16 burners with vaporiser tubes which had to be heated with hydrogen during the start, while the engine was spun by an air jet (c in the drawing) directed at the turbine. The compressed air and hydrogen had to come from workshop supplies. Once the engine was running, the vaporisers were kept hot by their own flames. To provide the required length for combustion, the compressor and turbine were moved apart and connected by a tubular shaft with bolted flanges.

Although the design thrust was 800–900 kg, the measured thrust at the start of testing in October 1938 was only about 400 kg. Nevertheless, the HeS 3 was hung under the He 118 V2 dive bomber and tested in the air on

many occasions from mid-May 1939 until the turbine burned out. The pilot was Wernher von Braun's Peenemünde test pilot Erich Warsitz, who kept in close touch with Heinkel's jet work at Rostock-Marienehe, and tested the He 176 (Chapter 12). The turbojet was managed by observer Walter Kunzel. Reasons for the poor performance included a generally cramped design to reduce overall diameter. Accordingly the engine was redesigned into the HeS 3B, with a larger compressor, three curved guide rings at the entry to the combustion chamber (B in the cutaway) and a new chamber exit with curved guide vanes (D) to direct the gas on to the turbine. This engine ran at the end of April 1939, reaching a static thrust of 450 kg (992 lb). Meanwhile, despite pressure from Gundermann to use two engines in the wing roots, Heinkel insisted on the He 178 having a single engine, even though the long inlet duct and jetpipe reduced static thrust to only 380 kg (834 lb). The engine weighed 360 kg (794 lb), and sfc was about 1.6.

On 24 August 1939 Warsitz began taxi trials with the He 178 V1, including a short hop. Early on the morning of Sunday 27 August, as Hitler's war machine moved to the Polish frontier, he made the first flight. The landing gears were locked down, with blanking plates over the main wells, a small bird was ingested as the aircraft began to move, and the Sun and mist made it difficult to find the airfield for the landing, but

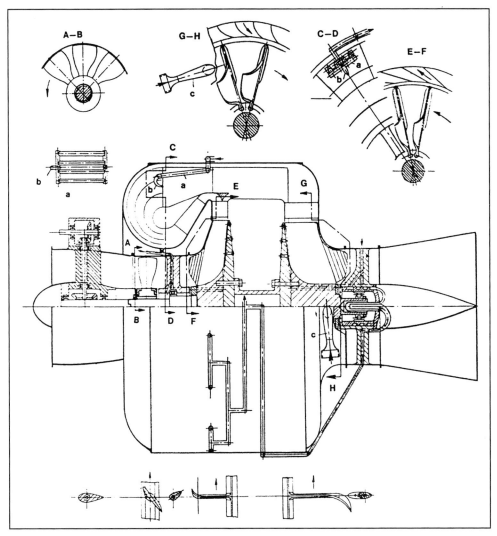

The completely redesigned HeS 3B, which powered the first jet aircraft in August 1939. The capital letters refer to the cross-sections; small letters, a, fuel pipes; b, flame stabilizer; c, compressed-air jet for starting.

The HeS 3B occupied the complete cross-section of the He 178 fuselage. This drawing shows the original aircraft design, slightly modified before it was built.

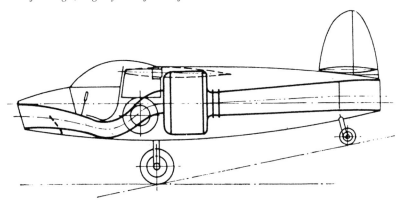

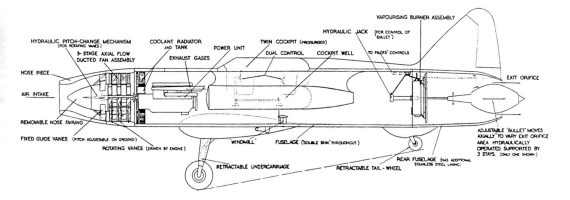

The Caproni-Campini N.1 was a piston-engined aircraft that looked like a jet.

this was the first flight of a turbojet aircraft. Several longer flights followed, some with the HeS 6 rated at 590 kg (1,300 lb), but the 178 accomplished little beyond proving the practicality of jet propulsion, triggering the start of design of the He 280 jet fighter. The 178 V1 was sent to the Air Museum in Berlin, where it was destroyed in an air raid. The V2 (second prototype) was never flown.

In 1931 American rocket pioneer Robert Goddard patented a constant-volume jet engine using a kind of rocket chamber fed with air/fuel mix through valves. Throughout the 1930s Italian Secondo Campini worked on clumsy piston-engined jet schemes. They were the sort of thing Whittle would have ended up with, had he not been inspired to replace the piston engine by a gas turbine. In 1938 Campini got Caproni to build the N.1, with a 900 hp Isotta-Fraschini 12-cylinder aero engine to drive a compressor with three stages of variable-pitch blading to blow air out of a tail nozzle with a variable central bullet. Its maximum speed was 330 km/h (205 mph),

This turbojet, intended to fit inside a wing, was the best of several Milo projects of 1933; the key is in the text.

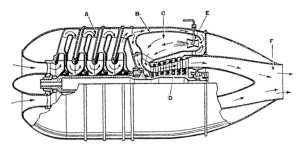

but at the cost of roughly doubled fuel consumption an 'afterburner' could be ignited to bring speed up to 370 km/h (230 mph). Rate of climb was 364 ft/min. Thus, all-round performance was much poorer than would have been obtained by an ordinary two-seater powered by the same engine driving a propeller.

On 30 November 1941 headlines around the world proclaimed 'the first jet flight'. The Italian Caproni-Campini N.1 had flown from Milan to Rome. But the average speed was 209 km/h (130 mph), and the 270-km (168-mile) flight could not be made without landing at Pisa to refuel! Ground staff at the Guidonia test centre 'did not hide their contempt'.

In 1933–36 French ramjet pioneer Leduc patented a turboprop, while Swedish patents were obtained by Lysholm, AB Ljungströms, and Milo AB for single-shaft gas turbines with a multi-stage axial turbine (D) driving a four-stage centrifugal compressor (A). Because of the differences in diameter, the folded annular combustion chamber (B, C and burner E) could surround the turbine. The basic idea was outstanding for its day. The drawing shows a Milo turbojet (nozzle at F) of Lysholm design; the Ljungström version was a turboprop, but no actual engine was built, neither did Brown Boveri build their turboprop, patented in 1939 and looking like Guillaume's 1921 patent plus a propeller gearbox.

We left Dr Griffith in 1930, unable to create his complex schemes. Not until 1936 did the RAE get permission to build 'Anne', an axial compressor with eight stages of free-vortex blading. Today its peak pressure ratio of 2.6 seems pathetic, but in October 1938 it was thought impressive. Next, in March 1937, a young RAE engineer, Hayne Constant, suggested a turboprop which led a month later to the start of work on an advanced machine with separate LP and HP axial spools. Only the HP was made; called the RAE B.10 and made by Metrovick, it first ran in December 1939. In the same year C. A. Parsons made 'Alice', with eight stages of RAE blading. Later axial machines appear in the next chapter. There was no sense of urgency; it was called long-term research.

Whittle's WU had terrified the BTH bystanders, so the Power Jets team had to move to the BTH Ladywood works, a most unsuitable derelict foundry, with sand everywhere. Testing quickly changed to kerosene, initially using vaporising burners. Pinned turbine blades were replaced by fir-tree roots; BTH failed to mill these, but Alfred Herbert succeeded by broaching.

On 23 August 1937 testing stopped for a major rebuild. The diffuser delivered air along ten pipes to the far end of the engine, where a 180° turn took it into a large combustion chamber, from which it passed forwards through the rebladed turbine, then being split into ten and turned through a second 180° to leave via ten jetpipes. The result looked clumsy but the engine was axi-symmetric, and offered heat transfer between the air and hot gas.

Testing resumed on 16 April 1938. On 6 May the WU ran for 1h 45min, reaching 13,000 rpm and measuring a thrust of 480 lb, when the turbine nozzle fouled the rotor. The intense friction quickly caused nine blades to fail, almost wrecking the engine. To Whittle's relief, neither BTH nor the Air Ministry said 'That's it, give up', and the Air Ministry even agreed to pay to rebuild the engine. Previously, the total Ministry input had been £1,000 for a report on testing,

Rotor and casing of the 'Anne' axial compressor of 1938. Diameter was 6 in. After 30 sec running it stripped its free-vortex blading because of a faulty oil seal; rebuilt, it did well until it was damaged by a bomb in August 1940.

plus £900 for testing the reconstructed engine.

For £2,000 the three BTH fitters, plus Whittle and Power Jets' sole employee, Victor Crompton, rebuilt the WU a second time. The radical change was to link the ten delivery pipes to ten smaller reverse-flow combustion chambers. The burners were at the back of the engine, and linking tubes connecting the flame tubes and cylindrical chamber casings ensured that all ten would unfailingly light-up on starting. There was thus only one 180° curve, the hot gas escaping through the new turbine in the original front-to-rear direction to a central jetpipe.

Testing resumed on 17 June 1939. Although there was to be prolonged tinkering with burners, Whittle now had a practical engine. On 30 June 1939 Dr Pye, by now Director of Scientific

The totally rearranged WU, which resumed testing on 16 April 1938.

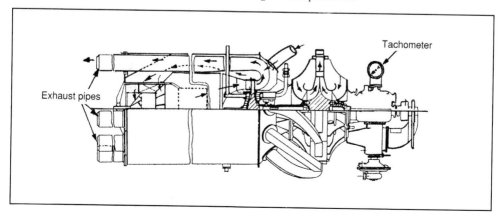

Research, witnessed a run to the highest speed then attained, 16,500 rpm. Previously a disbeliever, he immediately put in hand Air Ministry contracts for a flight engine and an aircraft in which to fly it. The resulting W.1 and E.28/39 are described in the next chapter.

In early 1939 Power Jets heard, via shareholder J. & G. Weir Ltd, a rumour that there were several German jet projects. In contrast, the German RLM was kept properly informed on Whittle's work, though it was not considered of major importance. Indeed, when in November 1939 Heinkel at last invited RLM and Luftwaffe staff to witness the He 178 flying, he was told; 'Your turbojet is not needed. We will win the war on piston engines.'

Nevertheless, unknown to Britain, there were by that time major turbojet projects with BMW, Bramo, and Junkers. BMW at Munich, led by Helmut Sachse, put Kurt Loehner in charge of an engine with a two-stage centrifugal compressor and a hot turbine with hollow blades. Bramo, at Berlin-Spandau, led by Bruno Bruckmann and Hermann Oestrich, looked to new ideas in order to stay in business, and had already flown a piston engine driving a ducted fan. During 1938 design began on a contrarotating axial jet, and also on a simpler axial jet which became the 109-003. Scorning what he perceived to be the slow-moving engine division, Junkers' farsighted airframe leader at Dessau, Herbert Wagner, secretly assigned Max Adolf Mueller the task of investigating jets and turboprops. A plethora of projects at Magdeburg led to the 109-006 axial turbojet, run in autumn

1938. Daimler-Benz alone remained aloof.

A key role was played by Hans A. Mauch, who in April 1938 became RLM head of rockets, quickly translated as 'head of jet engines', though the RLM Research Division already had a jet-propulsion department headed by Helmut Schelp. This was already financing Schmidt and Walther (Chapter 12). Supported by the DVL (aero research institute) and AVA (aerodynamic research), the RLM had by mid-1938 organised a comprehensive programme of jet development which increasingly involved airframes as well as engines. A major objective was to gain actual experience as quickly as possible with 'all types of engine'; another was to have all jet development carried out by experienced engine teams. Each engine project was given a number with the prefix 109-.

Thus, the work by Heinkel, an airframe firm, was to be handed to Daimler-Benz, and the Junkers airframe division jet was to be passed to the company's engine division. Mueller and his team resisted, and moved from Junkers to Heinkel. Heinkel also argued, and at the end of 1939 gained RLM support. The result was the HeS 8 (109-001), derived from the original von Ohain engines, plus axial jets brought by Mueller from Junkers, as described in the next chapter.

Meanwhile, Otto Mader, head of the huge Junkers engine division at Dessau, had told Dr Anselm Franz to study every possible kind of engine. One had an axial compressor driven by a two-stroke diesel of X-16 form, the air supercharging and scavenging the diesel and, with the

Layout of the WU on its third and final rebuild, with ten separate chambers. In true British fashion, this historic engine was scrapped in February 1941.

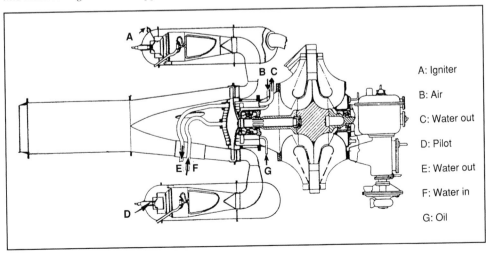

A: Igniter

B: Air

C: Water out

D: Pilot

E: Water out

F: Water in

G: Oil

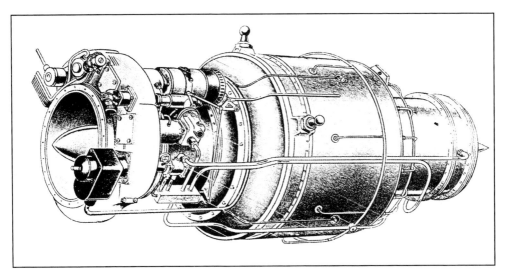

Perspective view of the HeS 8A (RLM designation 109-001), the ultimate form of von Ohain's centrifugal turbojet. The rotating assembly is pictured in Chapter 4.

exhaust, passing through combustion chambers to form propulsive jets. Franz soon decided that a simple axial turbojet would be much lighter, and in 1939 began the development of the 109-004, aimed at reaching production as soon as possible.

Thus, whereas Whittle, with a running engine, was still finding it desperately difficult to get anyone interested, in Germany the RLM could oversee the work of more than 2,000 engineers working on 12 jet and turboprop projects. In the closed society of the Soviet Union, Arkhip Mikhailovich Lyul'ka at the Kharkov aviation institute (KhAI) had by 1936 completed calculations for an axial turbojet. He overcame the usual sceptics, and, assisted by established engine designers I. F. Kozlov and P. S. Shyevchenko, was permitted in 1938 to open a small KB (design bureau) in Leningrad, where he began the design of the VRD-1. In Hungary, György Jendrassik had from 1932 been working on a 100 hp turboprop, and actually ran it in 1937.

11 The Second World War

'In its present state, and even considering the improvements possible when adopting the higher temperatures proposed for the immediate future, the gas-turbine engine could hardly be considered a feasible application to airplanes.'
COMMITTEE ON GAS TURBINES, US NATIONAL ACADEMY OF SCIENCES, 10 JUNE 1940

Whereas Germany started the war on 1 September 1939 with numerous diverse and fully funded programmes for turbojets, turboprops, pulsejets, and rockets, Great Britain declared war two days later not only in ignorance of all this, but without a single aircraft-engine firm showing any interest in jet propulsion.

Whittle had visited the major British aero-engine companies, but had been rebuffed. The reason was a combination of disbelief, unwillingness to spend a lot of money on research and development (especially at a time when they were stretched to the limit, and did not need extra projects), and hostility to new ideas. The hostility stemmed in part from the fact that the top engineers were respected as experts in their field, but knew almost nothing about gas turbines and resented the idea of having to learn a new technology in which an unknown called Whittle was far ahead of them.

The first established firm to be brought in was Metropolitan-Vickers, at Trafford Park, Manchester. With a background of axial steam turbines, it was well placed to take on the manufacture of Constant's complex 'Betty' B.10, as noted previously. The nine-stage compressor ran in December 1939, and though p.r. was only 1.2–2.0, the efficiency was a commendable 86–87 per cent. The turbine ran in May 1940 and the complete machine in October 1940. By this time the unwieldy 'Doris' D.11 had been tested, with a 17-stage compressor, 8-stage turbine and separate 5-stage power turbine to give 2,000 shp, but this was soon abandoned.

Instead, Whittle's influence prompted Metrovick to switch to a turbojet. In July 1940 Dr D. M. Smith, Dr I. S. Shannon, and K. Baumann, assisted by A. R. 'Taffy' Howell from the RAE, began work on 'Freda', the F.2 turbojet. This was to be a practical aero engine. It had a nine-stage compressor driven via a conical shaft by a two-stage turbine, with a remarkably compact annular combustion chamber. The F.2 ran in December 1941, and with a few changes passed a 25-hour Special Category test at 1,800 lb in November 1942, the compressor showing 88 per cent efficiency at a p.r. of 3.2. The third, further-modified F.2 flew from Baginton in the tail of Lancaster LL735 on 29 June 1943, and on 13 November 1943 at Farnborough two mounted in underwing nacelles powered the third Meteor, DG204/G. The test F.2 ran 210 h, including 90 h in flight, and stripped in good condition, but one of the Meteor engines exploded fatally on take-off on 1 April 1944.

Another F.2 to early standard was rebuilt with a ducted-fan augmentor, becoming the F.3. Immediately behind the turbine a second turbine was added with two pairs of contrarotating intermeshing blades with peripheral rings carrying fan blades in a surrounding duct. First tested in August 1943, this increased thrust from the original 2,400 lb to 4,600, with a 60 per cent reduction in sfc. Development of the F.2 continued, eventually doubling its original thrust to 3,500 lb with a 10-stage compressor and single-stage turbine as the F.2/4 in 1945. One F.2/4 was fitted with a two-stage open propfan to give an engine identical in layout to the American GE

Outstanding for its day, the Metrovick F.2 was developed by January 1943 to give 3,000 lb thrust at 7,700 rpm. In this group the Chief Engineer, Dr D. M. Smith, is the short man in the centre.

UDF of over 40 years later, but nobody was interested.

The company decided to give the engines the names of precious stones, and as the Beryl the F.2/4 reached 4,000 lb and powered the SRA.1 flying-boat fighter. Metrovick's final design, the F.9 Sapphire, was an outstanding large turbojet which in 1948 was handed to Armstrong Siddeley Motors (ASM) when the Ministry of Supply decreed that the Manchester firm should cease to develop aero engines.

After 1925 ASM had become an 'also ran' with its many piston engines, but in 1939 it manufactured to RAE design a counter-flow contrarotating turbocompressor to the Griffith formula, tested in 1940. The RAE continued

axial compressor development with the big 14-stage 'Sarah', and in November 1942 ASM received a contract to build a complete turbojet using it. The design team under Bill Saxton and A. Thomas planned the ASX with this compressor arranged back-to-front. Thus, the inlet was towards the back of the engine 'where deicing and filtering devices could easily be fitted'. On the front was nothing but a mass of accessories. Air from the compressor turned two 90° corners to return through 11 'long and accessible' combustion chambers. Although created quickly, and on test in April 1943, the ASX was complex and weighed 1,900 lb for an eventual thrust of 2,800 lb. It was tested in the bomb bay of the Universal Testbed Lancaster B.6 ND784/G, but

The F.2 was developed into turbofans and propfans, and also into an improved turbojet, the Beryl. Rated at 4,000 lb for a weight of 1,725 lb, it had a zero-stage on the compressor but a turbine redesigned with a single stage.

ASM wisely added a propeller reduction gear (Chapter 18).

We left Whittle beset by politics, jealousies and disbeliefs, but at least with a contract to build a W.1 flight engine. The much-rebuilt WU had by 1940 received Lubbock (Shell type) high-pressure atomizing burners, which at last promised good combustion. Ironically, these resembled burners briefly tested in 1936. Other improvements included curved guide vanes in the entry to the compressor, and turbine rotor blades held by fir-tree roots instead of tradition-al bulb fixings. In addition, the W.1 was made much lighter throughout, the compressor had 29 instead of 30 vanes to avoid resonance with the ten outlets, and the turbine had 72 instead of 66 blades. During construction by BTH various parts were deemed unairworthy, and together with spares these were built into the W.1X bench engine, first run on 14 December 1940. This was fitted into the first of the two Gloster E.28/39 aircraft, W4041/G, for taxiing trials at Brockworth. On 8 April 1941 Whittle did some himself at 15,000 rpm, and with the governor set at 16,000 Gloster test pilot P. E. G. 'Gerry' Sayer made three hops, each of about 900 ft, on the same day.

The W.1X assisted the detail design of the W.1, which incorporated a new control system (in which Dr Harry Ricardo participated) with a top speed governor and a barostatic control which reduced fuel pressure as a function of

Most of the valuable flying by the E.28/39s was done with the W.1A. This tested several new features already designed for the W.2, including curved guide vanes giving the air a pre-whirl in the direction of impeller rotation, and an air-cooled turbine. Rating was 1,450 lb.

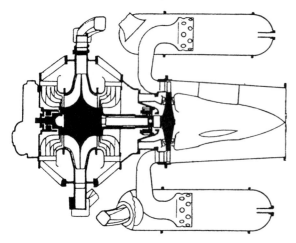

altitude. The W.1 first ran on 12 April 1941, and soon received 25-hour Special Category clear-ance at 850 lb thrust at 16,500 rpm. On 15 May 1941 it powered W4041/G on an uneventful first proper flight by Sayer from Cranwell. The installed weight was 623 lb. By 28 May the ini-tial flight programme was completed, after just under 40 h running of which 10 h 28 min was in flight. The engine stripped perfectly.

In October 1940 the official view of Whittle's engine had suddenly changed. Instead of being vague long-term research, it was recognised as an engine that could help win the war. Contracts were placed for 80 Gloster F.9/40 fighters per month, each powered by two W.2B engines to be made by BTH, Rover, and Vauxhall. This could all have happened, even though Power Jets was considered a mere laboratory, prevent-ed from competing with existing engine firms. But Beaverbrook's newly invented Ministry of Aircraft Production hit on a way of ensuring acrimony by telling Power Jets to design the W.2 for Rover to build in quantity, and telling Rover they could introduce modifications to facilitate production. BTH and Vauxhall dropped out.

From the start, Rover saw Power Jets as the people to beat, if not technically then politically. Their design changes ignored Whittle, and soon extended to redesigning the combustion cham-bers for straight-through flow. Just as it became fashionable to think centrifugal compressors were obsolete, so it soon became fashionable to think Whittle's reverse-flow combustor was obsolete (if you fly in a modern helicopter or turboprop you will probably be powered by both these obsolete ideas!).

To cut a long story short, Whittle's pioneer effort was so crippled that by November 1942 almost no progress was being made, and Power Jets and Rover were scarcely on speaking terms. Accordingly, Ernest (later Lord) Hives of Rolls-Royce took S. B. Wilks of Rover to dinner and said; 'Give us this jet job and we'll give you our tank-engine factory in Nottingham.' Thus, in a single sentence, Britain's vital jet programme was saved from near-extinction. But Rover had so delayed the programme that when the first F.9/40 at last flew it did so on engines of a total-ly different type. Whereas in December 1942 Rover logged 24 h testing the W.2B, in January 1943 Rolls-Royce logged nearly 400!

There were still problems with surging (com-pressor stall), blade failure, and combustion, but once Rolls-Royce took over the Rover works at Barnoldswick the project never looked back. The W.2B had first flown (at 1,250 lb) in the tail

The Rolls-Royce Welland I was the first turbojet in squadron service in the world, with RAF CRD Flight in May 1944 and No 616 Sqn on 12 July. Also called the W.2B/23, it at last incorporated the turbine blades with 5° twist which Whittle had been trying to get Rover to fit since 1941!

of Wellington II W5389/G in November 1942, and at 1,400 lb in the second E.28/39 in March 1943. Two (at 1,600 lb) at last powered an F.9/40 on 12 June 1943, 11 months after F.9/40 ground testing began. As the Rolls-Royce Welland I, a batch of 100 were delivered for the F.9/40 Meteor I, which in May 1944 became the first jet aircraft in the world to enter regular service. In July thrust was increased to 1,700 lb to help 616 Sqn catch flying bombs, in the first jet-v-jet engagements.

Using straight-through chambers, and many other changes including better fuel and oil systems, the Rolls-Royce B.37 went on test at 1,800 lb in August 1943. A batch of 500 rated at 2,000 lb and called Derwent I were built at Newcastle-under-Lyne from April 1944 to power the Meteor III. By this time, with proper effort being applied, the gas turbine was becoming understood, and development was not so much a desperate search for cures as a way to increase power and engine life.

Hives had such strategic vision that, unlike other British engine firms, Rolls-Royce was working on gas turbines in 1939. This was because Hives induced Dr A. A. Griffith to leave the Air Ministry and set up a small project office at the company's guest house, Duffield Bank. Hives just said 'Go on thinking'. Griffith had no inkling of practical mechanical design, but his assistant, Don Eyre, produced such beautiful drawings that for 21 years the Derby firm funded Griffith projects that were far ahead of their time.

The first project for a complete engine was the CR.1 (CR, contrarotating), schemed in 1940. This strongly resembled Rolls-Royce projects of 40 years later, being a turbofan with a BPR of 7. Air entered via ram inlets in the fan duct, turned 180° to be compressed by an axial compressor, continued through a second HP compressor, turned through another 180° in the combustion chamber in the front of the core fairing, and passed out via the multistage HP turbine and subsequent multistage LP turbine, the latter driving not only the inboard LP compressor but also the outboard fan blades.

Rolls-Royce was so captivated by such an advanced scheme that it actually tested the HP spool, beginning on 3 March 1942. Later it wasted more time and money on the even more advanced CR.2, finally giving up in 1944. By this time the engines Griffith despised, the simple Whittle turbojets, were in production. From the Derwent I, an outstanding development team at Barnoldswick under Dr Stanley Hooker produced the Mk II with 90° diffuser elbows, the Mk III to provide boundary-layer suction in the A.W.52 flying wing, and the Mk IV with a bigger compressor giving 2,450 lb.

In April 1944 Hooker visited GE in the USA and was startled to find not one, but two engines both rated at 4,000 lb. On his return he got his top team of Adrian Lombard, Fred Morley and Harry Pearson to scheme a completely new centrifugal turbojet handling 80 lb/s. The resulting RB.41 Nene was first run at midnight on 27 October 1944, and gave a thrust of 5,000 lb. In five months the small team had created the most powerful engine in the world, and it was simple, cheap, and weighed just 1,600 lb. By July 1945 a Nene was flying in a YP-80A Shooting Star, and subsequently Nenes were made in Canada and Australia, and under licence by Pratt &

Griffith's CR.1 contrarotating turbofan with double reverse flow and a BPR of 7 was roughly 60 years ahead of its time. Rolls-Royce was actually misguided enough to build and test the HP spool, shown enlarged in the second drawing. It never even approached its design performance.

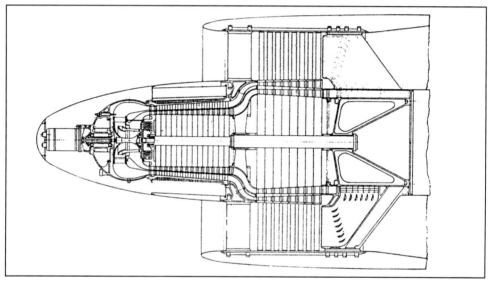

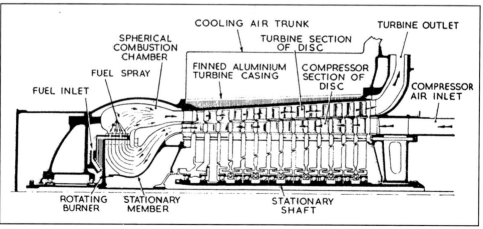

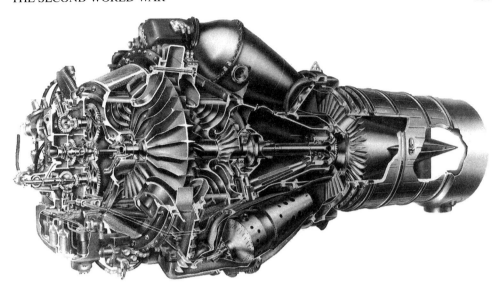

This cutaway of a Nene shows the auxiliary compressor amidships feeding cooling air to the turbine disc and main bearing. Also visible are the concentric curved vanes guiding air into the front and rear of the main compressor. Their importance was discovered on the very first run.

Whitney in the USA and Hispano-Suiza in France. Without a licence, the Soviet Union made 39,000, and in turn licensed a version to China which was in such demand in 1965 that it was transferred to a bigger factory, the last one coming off the line in 1979! In contrast, thinking centrifugal engines outdated, Britain almost ignored it, making a mere handful in 1950–54 for Attacker and Sea Hawk fighters for the Fleet Air Arm.

Even quicker than the Nene, the same team produced a version called Derwent V, scaled down to fit the nacelle of the Meteor. Rated at 2,650 lb, it ran on 7 June 1945, and at 3,500 lb powered a Meteor IV to a world speed record of 606 mph four months later. In 1943 Hooker began planning an advanced turboprop, the RB.39 Clyde, and in 1945 a Derwent II was quickly fitted with a reduction gear to drive a propeller; both are described in Chapter 17.

Whittle's small team, from early 1942 in improved premises at Whetstone near Leicester, demonstrated the way jet-engine design had progressed from being a vague art to a science – Whittle briefly thought 'an exact science' – in developing the W.2B. First, they lengthened the turbine blades from 2.455 to 2.73 in to produce the W.2B/500. This was designed in six months and ran on 2 September 1942, giving almost precisely the predicted performance (A in the table below). In July 1943 came the W.2/700, with turbine blades of Nimonic 80, an outstand-ing Ni-Cr alloy derived from material used in electric fires. The blades were made 0.3in longer at the trailing edge to match the gas expansion and allow mass flow to rise from 34.73 to 40.33 lb/s (B). Then came a redesign of the diffuser, from a long curve to two straight channels joined by a 90° corner with internal cascades (C). Finally came another increase in turbine-blade height, to 3.63 in, giving a mass flow of 47.15 lb/s (D). Compared with the W.1, thrust had been tripled without making the engine larger overall. Thus, all these versions could be tested in the E.28/39.

	A	B	C	D
rpm	16,750	16,750	16,750	16,750
Static thrust (lb)	1,755	1,850	2,040	2,487
Jetpipe temperature (°C)	606	620	597	647
Specific fuel consumption	1.13	1.12	1.07	1.05

Back in March 1936, the prospect of actually getting somewhere with his turbojet prompted Whittle to patent further ideas. For years he had recognised that propulsive efficiency for sub-sonic aircraft could be improved if his engine could somehow be 'geared down' to impart gentler acceleration to a greater airflow. The obvious answer was to increase the turbine power to 'drive a low-pressure compressor or fan capable of "breathing" far more air than the jet-engine itself, to provide a "cold jet" and also to "supercharge" the main engine'. The result was what we today call a turbofan, described in Patent 471,368 of 4 March 1936. On 2 March 1940 he was awarded patents for an aft fan (Chapter 6) and an aft fan plus core supercharger.

He called aft fans Thrust Augmentors, and by November 1942 Augmentors Nos 1, 2, and 3 had been designed, and two run on W.2/700 engines. Whittle also patented various forms of thrust boosting, as well as thrust spoilers which were tested on one of the several three-engined Wellingtons. In an extraordinary example of official long-range vision, a contract was signed on 8 October 1943 for four Miles E.24/43 aircraft (manufacturer's type number M.52) to explore the problems of flight at 1,000 mph at 36,000 ft. The first was to fly in January 1947, powered by a 2,487-lb W.2/700, with 18.5in nozzle, just sufficient for safe flight. Later M.52s were to go supersonic, powered by a W.2/700 plus No.4 Augmentor and with reheat in the fan duct and core, nozzle diameter being 37 in and take-off rating 7,000 lb.

Britain has an inexhaustible capacity for self-inflicted injury. In 1943, when Whittle was at the RAF Staff College, he planned the LR.1 turbofan for bombers having the range to attack Japan. It had 10 axial and 1 centrifugal compressor stages, a rating of 6,000 lb and a BPR of 3. Under pressure from the established engine firms, in April 1944 the government nationalized Power Jets and decreed it must never make any more engines, so the LR.1 was cancelled shortly before the first run. Another casualty of the same law was a 300 hp turboprop identical in layout to the Pratt & Whitney Canada PT6A, of which over 31,000 have been made. A third

disaster was that in January 1946 the M.52 was cancelled, the Ministry saying; '. . . the idea supersonic flight is just around the corner is quite erroneous . . . we have not the heart to ask pilots to fly it'.

The outrageous delay caused by Rover resulted in the F.9/40 making its first flight with engines of de Havilland manufacture. In 1940, though Whittle was still starved of funds, it was at last realised that the turbojet could be important to the war effort, and in January 1941 Tizard invited de Havilland Aircraft to produce a jet fighter and its engine. For the resulting H.1 Maj Frank Halford chose a configuration more like Whittle's 1930 patent than Whittle's own engines, with a single-sided centrifugal compressor, 16 slim straight-through chambers, and a single-stage turbine. The D.H.100 (later named Vampire) was to have this engine in a central nacelle, with an ideal short jetpipe, fed from inlets in the wing roots.

The H.1 ran on the bench in April 1942, reaching 3,100 lb thrust (a world record) in June. Halford's main reason for using a single-sided compressor was to take advantage of ram pressure in the inlet. The main penalty was that it increased the engine diameter. This was no problem in the D.H.100, but it was undesirable in a wing nacelle. As the H.1 would not fit in the E.28/39, two were air-tested in the fifth F.9/40 (DG206/G), which like other F.9/40s was lying un-engined. On 5 March 1943 it became the first Meteor to fly, the engines being cleared at a mere 2,000 lb. On 26 September 1943 an H.1, now named Goblin and cleared at 2,300 lb, powered the first D.H.100.

From 1943 de Havilland designed the D.H.106 Comet jetliner, and to power it produced the H.2 Ghost, effectively an H.1 scaled

The Power Jets LR.1 was the first turbofan to be designed. With staggering ineptitude, it was cancelled when the prototype was about to run.

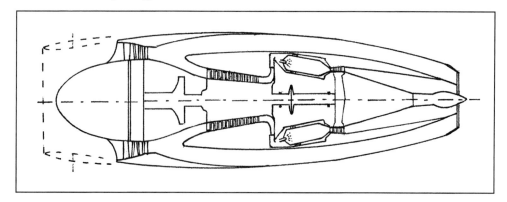

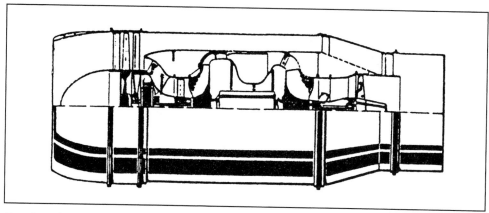

Based on the HeS 8 turbojet, the Heinkel S 10 was a far-sighted geared turbofan.

up from 63 to 88.5 lb/s. This first ran just after VJ-Day, on 2 September 1945. The H.3 turbo-prop of 500 hp was run but soon abandoned as having no market.

British wartime patents, based on work of the 1930s, included turboprop projects by Baynes-Muntz and Hewson-Locan. In 1942 Fairey patented a pressure-jet for driving the tip of a helicopter rotor (Chapter 18).

Meanwhile, the German engine firms saw Heinkel in the same light as British firms saw Whittle. The RLM tried to make him stick to airframes, but from 1939 did help fund Heinkel gas-turbine work, which was expanded in 1941 by taking over the Hirth factory at Stuttgart-Zuffenhausen. The early von Ohain engines were quickly tapered off, and even the advanced 109-006 designed by Max Adolf Müller was abandoned, though some work was done on the S10. This was a practical turbofan derived from the S8 turbojet, with a front fan driven via a reduction gear. Three were built and run in 1940, but from 1942 all effort was applied to the big 109-011 to be rated at 2,640 lb, rising later to 3,520 lb. This engine was regarded as one of the most important for the future Luftwaffe, but though five were run in 1943 and five more in 1944 it never flew except under a Ju 88 testbed.

In 1943 Daimler-Benz used the He S 011 turbojet as the basis for the 021 turboprop (for a long-range Ar 234). This February 1945 drawing shows the final stage reached.

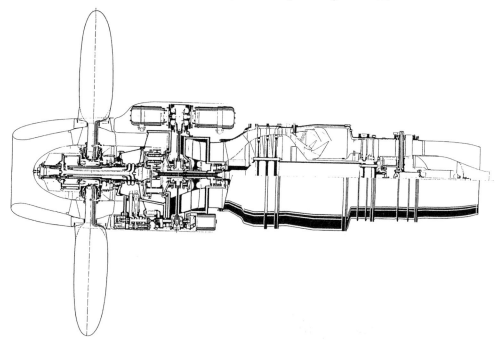

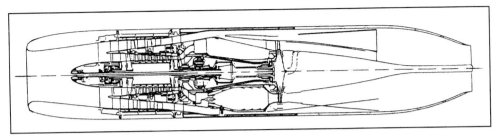

The DB 109-007 was a remarkable turbofan with fan blading on the outside of the compressor casing which rotated in opposition to the compressor rotor. The drawing shows how the turbine blades passed alternately through very hot gas (lower section) and bypass air (upper section).

The 021 turboprop version was taken over by Daimler-Benz. At the collapse in 1945 this had a two-stage turbine driving an axial inducer, diagonal plus three-axial compressor, and reduction gear to a propeller with a ducted spinner.

As already noted, Daimler-Benz stood out as the German engine company that did not wish to get involved in gas turbines. The RLM thus applied pressure, and reluctantly the giant of Stuttgart embarked on the ambitious 109-007 ducted fan, headed by Prof Karl Leist, who had headed turbosuperchargers at the DVL. The 007 had contrarotating spools, driven through a gearbox by a single highly loaded turbine. One spool had nine stages of compressor blading and the other eight interleaved compressor stages and three (the drawing shows two) stages of projecting fan blades in a bypass duct. Design thrust was 1.4 t (3,080 lb) at 560 mph, but it never came near this figure and never flew. The company never built a 109-021 turboprop, but

drew several forms of P.100 axial turbojet with ratings up to 13 t (28,660 lb).

Thus, by far the most important jet firms were BMW/Bramo and Junkers. The Bramo 109-002 was dropped in 1942, but the relatively simple 003 was pressed ahead, though initial results were poor. Despite a host of problems, two 003s were flown on the first Me 262 on 25 March 1942; both quickly failed, and the circuit was completed on the Jumo 210 piston engine in the nose which was needed to enable the aircraft to take off. The chief designer at Berlin-Spandau, Hans Rosskopf, produced what amounted to a new and larger engine, the 003A, first run in December 1942 and reaching 500 kg. After intensive development an 003A-0 flew under a Ju 88 on 17 October 1943, and by January 1944 was urgently being cleared for production at 800 kg (1,764 lb) with a life of 50 h. In August 1944 100 were delivered, and the total by the end of 1944 exceeded 3,500. Few of these reached airframes, though the He 162A was to

When the BMW 109-003 was an active programme there was no time to make a sectioned display exhibit. This 003E-1 was constructed in the late 1950s. The E was the final production version for the He 162, similar to the A-1 but with a 30-second boost rating of 920 kg.

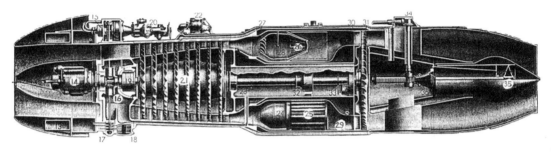

*A production Jumo 109-004B: **1**, nose cowl; **2**, oil tank; **3**, inlet case; **4**, auxiliary gearbox; **5**, compressor casing; **6**, servomotor; **7**, igniter box; **8**, throttle lever; **9**, outer case; **10**, attachments; **11**, bullet shaft from 6; **12**, jetpipe case; **13**, starter fuel tank; **14**, Riedel starter; **15**, fuel pump; **16**, tower shafts; **17**, oil pump; **18**, oil filter; **19**, front bearing; **20**, speed governor; **21**, compressor rotor; **22**, fuel filter; **23**, rear bearing; **24**, flame tube; **25**, muffle; **26**, turbulence bars; **27**, combustion chamber casing; **28**, burner; **29**, turbine entry; **30**, nozzle guide vanes; **31**, turbine rotor; **32**, forward turbine bearing; **33**, rear bearing and scavenge pump; **34**, bullet drive gears; **35**, bullet; **36**, bullet support ring.*

be built at the rate of 2,000 per month.

Partly because of its smaller diameter, the 003 was considered to have greater potential than the rival 004B. The 003C had an improved compressor, the 003D was largely redesigned, and the 003R was an 003A combined with a BMW 718 rocket. In 1941 development began of the 018 turbojet and 028 turboprop, both with 12-stage compressor and 24-burner annular chamber. The 018 had a 3-stage turbine and the turboprop four stages. Allied troops reached Munich-Allach as the first 018 was being installed on the testbed.

We left the Junkers engine division at Dessau acrimoniously taking over the airframe division's jet projects at Magdeburg, with Wagner and Müller both walking out. In August 1939 the RLM awarded a contract for a completely new axial jet, the 109-004, the only stipulation being a thrust at 900 km/h of 600 kg (1,320 lb). Designed to run on diesel oil, the 004 had a simple eight-stage compressor of about 3.0 p.r., six flame tubes in an annular chamber, a single-stage turbine, and rack/pinion drive to an adjustable nozzle bullet. Under Anselm Franz, development

was aimed at reliability rather than performance. In January 1942 the fifth 004A-0 reached 2,205 lb (normal rating 1,848 lb) for a weight of 1,870 lb, on 15 March 1942 an engine began flight test under a Bf 110, and on 18 July 1942 two provided sole power for the Me 262 V2.

By early 1943 the 004B was on test, made largely of sheet metal and with strategic alloys halved, rating remaining 1,848 lb but weight being reduced to 1,650 lb. After many further problems had been solved the 004B-1 went into large-scale production in late March 1944, and about 6,000 were built by VE-Day. Standard rating was 900 kg (1,984 lb) at 8,700 rpm; basic length was 3,865 mm (152in) and diameter 806 mm (31.7in), dry weight being about 720 kg (1,587 lb). Features included a spigoted-disc light-alloy compressor with wrapped-sheet stators and solid rotor blades held in dovetails by grub screws, an annular combustion chamber of aluminised mild steel housing six flame tubes with upstream swirl cascades, upstream injection and downstream baffles, and a single-stage turbine with 61 Ni-Cr steel rotor blades soldered and pegged into a disc of austenitic steel. The starter was a Riedel flat-twin two-stroke inside the nose bullet, itself started by a pull cord or electrically by the pilot, and running on petrol from a 3-litre tank. Rated at 10 hp at 10,000 rpm, these became popular post-war for Go-karts.

By late 1944 Junkers had perfected the manufacture of hollow turbine rotor blades by deep drawing. These increased reliability and were easier to make using less scarce material, and entered production in the 004B-4 at the war's end. The D-4, with a minor combustor change, was about to enter production at 1,050 kg (2,315 lb). Like BMW, Junkers also had contracts for a much bigger turbojet, the 012, and the 022 turboprop version, but made slow progress.

In the United States there was little prewar interest. Rudolph Birmann, a Swiss whose 1922 thesis described a gas turbine, formed Turbo Engineering Corporation (TEC) in 1937, receiving a US Navy contract for advanced turbochargers, followed in October 1942 by a turbojet of 1,100 lb thrust (envisaged as a booster only). Predictably, TEC was too small, and in 1944 had to drop the jet and concentrate on the turbochargers. In August 1939 Northrop's newly independent company began working on

gas turbines at the insistence of its Czech research head, Vladimir Pavlecka. In early 1941 Gen Henry H. Arnold, Chief of the Army Air Corps, wrote to the National Advisory Committee for Aeronautics (NACA) recommending a study of jet propulsion, by which he meant rockets. In March 1941 the Special Committee on Jet Propulsion was set up under William F. Durand, who had been Moss's professor at Cornell and Chairman of the NACA in the First World War. They quickly began to study a wide spectrum of possible engines, and in July 1941 awarded contracts to three steam-turbine firms, Westinghouse to build a turbojet, Allis-Chalmers a ducted fan, and General Electric (GE) a turboprop.

Ten days after writing to the NACA, Gen Arnold was on a fact-finding trip to Britain, where he was told that Whittle's engine was soon to fly. GE's D. Roy Shoults, a turbosupercharger expert on liaison in London, already knew about Whittle's work, and Arnold quickly arranged for a W.1-type engine to be made by GE. For a total licence fee of $800,000 the USA was given full information on Whittle's various projects, and permission to use it for civil as well as military purposes. They also received

Cross-section of a General Electric I-A: A, inlets; B, impeller; C, diffuser; D, combustor inlet; E, outer case; F, reverse-flow combustor; G, fuel nozzles; H, spark plug; J, turbine nozzle ring; K, turbine; L, shaft; M, jetpipe.

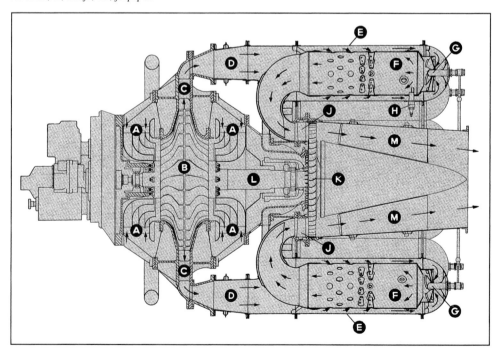

Power Jets' only bench engine, the W.1X, which with a support team was flown to GE's factory at Lynn, Massachusetts, on 1 October 1941.

The result was the General Electric I, first run on 18 April 1942. On 2 October 1942 a pair of improved I-A engines, each rated at 1,250 lb, took the first Bell XP-59A Airacomet into the sky. By this time the improved I-16 was already running at 1,400 lb, leading to the delivery of 241 I-16s to the Army for P-59As and Navy FR-1s, the designation reflecting the improved rating of 1,600 lb. In early 1943 the Army asked GE to consider an engine of 4,000 lb thrust. The result was the I-40, created by a group under Dale Streid and first run on 9 January 1944. This important engine, which paralleled British work in switching to 14 straight-through combustion chambers, went into production at GE's Syracuse plant as the J33, but at the insistence of the Army Air Force the engine was also licensed to Allison Division of General Motors, thus bringing in an established aero-engine supplier. In September 1945 Allison took over the J33 completely.

One of the chief applications of the J33 was the Lockheed P-80 Shooting Star. Lockheed had tried to produce a turbojet in 1940. Research engineer Nathan Price designed the startlingly advanced L-1000 with two-spool axial compressors aimed at a p.r. of 17 plus an afterburner, but when President Robert E. Gross took the brochures for the L-1000 and the L-133 canard fighter to be powered by it to Wright Field in 1942 he was told 'Forget it, keep building P-38s'. But in 1943 the Army did support the engine as a long-range project called XJ37. Like so many projects, it combined great promise with complexity and severe problems, so in October 1945 Lockheed handed the whole effort, including Price, to Menasco. This company in turn passed it to Wright, which eventu-

ally gave up.

Also in 1943, Lockheed's Chief Research Engineer, C. L. 'Kelly' Johnson, was told that the Army and Navy both wanted to have aircraft powered by Halford's H-1 (Goblin), and Allis-Chalmers was given a contract to make the 3,000 lb engine under licence as the J36. Johnson quickly designed the XP-80 and had it at Lake Muroc ready for ground running on 15 November 1943. Guy Bristow, the de Havilland rep, was horrified when he saw the flimsy inlet ducts, and said 'the sheet gauge ought to be thicker'. He was right; the British engine sucked the duct into its compressor. Generously, de Havilland took the only available engine out of the second prototype Vampire and sent it to Lockheed. It enabled XP-80 testing to begin on 8 January 1944. But Allis-Chalmers failed to produce J36s, and the British engine was replaced by the J33.

Meanwhile, GE's turboprop requested by the Durand Committee was taking shape at the steam-turbine works at Schenectady (Chapter 17). Its gas generator first ran on 15 May 1943, and this made the engineers think; 'Why not an axial turbojet?' After agonising doubts about diluting their jet effort, it was decided that the greater potential of the axial made it a candidate for the future (though probably not for the Second World War), and the Army Air Force agreed. Accordingly, the prototype TG-180 ran on the testbed on 21 April 1944. It had an 11-stage compressor handling 75 lb/s at a p.r. of 5, eight tubular chambers and a single-stage turbine. Static thrust, like that of the I-40, was 4,000 lb. Coming from nowhere, GE had produced two quite different turbojets, each more powerful than anything in Britain or Germany. Surprisingly, the TG-180 first flew in a single-engined fighter, the XP-84, on 28 February 1946.

Compared with the I-40, the TG-180 was

General Electric's TG-180 became the Allison J35.

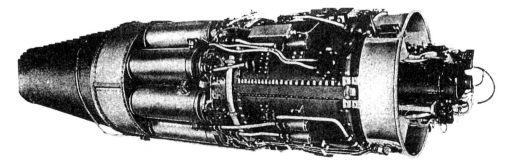

The Westinghouse J30 was the turbojet of 19 in diameter contracted for on the day following Pearl Harbor. It was the first American-designed turbojet to run.

slimmer (37.5 in against 48) and offered fractionally better fuel economy (1.075 against 1.185), but for the same thrust was heavier (2,335 lb against 1,820), considerably more complicated and expensive, and much more difficult to develop. GE eventually delivered 140, with Army/Navy designation J35, and these powered a wide range of fighter, bomber, and research prototypes. For series manufacture the USAAF brought in the Chevrolet Division of General Motors at Tonawanda, NY, but after delivering small batches this firm withdrew to resume car production, and in September 1946 the J35, like the J33, was assigned wholly to Allison for production and further development.

The Navy's early attitude was secretive. The Northrop and TEC projects were not disclosed even to other Government agencies, and when contracts were awarded to Allis-Chalmers for a ducted fan in October 1941, and to Westinghouse for a turbojet on 8 December (the day after Pearl Harbor), the recipients were specifically instructed not to communicate with any other company. Allis-Chalmers got nowhere, but Westinghouse had so much steam-turbine experience that R. P. Kroon's team at South Philadelphia flew their engine before the end of the war. Called the 19A because its diameter was 19in, this neat engine was seen mainly as a booster, not a primary powerplant. It had a six-stage compressor, an annular combustor, and a single-stage turbine, and first ran on 19 March 1943. A prototype rated at 1,200 lb flew slung under an FG-1 Corsair in January 1944.

To get Pratt & Whitney into gas turbines, the Navy awarded the company a contract to build 130 of the 261 Westinghouse 19XB engines, rated at 1,600 lb. Designated J30, its first major

application was to power the twin-jet McDonnell FD-1 (later FH-1) Phantom, the prototype of which made its first flight on 25 January 1945 on one engine (the second had not then been delivered)! Westinghouse also produced axial turbojets of 9.5 and 24in diameter, the 24in being built in large numbers as the J34. First run in 1944, this had an 11-stage compressor, initially handling 50 lb/s with a p.r. of 3.65 and later 55 lb/s at 4.35, and a combustion chamber with 24 downstream burners round an inner flame tube and 36 round the outer. Rated at 3,000 (later 3,500) lb at 12,500 rpm, the J34 typically weighed 1,220 lb and was an important Navy engine until after 1960. Sadly, it did not give Westinghouse the experience it was later to need.

Apart from making Westinghouse turbojets, Pratt & Whitney had dabbled in gas turbines from 1939; Chief Engineer L. S. 'Luke' Hobbs told the author he mistakenly dismissed such work as uncompetitive. But in October 1940 Jerome C. Hunsaker of MIT told Hobbs about his colleague Andrew Kalitinsky's work using a two-stroke diesel to drive a compressor and provide gas for a turboprop. Hobbs had the numbers checked and launched the PT-1 (Chapter 18). He also began project design of a turbojet immediately after the war. The other engine giant, Wright Aeronautical, remained aloof until late 1945, when it began work on a turboprop.

In the Soviet Union the greatest pioneer was Arkhip Mikhailovich Lyul'ka. Like Whittle, he encountered disbelief when in 1936 at Kharkov Aviation Institute (KhAI) he produced calculations describing an axial turbojet. However, Russians tend to have open minds and accept that 2 + 2 probably equals 4, so with established engine designers I. F. Kozlov and P. S. Shyevchenko he was allowed in 1938 to start work on the VRD-1 at the Leningrad plant named for S. M. Kirov. With an eight-stage compressor and 700°C turbine gas temperature, the VRD-1 was to weigh 500 kg and give a thrust of 600 kg (1,323 lb), but before it could run Germany invaded, and Lyul'ka was sent elsewhere to do more urgent work (including the BI-1, next chapter).

In 1942 he returned to the besieged city and began work on two engines to give 1300 kg, one centrifugal and the other axial. After running the VRD-2 research engine rated at 700 kg, he went ahead with the VRD-3, later called S-18 and finally TR-1, with an improved eight-stage compressor and annular combustor. It weighed 885 kg, and in 1944 reached its design thrust of 1,300 kg (2,866 lb) at 6,950 rpm. Among its

Little known outside the Russian Federation, the Lyul'ka TR-1 was an outstanding axial turbojet first run in December 1943. This example is displayed sectioned at Monino.

applications were the Il-22 and Su-10 four-jet bombers. Subsequently Lyul'ka became famous.

In Canada in 1944 the government opened a laboratory at Malton, Toronto, called Turbo Research, to study gas turbines. After the war this became a division of Avro Canada (Chapter 13).

In Japan the Navy developed various piston/compressor jet units, notably the Tsu-11 of 200 kg (441 lb) thrust, but in 1941 work began at the Navy Yokosuka plant, helped by outside firms, on the Ne-00 turbojet, which in 1943 was tested slung under a Ki-48-II bomber. This led to the 320 kg thrust Ne-12, but in 1944 all effort was switched to the Ne-20, based on photographs sent from Germany of the BMW 109-003A (it had been hoped to receive actual engines, sent by U-boat). Rated at 1,050 lb, the Ne-20 first ran in March 1945, and this engine was used in the twin-jet Navy J8N1 attack aircraft which made one flight on 7 August 1945.

In France such pioneers as Leduc (next chapter) and Rateau could do little during the occupation, but in 1941 a team at Cie Electro-Mécanique began work on a turboprop, the TGA 1bis, TGA meaning *Turbine à Gaz d'Aviation* but being represented to the German authorities as *Turbo-Groupe d'Autorail*. Using Brown Boveri technology, the company was ready to build as soon as Paris was liberated. Famed for superchargers and turbocompressors, Josef Szydlowski's Société Turbomeca moved in June 1940 to Bordes, Pyrenees, and a year later began work on a small gas turbine to give shaft power or jet thrust. This, too, had to wait until after the war.

Never seen before, this cutaway of the first turbojet tested by Arkhip Lyul'ka shows the relatively enormous size of the combustion chamber in 1940. The engine was designated RD-1, rated at 500 kg (1,102 lb).

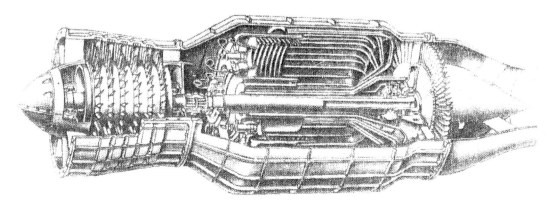

12 Rockets and ramjets

The point was made earlier that the chief reason for the lack of success with aeroplanes of the 19th century was the absence of a suitable propulsion system, and that this could have been provided quite simply by solid-propellant rockets. So far as the author knows, nobody actually did this until 11 June 1928, when Fritz Stamer flew a tailless glider – the *Ente* (Duck), designed by Alexander M. Lippisch – about 4,300 ft in 70 sec. A third rocket glider gained much publicity by covering about a mile at around 95 mph on 30 September 1929. Piloted by Max Valier, the Opel-Hatry

Assisted-take-off rockets are usually of the simple solid-propellant type (upper drawing); the lower diagram shows major elements of a throttleable liquid-propellant rocket engine (Pratt & Whitney).

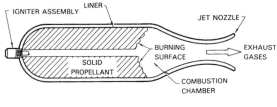

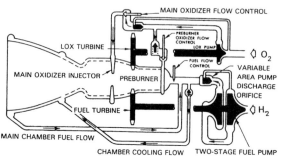

Rak I was powered by four groups of four Sander cordite motors each giving 55 lb thrust, fired in sequence.

These experiments made no contribution to aviation, but this chapter is needed to cover the many forms of jet engine that do not rely on a gas turbine. These are dealt with in groups: rockets, then ramjets and pulsejets; helicopter pressure jets are in Chapter 18.

To continue with rockets, Germany, the location for the 1928–29 flights, was also the next country to build rocket aircraft. In the first experiments, Wernher von Braun and Walter Dornberger – who later masterminded the A.4 'V2' missile – tested two different liquid-propellant rockets in two prototypes of the He 112 fighter; both exploded on the ground as soon as the rocket was fired. In April 1937 Heinkel rather less eagerly donated yet another He 112, the V7, for testing with a von Braun lox (liquid oxygen)/alcohol motor. First flown in late April 1937, this aircraft made several flights before it too was destroyed.

Rather scorned by the mainstream designers, BMW developed small rocket engines from 1937, culminating during the war in the 109-718. This was a powerful bipropellant unit packaged with its pressurized tankage to fit above a turbojet. Its rating was 2,700 lb for three minutes. It powered the Me 262C family, the engines being BMW 003Rs (003B plus rocket).

In May 1937 Heinkel got one of his designers, Hans Regner, to plan a single-seater to test the R 1 liquid-propellant engine, which appeared rather less dangerous. It was the first properly engineered aircraft rocket to be developed by HWK (Helmut Walter Werke of Kiel). It operated on *T-stoff* (concentrated hydrogen peroxide), which on being passed through calcium permanganate catalyst instantly decomposed into superheated steam and oxygen. It had a degree of pilot control up to a maximum thrust of about 600 lb. Erich Warsitz flew the resulting He 176 at Peenemünde-Ost on 20 June 1939. Although it is sometimes represented in artwork as a bullet-like high-speed aircraft, the 176 actually had fixed tricycle landing gear and an open cockpit.

In 1938 HWK had begun working with Lippisch's team on the DFS 194 tailless testbed, which was audaciously intended to lead to a short-range interceptor with unmatched performance. In January 1939 Lippisch's team was transferred from DFS to Messerschmitt AG. The DFS 194 was flown with an R I engine in 1940, followed in 1941 by the first two Me 163 prototypes with the more powerful R II engine. The permanganate tended to clog the injector nozzles, and HWK developed a larger and better engine, the 109-509, fed with *T-stoff* and *C-stoff*, the latter being hydrazine hydrate and methyl alcohol. When the two liquids were mixed, the reaction was violent enough to be called an explosion.

First flown under power in August 1943, the Me 163B Komet was the only reusable manned rocket aircraft to go into production, and was certainly more dangerous to its pilots than to its enemies. About 364 were delivered, becoming operational from late June 1944. The 109-509A engine had a thrust range of 441-3,748 lb at sea level, and about 10 per cent more at 30,000 ft. Late in the war HWK began delivering the 509C engine, for the totally redesigned Ju 248 (Me 263), in which an auxiliary chamber rated at 662 lb enabled flight endurance to be extended from 8 to 12 minutes. HWK 509A engines were also used in the ultra-high-flying DFS 228 reconnaissance aircraft, and the DFS 346 supersonic research aircraft which was flown briefly by the Russians after the war.

In fact, the Soviet manned-aircraft rocket programme in 1924–60 was easily the biggest and most diverse in the world. The OR-2 lox/kerosene engine was tested in 1933 and the BICh-11 tailless glider was also flown, but the OR-2 never powered the BICh-11 as had been intended. Of a host of experimenters, the names of S. P. Korolyev, L. S. Dushkin, and V. P. Glushko stand out, the last two as leaders of engine design teams. Glushko used several propellant combinations, notably nitrogen peroxide and toluene methylbenzene, and achieved a

Block diagram of the D-1A engine of the BI-1, which had an installed weight of 48 kg (106 lb): **1**, *air bottles (11);* **2**, *kerosene filter;* **3**, *air collector;* **4**, *air filter;* **5**, *line valves;* **6**, *pressure reducer;* **7**, *non-return valves;* **8**, *air supply pressure;* **9**, *kerosene pressure;* **10**, *crash-landing shut-off valves;* **11**, *RFNA tanks;* **12**, *acid filter;* **13**, *kerosene tank;* **14**, *aux kerosene tank;* **15**, *nitrogen consumption sensor;* **16**, *kerosene consumption sensor;* **17**, *acid/nitrogen drains;* **18**, *kerosene drains (4);* **19**, *stop valves (3);* **20**, *filters (3);* **21**, *combustion chamber;* **22**, *venturi;* **23**, *kerosene thermometer;* **24**, *thermo-relay;* **25**, *throttle valve;* **26**, *chamber pressure;* **27**, *power lever;* **28**, *three-way air valve;* **29**, *starting valve;* **30**, *T-50 reducing valve.*

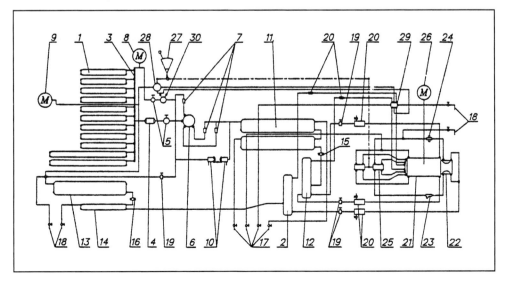

flight under power in the RP-318-1, a converted SK-9 sailplane, on 28 February 1940.

Dushkin managed the development of 17 different engines, all burning RFNA (red fuming nitric acid) and kerosene or petrol. One of the major programmes was the Berezniak-Isayev BI-1 fighter, powered by a D-1A engine rated at 1,100 kg (2,425 lb) at sea level. Gliding trials with the first of five BI-1s began on 10 September 1941, followed by powered flights from 15 May 1942, but these ran into compressibility problems causing structural failure. This put a blight on rocket fighters, curtailing work on the Tikhonravov 302 and Polikarpov Malyutka, but Dushkin did manage to see his booster engines, notably the RD-1KhZ, flown in the tail of a variety of piston-engined aircraft. His RD-2M-3V, with two chambers with sea-level ratings of 3,200 and 882 lb, powered the MiG I-270, derived from the Me 263.

In California on 22 June 1944 a Northrop MX-334 prone-pilot flying wing first flew powered by its tiny (200 lb thrust) Aerojet engine running on RFNA and aniline. Subsequently Aerojet-General concentrated on assisted-take-off motors and engines for unmanned vehicles. In Japan a three-barrel solid-propellant motor powered the MXY7 Ohka suicide attack aircraft, first flown under power in November 1944 and used operationally.

In early 1944 the USAAF and NACA decided to fund aircraft to fly faster than sound, and quickly decided on rocket propulsion. The immediate result was three Bell XS-1 (later X-1) aircraft; No 1 (46-062) made a gliding flight on 25 January 1946, No 2 made the first flight under power on 9 December 1946, and No 1 exceeded the speed of sound on 14 October 1947. All were powered by a Reaction Motors 6000C4 (LR11) engine with four chambers, each with a sea-level rating of 1,500 lb, which could be fired consecutively or together. Propellants were lox and ethyl alcohol, and failure to develop the turbopump so curtailed endurance that the X-1s had to be air-launched. The No 3 aircraft, which flew just once (on 20 July 1951) did have turbopump feed, but was destroyed in an explosion during preparations for a second flight. This was one of a series of catastrophic explosions of X-1 and X-2 aircraft caused by the rash use of sealing gaskets made of Ulmer leather in contact with lox. Generally similar engines powered the X-1A, B, and D, which had greater tankage and longer endurance, as well as the Republic XF-91, which also had a J47 turbojet. A developed version powered the X-1E and three Navy Douglas D-558-II Skyrockets. The latter were designed also to have a turbojet, but this was later removed and the rocket tankage doubled.

Bell later built two X-2 aircraft, powered by the Curtiss-Wright XLR25 engine. Using the same lox/alcohol mix, this had one small and one large chamber to give thrust variable at sea level from 2,500 to 15,000 lb. Both chambers were later fitted with extended divergent nozzles to expand the jet to higher velocity at high altitude.

The Skyrocket's designer, Ed Heinemann, naturally planned a successor, the D-558-3. Powered by a Reaction Motors XLR30 engine with a sea-level thrust of 50,000 lb, operating on lox and ammonia, this would have had fantastic hypersonic performance in the atmosphere and in space. It was never built, its place being taken by the North American X-15, funded by NASA with the USAF and Navy. The first of three X-15s made a gliding flight on 8 June 1959. The No 2 made the first powered flight on 17 September of that year, with two XLR11 engines similar to those of the X-1s. On 15 November 1960 the No 2 aircraft flew with the Reaction Motors XLR99, very like the XLR30 that would have powered the D-558-3. Its single chamber was controllable from 25,000 lb to 50,000 lb at sea level (57,850 lb at high altitude), fed by a peroxide-powered turbopump with anhydrous ammonia and lox at the rate of 13,000 lb/min. This gave a full-power endurance of 85 seconds, but the No 2 aircraft (56-6671) was rebuilt as the X-15A-2 with about 70 per cent greater endurance. Peak performance included 4,534 mph, Mach 6.71, and 354,200 ft, figures never approached by any other aircraft.

In Britain, Armstrong Siddeley Motors developed controllable engines operating on lox/alcohol, while de Havilland Engines concentrated on high-test peroxide (HTP), later adding kerosene to burn in the freed oxygen. In France, SEP selected RFNA as oxidant and alcohol, petrol, or furaline (turpentine) as fuel for a range of engines which powered the Trident interceptor and served as boosters for early Mirage IIIs. Further details are given in the PSL book *Faster than Sound*.

Patents for ramjets – also called athodyds (from aero-thermodynamic duct) or 'flying stovepipes' – extend back to early in this century, Lorin's being most promising. Another pioneer was a Hungarian, Fon, who in 1928 proposed variations on Lorin. For dogged persistence few pioneers equalled René Leduc, who began bench tests in 1929, designed the

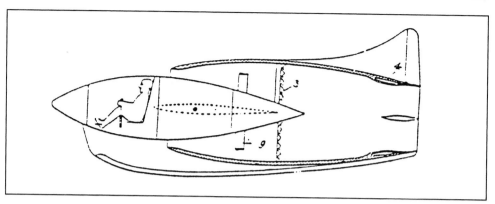

Drawing from Leduc's first patent of 1933, showing the cockpit housed in the centrebody in the inlet to the integral ramjet fuselage.

O.10 aircraft in 1935–37, and completed it in 1947. His idea was to confine the pilot in a centrebody in the circular inlet to a giant ramjet forming the entire fuselage. The O.10 made its first flight under power on 21 April 1949. Several further Leduc aircraft followed, culminating in the O.22, which had an Atar turbojet to enable it to take off by itself. It flew on Atar power on 26 December 1956, but was abandoned before the ramjet could be brought into action.

In wartime Germany various ramjets were tested on the ground or on testbed aircraft, the largest being a formidable unit developed by Dr E. Sänger of the DFS, which was flown mounted high above a Do 217E. The war ended before any of the numerous ramjet aircraft projects could be built. In the USA the champion of ramjets was Roy E. Marquardt, but though his subsonic ramjets were tested on many aircraft, all production versions, and all his supersonic engines, found unmanned applications. Smaller Marquardt ramjets were used to power helicopters (Chapter 18).

For over 40 years one of the favoured propulsion systems for hypersonic aircraft (those designed for Mach numbers of 4 or more) has been the external-burning ramjet (EBRJ). In this, the airflow is slowed in a supersonic venturi formed by a flat angular wedge on the underside of the aircraft. Fuel is injected, and made to burn stably in the supersonic airflow, the very hot gas finally accelerating to perhaps Mach 6 in a wedge diffuser. The example shown has upper and lower walls to the duct, fuel being injected at X. No full-scale EBRJ has yet been tested, far less flown.

Helicopters were a fertile field for proponents of small rockets, and also of pulsejets, of which many kinds have been used for fixed-wing

A typical hypersonic external-burning ramjet (EBRJ), in this case with a duct formed between upper and lower walls (the latter usually absent).

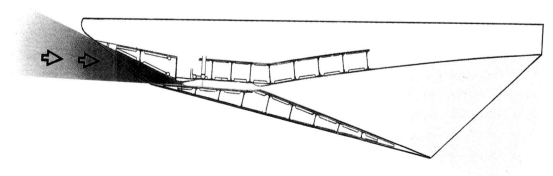

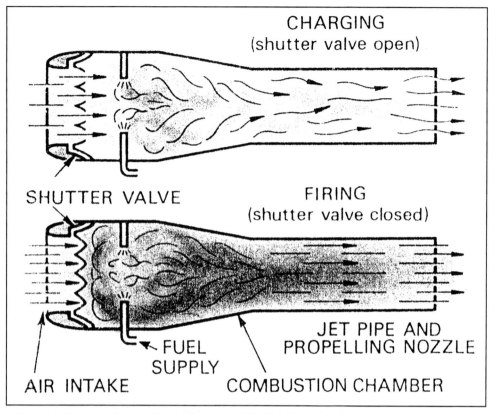

CHARGING
(shutter valve open)

SHUTTER VALVE

FIRING
(shutter valve closed)

JET PIPE AND
PROPELLING NOZZLE

FUEL
SUPPLY

AIR INTAKE   COMBUSTION CHAMBER

Schematic diagram of a pulsejet similar to the 109-014 flying-bomb engine.

propulsion. One of the earliest experimenters was Karavodine, who tried to drive a turbine with gas discharged intermittently from a long tube with a unidirectional inlet valve. More effective was the resonant pulsejet patented by Paul Schmidt in Germany in 1931. This eventually led to the Argus 109-014, with 63 spring-steel flap valves and nine injectors for petrol at the inlet to a duct just over 11 ft long. Resonant frequency was typically 45 Hz, or 2,700 pulses/min, and thrust at 360 mph was 600–650 lb. Some 31,000 were delivered, most for Fi 103 ('V1') flying bombs, but including the Reichenberg piloted version and the Me 328 twin-jet attack aircraft.

Many inventors have patented valveless resonant pulsejets. Some of the first to be flown (from 30 November 1950 on an SA 104 Emouchet glider) were developed by SNECMA in France. These simple but carefully profiled tubes initially had a forward-facing inlet, in front of which was a U-tube to turn gas expelled from the nozzle through 180° to give thrust. Later versions, the Ecrevisse (Crayfish) family, were completely in the form of a U with the inlet and nozzle facing aft. They resonated at about 60 Hz and gave from 22 to 331 lb thrust, depending on size. In the USA for many years E. M. Gluhareff worked on resonant pulsejets fuelled by propane and available with or without a ram-type inlet which reduced static thrust but increased it in flight. None of these devices found a really useful market.

13 The post-war era to 1960

The end of the Second World War left air forces everywhere equipped with piston-engined aircraft which, in the main, suddenly appeared obsolete. The only factories that had produced gas-turbine engines in large numbers, at Munich and Magdeburg, had been first bombed and then looted. So most countries went into overdrive to build a national capability in the technology and production of such engines, and the aircraft to be powered by them.

There was one startling exception. Great Britain had elected a new government which regarded defence expenditure as anathema. It showed no interest in the national lead in the new species of aero engine, and the only thing it did with the world-beating Nene was to ship some to the Russians in September 1946, in the naive belief that this would make the Russians less unfriendly. Maj-Gen Vladimir Klimov, who had been trying to produce a copy of the Nene, was staggered that Britain should do such a thing. He quickly organised production of a Soviet version, called RD-45 after the number (GAZ-45) of the Moscow factory, and had it in production in August 1947. Soon factories at Ufa and Kuibyshyev came on stream, and altogether 39,000 engines based on the Nene were delivered, most of them VK-1s rated at 2,700 kg (5,952 lb) and afterburning VK-1Fs rated at 3,380 kg (7,450 lb). As noted earlier, various VK-1 versions were licensed from Moscow to China, where a further 8,500 were built. When Dr Stanley Hooker visited ShenYang in 1972, he was astonished to find the WP-5D version still in production, and convulsed his hosts by saying; 'The Russians made a good copy of the Nene, they even copied my mistakes!'

Apart from the Nene, which at a stroke solved propulsion problems for such important aircraft as the MiG-15, Il-28, and Tu-14, the Soviet Union also naturally did what it could with German technology. For three years substantial numbers of engines based on the Jumo 004 were produced as the RD-10, and others based on the BMW 003 as the RD-20, but the belief widely held in the West for 20 years after the war that almost all Soviet gas turbines were based on German designs was founded on ignorance. On the other hand, considerable use was made of German engineers, notably in the design of the first turboprops (Chapter 18).

The chief engine design bureaux were headed by Klimov and by A. A. Mikulin, N. D. Kuznetsov, A. D. Shvetsov, A. M. Lyul'ka and V. A. Dobrynin. Klimov failed in an attempt to enlarge the basic Nene design (as the VK-5), and his successor, S. P. Isotov, concentrated on shaft-drive engines. Mikulin's first important gas turbine was the AM-3, a large single-shaft turbojet designed for reliability rather than performance. It had a diameter of 1.4 m (over 55 in) and an eight-stage compressor handling 298 lb/s. It began in 1950 at 14,900 lb, and over 9,000 were later produced at 19,180–20,945 lb as the RD-3 family, and in China as the WP-8, mainly for the Tu-16 (Chinese H-6) and Tu-104. The RD-3M was also adopted as temporary engine of the M-4 heavy bomber, predecessor of the 3M 'Bison' powered by the VD-7 (see later). The latter proved unreliable, and many 3M were re-engined with the RD-3M-500A, reducing thrust from 11 to 9.5 tonnes and considerably shortening the range.

The next mass-production engine was the AM-9, a slim (813 mm, 32 in) turbojet with a nine-stage compressor, a can-annular combustor, and a two-stage turbine. In 1956 Mikulin was replaced by his deputy, S. K. Tumanskii, who, starting with the RD-9 (AM-9 redesignated), saw his engines produced in greater quant-

Big, simple and reliable, the Mikulin RD-3M-500 was built in large numbers.

ity than those of any other engine designer since 1945, a total of some 72,000. Even the original RD-9 was still in production as the WP-6A in China until 1990, rated at up to 8,929 lb with afterburner. Next came the R-11, the most numerous of all. First run in 1953, this simple two-shaft turbojet has a three-stage LP compressor and three-stage HP compressor, the first LP stage being overhung ahead of the front bearing without inlet guide vanes, then a bold

A typical fighter engine of the immediate postwar era, the RD-9BF was licensed to China as the WP-6A: **1**, *variable inlet vane;* **2**, *stators;* **3**, *case;* **4**, *air-bleed band;* **5**, *bleed actuator;* **6**, *rear load-relief cavity;* **7**, *centre bearing;* **8**, *starting igniter;* **9**, *Stage 1 nozzle;* **10**, *turbine rotor;* **11**, *Stage 2 nozzle;* **12**, *quick-release ring;* **13**, *afterburner core;* **14**, *quick-release ring;* **15**, *front flange;* **16**, *flameholder gutters;* **17**, *shroud;* **18**, *bracket;* **19**, *nozzle actuator;* **20**, *flap hinge ring;* **21**, *piston heat shield;* **22**, *nozzle drive ring;* **23**, *copper plate;* **24**, *nozzle flap;* **25**, *centring pin;* **26**, *actuator casing;* **27**, *clamping strip;* **28**, *rear bearing;* **29**, *oil jet;* **30**, *flame tube (12);* **31**, *rotor blade;* **32**, *front load-relief cavity;* **33**, *front bearing;* **34**, *front frame with four inlet struts.*

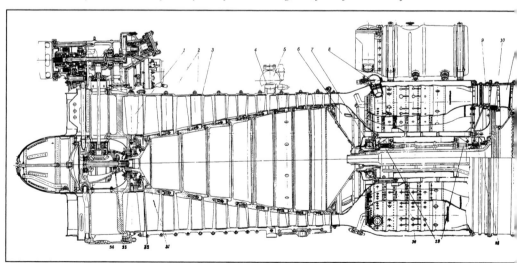

innovation. Early versions were rated at about 8,600 lb dry and 11,240 lb with afterburner, but later models have afterburning thrust up to 13,668 lb. Some versions for the MiG-21 are equipped with large bleed manifolds for flap blowing. From it were developed such mass-production engines as the Gavrilov R-13 and R-25 and the Khachaturov R-27 and R-29, the names being the chief designers of those particular engines.

A. D. Shvetsov was near retirement in 1945, and left the new field of gas turbines to P. A. Solovyev, who succeeded him as General Constructor in 1953. His engines are covered in later chapters.

Arkhip Lyul'ka would soon have given the Soviet Union outstanding engines without any need for German assistance. By 1950 his VRD-5 (TR-3) had reached its design thrust of 4,5 t (9,921 lb), and a small batch rated at 10,141 lb with automatic single-lever control were delivered for various prototypes. In the same year he became a General Constructor, so the VRD-5 became the AL-5. In 1952 the AL-7 was on test, and this outstanding engine, with supersonic flow through the widely separated first two stages of the nine-stage compressor handling 251 lb/s, was produced in large numbers with ratings up to 15,432 lb, or 22,282 with afterburner.

V. A. Dobrynin produced large and complex piston engines, but made the transition to gas turbines so well that the VD-5 was the most powerful engine of 1954 (13 t, 28,660 lb, dry). The smaller VD-7 was the most powerful production engine. First run in 1954, it went into production two years later for the 3M rated at 24,250 lb., and with remarkable fuel economy. Sadly, it was troublesome, and after some bombers had been re-engined as 3MS with the older RD-3M-500A the VD-7B was provided, derated to 20,945 lb but with good reliability and economy. Dobrynin's successor, P. A. Koliesov, developed the VD-7M for supersonic applications, with an afterburner giving 35,275 lb. In 1958–59 this powered Tu-105 and M-50 prototypes, and in 1994 the production Tu-22 was still in service with the RD-7M-2, rated at 36,376 lb.

The USA was another country glad to make use of the Rolls-Royce Nene. At the instigation of the Navy, Pratt & Whitney took a licence and from mid-1948 delivered large numbers of an Americanised version designated J42, followed by the J48 with airflow increased from 80 to 130 lb/s and rated up to 8,750 lb with a simple on/off afterburner. P&W had already produced the T34 turboprop, and via the two-shaft T45 began working on prototypes of various forms of JT3 turbojet. The Air Force called this the J57, and believed it could be the high-compression engine that they desperately needed, with unprecedented fuel economy for long-range bombers and high afterburning thrust for supersonic fighters. Prototypes were disappointing, and in February 1949 Andy Willgoos, who had been chief designer from the company's formation in 1925, completely revised the engine with a 'wasp waisted' exterior. A month later he died after clearing snow from his drive, but the work was done.

The resulting JT3 was probably the most important postwar jet engine. As the J57 it enabled the vital B-52 to be a jet, not a slower turboprop. With afterburner it thrust such fighters as the F-100, F-101, and F-102 faster than sound. Unexpectedly, it also had to replace the unsuccessful J40 in the principal fighters and bombers of the Navy. As the commercial JT3 it powered the 707 and DC-8 which, far more than the faltering Comet, opened the jet age on the world's airlines. P&W made 15,024, and Ford a further 6,202. It was logical to build an enlarged version, the J75, which powered Mach 2 fighters and, as the JT4, the heavier intercontinental versions of 707 and DC-8. In 1959, in a brilliant move, the first three LP stages of the JT3 were replaced by two fan stages handling 450 lb/s, with a BPR of 1.5. Not only did this dramatically improve thrust, fuel economy, and noise, incidentally staving off competition from the British Conway, but operators could convert their JT3C turbojets into JT3D turbofans with a simple field kit.

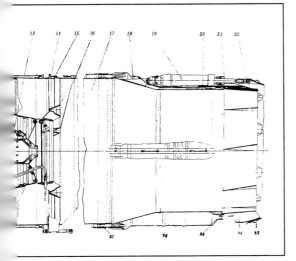

Pratt & Whitney X-176, the original prototype of the JT3.

In 1954, Westinghouse having lost the Navy's confidence, Pratt & Whitney was asked to produce a JT3 scaled down to 7,500 lb. This was produced as the J52 for cruise missiles, replaced the J65 in the A-4 Skyhawk, and powered all versions of the A-6 Intruder. The LP compressor was then redesigned with two fan and six compressor stages, driven by a turbine with two extra stages for a total of three. The result was the JT8D, one of the first gas turbines developed without military backing, as explained later.

During the 1950s the Hartford giant also had major secret programmes. The first, launched in 1951, was to provide propulsion for USAF Weapon System 125A, the nuclear-powered bomber (NPB). It was accepted that this would probably be subsonic, but it was expected to compensate by having flight endurance limited only by the available food and drink for the crew, so that any target in the world could be approached at any time from any direction. The first scheme was for a monster ducted fan, 67 ft 7 in long with the usual combustion chamber

From the same aspect, an early JT3P or J57, showing the oil tank and accessory group tucked into the 'wasp waist'. It was the most important US engine of the 1950s.

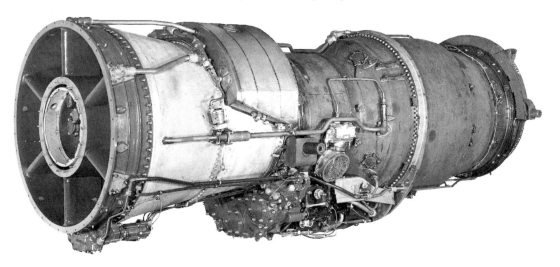

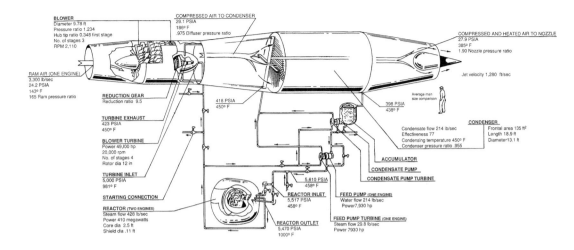

Schematic diagram of the huge ducted-fan turbojet proposed by Pratt & Whitney for the WS-125 Nuclear-Powered Bomber. The reactor delivered high-pressure steam which drove the fan turbine and was then condensed back for a repeat cycle.

replaced by a heat exchanger with indirect-cycle heating from a pressurized-water reactor (PWR). In 1953 this was replaced by molten-salt heat transfer to six J91 turbojets. There were several J91 versions, two of which, with company designation JT9A-20, were built and run in 1957 using kerosene fuel and high-energy HEF-3 (described later) in the afterburner, to give thrusts with water injection up to 44,000 lb. The JT9As had only distant kinship with the JT9D which made the 747 possible.

The almost equally large Project Suntan began in 1955. This was a supersonic turbojet running on liquid hydrogen to power a successor to the Lockheed U-2 at over 100,000 ft. The problems were severe, but on 11 September 1957 the engine – known only as '304' because of its highly classified status – ran on gaseous hydrogen. When it ran on liquid hydrogen an incredible heat exchanger heated the fuel from 18°K (minus 255°C) to over 1,000°K in a split second, expanding the gas through a small 18-stage turbine giving 12,000 hp. Downstream some hydrogen was burned, to give a jet at 1,500°K.

In 1958 the problems of such intensely cold fuel finally prompted a switch to hydrocarbon fuel, but no ordinary fuel would do. To power 'Kelly' Johnson's A-12/YF-12/SR-71 Black-birds, Pratt & Whitney had to develop the J58 (JT11D-20B). This remarkable single-spool engine delivered much of its airflow through six bypass pipes to feed the afterburner, which, at

Mach 3.35, provided almost all of the thrust. It used special fuel, special oil and many special techniques. In contrast, P&W also produced (originally in Canada) the small J60 (JT12A) turbojet of 3,000 lb thrust for missiles and biz-jets. It led to the T73 helicopter engine.

Thus, coming from behind, the established aero-engine firm of Pratt & Whitney got to the point where for many years it has claimed to be 'the world's largest producer of gas-turbine engines', despite the head start of GE, which until 1942 had never made an aero engine. Having been pressured into handing their first two production turbojets to Allison, GE continued with a redesigned TG-180 (J35) called

The Hartford engineers were probably better fitted to tackle the highly secret Project Suntan engine than any other team anywhere, but the liquid-hydrogen pump and heat exchanger were new challenges. The 23,500-rpm pump had oil lubrication! The heat exchanger, of 71.6 in diameter, transferred heat at a rate that would 'heat 700 six-room houses'.

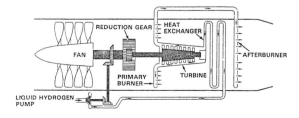

TG-190. First run on 21 June 1947, this appeared at first to be a backward step, because despite having an extra (12th) stage on the compressor it handled less air (92 lb/s) at lower p.r. (5.0 instead of 5.5) and gave less thrust (5,200 lb) despite being heavier (around 2,500 lb). Such an assessment would have been misleading, and as the J47 the TG-190 became by far the most important engine of the USAF, Studebaker and Packard helping to produce 36,500 by 1956. The biggest application was in the B-47, followed by the F-86, many of which (F-86D and K) had an afterburner and gave up to 7,650 lb.

Next came the awesome J53, intended for supersonic fighters, bombers, and cruise missiles, but despite giving 21,000 lb without afterburner its applications stayed on paper. So GE scaled the 53 back to J47 frame size to produce the J73, giving 9,200 lb dry for the F-86H. In parallel, GE worked on alternatives to P&W's proposals for an NPB. After false starts, project director Bruckmann (wartime head of BMW gas turbines) boldly decided to design the world's biggest turbojet, the 41ft XJ87, to be used in a twin installation with the 300 lb/s airflow of each engine split into two, the inboard half passing through a reactor core with enriched U-235 fuel moderated by zirconium hydride. Beryllium oxide and other ceramic fuel elements were tested. The complete system was called the X-211, and by the time the pro-

gramme was terminated in March 1961 a single in-line engine, representing half the X-211, had achieved a thrust of 27,370 lb at a TET of 982°C.

Fortunately, GE had also created a superb successor to the J47 to fill the expanding new plant at Evendale outside Cincinnati, Ohio. In March 1952 it was decided to go ahead with a totally new turbojet offering fuel economy at Mach 0.9 and high thrust for Mach 2. The key was increased pressure ratio, and the choice was between a two-shaft 'split compressor' design, as adopted by P&W and Bristol, or the completely new single spool with variable stators, which was already in preliminary test by GE's Gerhard Neumann. After intense research and argument (who knows how far the decision was influenced by a wish not to copy the competition?), the choice fell in November 1952 on the variable stator. The result, via the GOL-1590 demonstrator, was to become famous as the J79, first run on 8 June 1954, flown under a B-45 in May 1955, and tested in an XF4D Skyray from 8 December 1955. Probably the world's first fully engineered Mach-2 propulsion system was seen in the XB-58 bomber, first flown in November 1956. Altogether just under 17,000 J79s were built, 3,249 by licensees. They set 46 world records, and until Concorde had more Mach 2 time than all other engines combined (said GE).

Build-up of the nuclear-power X-211 at GE's Evendale plant.

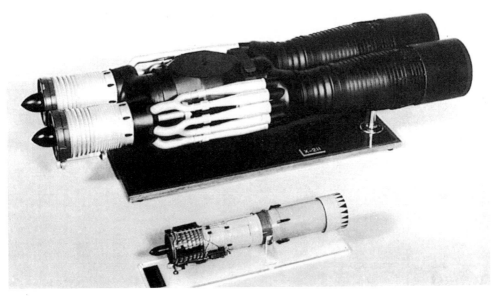

A model of the giant nuclear X-211 to the same scale as a J79 with afterburner.

Small numbers were sold of the CJ-805 commercial version, and the vastly improved aft-fan version also failed to find a rewarding market. Another programme that failed to find the expected market began in 1955 as the chemically powered bomber (CPB), the alternative to the NPB. Whereas it was accepted that the NPB would be subsonic, the CPB was to be able to cross enemy territory at the highest speed possible; eventually the XB-70 was designed to cruise at Mach 3. Again coming from behind, GE won the giant propulsion contract with a variable-stator engine bigger than the J79, called the YJ93. A key feature was the use of so-called 'Zip' high-energy fuel, such as HEF-3 based on ethyl borane. Different versions of J93, both in the 27,000 lb class, were produced for the B-70 and for the companion F-108 Rapier interceptor. The problems with Zip fuel were such that in 1958 it was confined to the afterburner only, and in August 1959 it was abandoned entirely, after enormous expenditure on plants to produce it. Neither the bomber nor the interceptor went into production.

On the other hand, in October 1953 GE had formed a Small Aircraft Engine Department at the older plant at Lynn, outside Boston, and this quickly produced a diminutive turboshaft engine, the T58, and then a small turbojet, the J85. The latter was designed as a short-life missile engine, but soon became much more important as an engine for fighters, trainers, and the new species of bizjets. Rated from 2,850 lb up

to 5,000 lb with afterburner, more than 16,000 were produced, ending in 1988.

Other producers of small jets included Fairchild, with the monocoque-construction J44, and Continental, which took a licence from Turbomeca and made a useful quantity of small Marboré II turbojets as the J69, virtually all for USAF T-37 trainers. Rather surprisingly, Allison did little to develop the GE-designed J33 and J35 jets, and achieved limited success with the more powerful J71, preferring to concentrate on turboprops. Another giant of the piston age, Wright, failed to move fast enough and then, in 1950, tried to make up for lost time by building the British Sapphire and Olympus under licence as the J65 and J67. The latter never sold, but, years later than planned because of extensive 'Americanization', the J65 did become an important engine in such aircraft as the USAF F-84F and B-57, the Navy FJ-3 and -4, F11F, and early versions of A4D (A-4).

This leaves the steam-turbine and electronics giant, Westinghouse. Having achieved a valuable product-line with the Navy J34 axial turbojet, typically rated at 3,500 lb, the Aviation Gas Turbine Division relocated at Kansas City and continued with the J46 and J40, again for the Navy. The former was a modestly enlarged J34, and a mere handful were sold for the F7U-3 Cutlass. In contrast, the J40, of 40 in diameter, was planned as the engine of the next generation of Navy fighters and bombers, including the A3D Skywarrior, F4D Skyray, and F3H Demon.

Intended to be rated at 7,500–8,000 lb dry, and 10,500–12,000 lb with afterburner, the J40 proved too difficult for the engineers to develop. In panic moves, the aircraft had to be redesigned for the J57 and J71. Desperately, in June 1953 Westinghouse signed a 10-year agreement with Rolls-Royce, but by fiddling with the British technology ground to a halt.

Rolls-Royce itself was in most respects the world leader in aero gas turbines in the immediate post-war years, perhaps its only mistake being not to make use of the Clyde turboprop. General Manager Hives underestimated its market, instead having a fixation on the AJ.65 ('axial jet, 6,500 lb'), later called Avon, as 'the Merlin of the jet age'. In most respects it was little advance over the Nene, and early versions were no more powerful or fuel-efficient than the Nene, much heavier, far more expensive, and many times more troublesome. Unexpectedly, the design team at Barnoldswick, led by brilliant aerodynamicist Dr Stanley Hooker, found the axial compressor a desperately difficult beast to tame. In 1949 the turbine was redesigned to have two stages, which not only increased power but surprisingly reduced weight. In the same year a compressor with five variable stators was tested, five years ahead of GE, but in the end the Avon was cajoled into behaving with just variable inlet vanes and large bleed valves. Eventually, with a totally redesigned compressor, the 200-series Avons went beyond 10,000 lb, and with cooled blades 11,250 lb. By 1960 fighter Avons with afterburner had reached 17,000 lb, and highly efficient airline versions with a 17-stage compressor driven by a three-stage turbine gave 12,600 lb.

Rolls-Royce pioneered the turbofan and jet-lift turbojets. In contrast, the British leader with axial jets, Metrovick, was required to hand over its work to Armstrong Siddeley Motors (ASM), of Coventry, which accordingly took over the F.9 Sapphire in 1947. Blessed with an outstanding 13-stage compressor, this was type-tested at 7,500 lb in November 1951 and 8,300 lb in May 1952, for a weight of 2,550 lb. Its first application was the Hawker Hunter fighter. Although the Sapphire was a Hawker Siddeley Group product, few Hunters had Sapphires (Mks 2 and 5) and they were progressively phased out of service. Most Hunters had Avons, which flamed out when the guns were fired, causing the RAF's new fighter to be restricted to 25,000 ft and below, and later to having the fuel supply suddenly reduced as the gun trigger was depressed! Subsequently, 200-series Sapphires, with greater airflow, powered the Victor

bomber/tanker at 11,050 lb and the Javelin all-weather fighter at 13,390 lb with a simple afterburner.

Apart from turboprops, ASM's main effort centred on the Viper, a small turbojet rated at 1,640 lb and intended for short-life applications such as missiles and targets. It had a seven-stage compressor handling 32 lb/s and a fully annular combustion chamber with vaporising burners. By 1952 chief engineer John Marlow had qualified a version for long-life manned applications at the same thrust, and the Viper continued to be developed for the next 40 years. The 200-series increased mass flow to 44 lb/s, and thrust to 2,500 lb. The 500-series added a zero stage (an extra stage on the front of the compressor) to raise mass flow to 52.8 lb/s and thrust to over 3,000 lb. The 600-series introduced a two-stage turbine, and with airflow of 58.4 lb/s thrust climbed to 4,360 lb, or over 5,000 with a simple afterburner. Few engines have been so transformed over so long a period.

Formed in 1944, with Halford as chairman, the de Havilland Engine Co had originally designed the H.2 Ghost to power the aircraft company's Comet jetliner. This had four engines mounted between the spars at the roots of the wing, fed from inlets in the leading edge. For tropical take-offs it was proposed to fit a Sprite booster rocket between each pair of jet-pipes. Similar engines were made in Sweden for the J29 fighter, Flygmotor later adding an afterburner of their own design. de Havilland then developed the Vampire into the thin-wing Venom and Sea Venom, powered by a Ghost with a bifurcated inlet.

To meet what seemed an obvious need for a very powerful engine for Mach 2 aircraft, DH Engines then produced the H.4 Gyron, with a seven-stage compressor of 46in diameter handling 320 lb/s. First run on 3 January 1953, it was tested at 20,000 lb, and over 27,000 lb with a long afterburner, but was thrown on the scrap heap by the nonsensical decree by the 1957 government that the RAF would never need any new types of fighters or bombers. During the 1950s the company had recognised that the high-test hydrogen peroxide (HTP) used in its rocket engines could form a very potent fuel not only for starters but also for accessory drives that would maintain full power up to 70,000 ft or more. They envisaged fighters powered by a mix of Gyrons and Spectre rocket engines, and Gyrons were fitted with 600 hp HTP starters.

To power the NA.39, which became the Buccaneer, the H.6 Gyron Junior was designed to a linear scale of 0.45, giving a thrust of

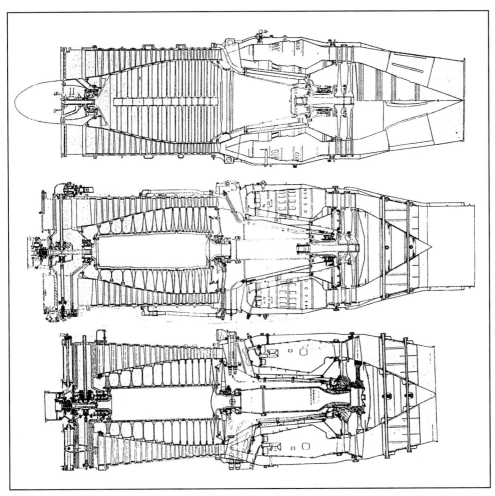

Development of an axial turbojet: top, Metrovick F.9 (1946) with welded drum compressor and only two bearings; centre, Sapphire 100 series (1950) with disc compressor and centre bearing; bottom, Sapphire 200 series (1953) with major redesign including flared turbine.

11,000 lb not allowing for an enormous boundary-layer control bleed. For the Bristol 188 a supersonic afterburning version was rated at 14,000 lb, but the de Havilland jet-engine business had almost disappeared when the firm became part of Bristol Siddeley in 1961.

Sir Roy Fedden, who until 1942 had been chief engineer of the Engine Department at Bristol, started his own company in 1946 to make an unconventional family car, a piston aero engine, and a turboprop of 1,425 h.p. The last, named Cotswold, had a ducted spinner of the same 27in diameter as the 10-stage compressor, tubular chambers, and 2-stage turbine. It had to be abandoned when Ministry funding was cancelled because of lobbying by rivals

whose engines could not come near the Cotswold's fuel economy (s.f.c., 0.67).

In France, Auguste Rateau, with René Anxionnaz, spent the Second World War designing the SRA.1 turbojet with a 16-stage compressor driven by a 2-stage turbine. The first four stages were oversized, and the excess air was fed through bypass ducts. This was run at La Courneuve, north-east Paris, in 1947, and led to the SRA.101 Savoie with a 10-stage compressor and no bypass flow, but these efforts came to nothing. Another French group which worked under difficulties during the war was SOCEMA. In 1945, the TGA.1 turboprop being on test, work began on the TGAR.1008 turbojet, a simple engine with an 8-stage compressor of

3.7 p.r., handling 62 lb/s, 20 flame tubes (10 small inners and 10 large outers) in an annular combustor, a single-stage turbine, and a variable nozzle bullet. This engine developed up to 4,620 lb on the bench but never flew.

In contrast, another simple Germanic turbojet eventually took over from the Hispano-Suiza-built RR Nene as standard engine of French combat aircraft. The chief designer was Dr Hermann Oestrich of BMW, who in May 1945 escaped from Germany and set up a small office at Rickenbach, Switzerland. The company gave its office a French name, the Atelier Technique Aéronautique Rickenbach, or Atar. In December 1945 it received a French government contract, moved to Decize (Nièvre) and restyled itself 'Aeroplanes G. Voisin, Groupe O'. From this shaky beginning came the first Atar, based entirely on BMW technology and actually manufactured by SNECMA. This was the giant nationalized group that embraced the former Gnome-Rhône, Renault, Regnier, SECM, and GEHL firms.

First run in March 1948 at what was to become a vast engine-test site at Melun-Villaroche, south-east of Paris, the Atar began as a simple turbojet with a 7-stage compressor (p.r. 4.2 at 8,050 rpm), 20 burners in an annular chamber, a single-stage turbine, and a variable nozzle bullet. It flew at a thrust of 3,307 lb in a B-26 Marauder, and at 3,750 lb on a pylon above a Languedoc. Meteor RA491 tested Atar 101B2 engines at 5,181 lb in 1949, and in early 1953 the Dassault Mystère IIC was in production, powered by an Atar 101C rated at 6,052 lb at 8,500 rpm. The 101D had a larger turbine,

and the D2A replaced the nozzle bullet by twin eyelids. On 3 August 1954 a D2A, in a SFEC-MAS Gerfaut, became the first turbojet to exceed Mach 1 on the level without afterburner. With afterburner and much larger nozzle eyelids the D became the F, rated at 8,375 lb. The 101E added a zero stage, increasing airflow 15 per cent to 132 lb/s, and though designed for 7,275 lb the 400 production E5 engines for Vautours actually produced 8,000 lb. With afterburner the E became the Atar 101G, the G2 version for the Super Mystère B2 being rated at 7,341 lb dry and 9,502 with reheat. In 1957 a 101G became the first European turbojet to exceed Mach 2, in the prototype Mirage III. Thus, by concentrating on the assets of high thrust per unit frontal area, and low cost, SNECMA kept an obsolete design in production almost 10 years.

Moreover, in 1953 the company began a near-total redesign which in 1960 went into production as the Atar 8. A new compressor increased airflow and p.r. by 20 per cent, to give thrust of 9,502 lb dry. With a large flap-blowing bleed, this engine powered the Etendard, while the corresponding Atar 09B3, rated at 9,193 lb dry and 12,963 lb with afterburner, powered the Mirage IIIC. This takes a remarkable success story to beyond 1960. In contrast, neither the big SNEC-MA Vulcain, run at up to 13,228 lb dry in 1954–55, nor the trio of small turbojets ordered for light fighters in 1953 – SNECMA Vesta, Hispano-Suiza R.800, and Turbomeca Gabizo – all rated at 2,425–3,968 lb, found any market.

On the other hand, the little Turbomeca company started by Josef Szydlowski in 1938 to make better superchargers, went from strength

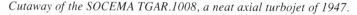

Cutaway of the SOCEMA TGAR.1008, a neat axial turbojet of 1947.

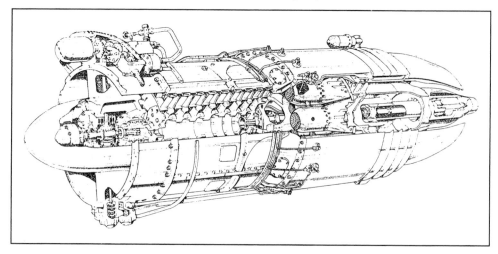

to strength. Small shaft turbines were run in 1947, followed in 1948 by a tiny turbojet, the Piméné. Rated at 221 lb thrust, it had a simple centrifugal compressor driven by a blisk turbine, and an unusual feature repeated in all subsequent Turbomeca engines was that the fuel was injected centrifugally from a perforated ring on the main drive shaft. A Piméné first flew, mounted above a sailplane, on 14 July 1949. Subsequently Turbomeca produced more than 50 types of small engine, including a few turbojets, of which by far the most important was the Marboré. A Piméné scaled up to 662 lb thrust, this flew on a Gemaux twin-fuselage sailplane on 16 June 1951, and the Marboré II, rated at 882 lb, was built by the thousand for many aircraft including the Magister trainer and, made by Continental, the T-37.

The former Turbo Research team at Avro Canada moved in 1946 into the field of original engine design, starting with a proof-of-concept axial turbojet, the Chinook, of 2,600 lb thrust. This was then scaled up to produce the Orenda, with a 10-stage compressor handling 90–106 lb/s. By 1958 3,794 Orendas rated from 6,000 to 7,275 lb had been delivered to power the CF-100 and Canadian-built Sabres. In 1953 work began on what became the Iroquois, an outstanding supersonic afterburning turbojet to power the great CF-105 Arrow. One of the first engines made principally of titanium, its two-spool compressor handled 300 lb/s and the accessories were protected from the hurricane surrounding airflow by a fireproof box. Its weight was 4,120 lb and take-off rating 27,000 lb. Flight testing began in 1956 with an engine mounted on the right side of the tail of a B-47, but the entire CF-105 programme was cancelled in February 1959.

Finally, the Swedes followed up their pioneer gas-turbine work by developing the STAL (Ljungström) Skuten axial turbojet of 3,200 lb thrust, run in 1949, which led to the excellent Dovern of 7,275 lb of two years later. Sadly it was recognised that perhaps the costs of doing their own engines were not justified, and the Saab-32 Lansen and J35 Draken went into production with licence-built RR Avons.

14 Powered lift

Before getting into this chapter, it is necessary to go back over the concepts of power and thrust. Thrust is easily comprehended; it is a force, and it is present whether or not the point of application moves (in which case useful work is done) or is stationary (in which case the work and power are zero). Power is rate of doing work, so in dimensions it is equal to thrust multiplied by the speed of the point of application measured along the axis of the force. If a thrust is used to accelerate a mass, the power is numerically equal to the mass multiplied by the acceleration, so it can accelerate a

Today's designer can choose from a complete spectrum of V/STOL aircraft, ranging from the helicopter to the vectored-thrust jet (a rocket would be off the picture to top right). The vertical axis, here shown as disc loading (a jet nozzle counts as a disc), could equally be fuel consumption per unit weight. Thus, the helicopter is brilliant in hovering economy but unimpressive going from A to B; the vectored jet is fast and efficient in wingborne flight, but unimpressive when hovering.

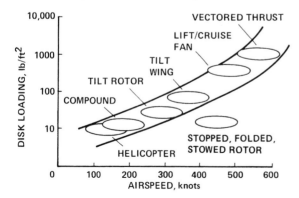

small mass violently or a heavy mass gently (ignoring friction).

Early aeroplanes used the power of the engine (say, 100 hp) to drive a propeller to accelerate a large airflow to the rear to generate a thrust (say, 300 lb). This in turn accelerated the whole aircraft horizontally until the lift from the wings equalled the total weight, which might be 3,000 lb. It would then continue the acceleration to a speed at which the drag was the same as the thrust. There was absolutely no way the engine/propeller could have generated a thrust of 3,000 lb, so vertical take-off and landing (VTOL) would not have been possible. On the other hand, the same engine geared down to a giant propeller (say, 40 ft diameter) might have been able to achieve VTOL, and with a great deal of refinement this led to the helicopter.

One seldom gets anything for nothing. By using a large wing liberally endowed with slats and flaps, to raise maximum lift coefficient from the usual 1.8–2.2 up to 3.8–4.5, it is possible with unchanged engine power to achieve such short take-off and landing (STOL) performance that it is possible to hover in a brisk breeze. The penalty is a rather clumsy aeroplane and poor flight performance. These limitations were dramatically altered by the introduction of modern gas turbines. Suddenly, designers had so much power (or thrust) available that VTOL came within their grasp.

During the Second World War, German designers schemed various VTOLs, but could not spare the effort to build them. One of the more radical was the Focke-Wulf *Triebflügel* (thrust-wing). This stood on its tail while the pilot fired small rockets on the tips of three wings projecting radially like a helicopter rotor. The thrust made the wings spin rapidly round the fuselage until the tip speed was high enough for tip ramjets to be lit. The aircraft then took off, tilted over into forward flight, and engaged the enemy. The landing, reversing the take-off procedure (without needing the rockets, of course) would have been tricky in the extreme.

One project that actually got built was the Ba 349 target-defence interceptor, which was just a manned rocket, fired vertically upwards. After engaging bombers in controlled wingborne flight, it separated into parts, the pilot and rocket engine descending on separate parachutes and the rest of the airframe being allowed to crash to the ground. In the decade following the war, Britain and France were among countries which looked carefully at short-duration target-defence interceptors, including possible VTOL schemes.

Predictably, once they had realized that gas turbines made VTOL possible, the USA explored everything they could think of. It was soon obvious that the possible ways of using gas-turbine engines for V/STOL aircraft are many and diverse. This chapter considers the subject idea by idea. It is logical to begin with jet engines pointing downwards, either by vectoring the whole aircraft to point to the sky or by vectoring the engine(s) or nozzles.

A pioneer of jet lift was Britain's Fairey Aviation. In 1946 the Air Ministry, impressed by the Ba 349 (!) proposed a project to develop a VTOL jet fighter, and Ministry of Supply specification E.10/47 was issued. Fairey began with delta-wing models launched up an almost vertical ramp by twin rocket engines, with swivelling nozzles to provide control. Next came the manned aircraft, the FD.1. This was a small delta, powered by a Rolls-Royce Derwent, to be launched almost vertically by four solid rocket motors. Control during the launch phase was to be provided by four swivelling nozzles at the tail, either fed by the main engine or using additional small rockets. Fortunately the idea was dropped, and the FD.1 flew only from runways.

At roughly the same time, Dr A. A. Griffith of Rolls-Royce began serious study of 'flat-riser' VTOL aircraft, so called because they do not need to point to the sky. He proposed the use of batteries of simple turbojets of high thrust/weight ratio fixed in the near-vertical attitude and used to lift the aircraft during VTOL. The idea naturally appealed because his employer could see a market for thousands of extra engines, even if they were individually of low cost. To explore the concept the company built two Thrust-Measuring Rigs (TMRs), better known as Flying Bedsteads.

Each comprised two Nene turbojets pointing towards each other. One jetpipe was angled 90°

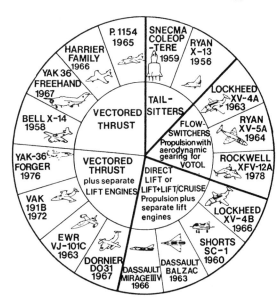

This diagram shows pioneer types of jet lift; it does not include helicopters, tilt-rotors, or stopped rotors (British Aerospace).

down at the centre of gravity (CG); the other was split to feed two half-size pipes evenly disposed on each side. Thus, should either engine fail, though the beast would come down fast, at least it would remain upright. Large air-bleed pipes fed control jets at front and rear, for pitch, and smaller lateral jets for roll. The pitch jets could swivel, for yaw (pointing) control, all governed by the pilot sitting in a non-ejection seat!

The device was obviously very dangerous to fly, and the original project pilot chickened out. The task was assigned to a pilot at the RAE Farnborough, Sqn Ldr Ron Harvey. He approached the challenge with enthusiasm, and on 3 July 1953 began hovering in a gantry with loose tethers and using autostabilization. Before long he was making free flights in full manual control, with autostab switched off. Apart from the probable lethality of any mechanical failure or pilot error, the main hazard was that, if the rig was landed when moving (even at 5 mph) it would roll over and kill the pilot. Harvey recommended the addition of a crash pylon to protect the pilot, but his 23 hazardous flights were made without one. Then a pylon was added, but the pilot Harvey handed over to crashed and was killed.

In April 1947 the US Navy awarded a contract to Ryan to investigate flight control at zero airspeed, when absence of relative wind makes

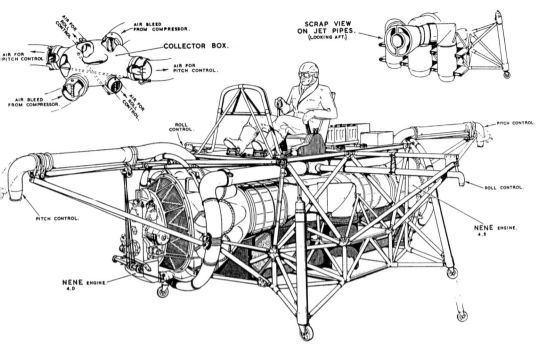

Crude it may have been, but the 'Bedstead' was the first to explore the problems of stability and control when supported only on a jet of hot gas (in this case, one big jet and two half-size ones) (Rolls-Royce).

ordinary control surfaces useless. More than 80 ideas were studied, and many were tested on a tethered rig lifted by a vertical Allison J33, which lifted off on 31 May 1951. Then a simulated delta wing was added, and finally a tilting seat and instruments, so that on 24 November 1953 Ryan chief test pilot Pete Girard began manned hovering, just five months after Harvey's first lift-off in the Bedstead.

Eventually, funded by the USAF, Ryan built two X-13 Vertijets. These were the smallest tail-less delta aircraft that could be wrapped round a Rolls-Royce Avon of 10,000 lb thrust. This engine had nozzle deflectors to vector the jet up/down or left/right through small angles, while bleed-air jets were fitted at the tips of the wings and fin. For translational (forwards) flight there were full-span elevons and a large rudder. The first X-13 made a conventional take-off on 10 December 1955. The first hover, pointing skywards on a temporary four-wheel 'landing gear', was made on 28 May 1956. Later testing was done with no landing gear, the aircraft hanging like a moth from a large truck-mounted

The second X-13 Vertijet about to unhook from its hinged platform.

platform which could be pivoted up through 90°. The pilot seat could pivot forward through 50° to give a less-uncomfortable attitude with the aircraft upright. On 12 April 1957 Girard made the world's first flight starting as a VTO (unhooking from the platform), accelerating through a forward transition to fly at high-subsonic speeds, then making a decelerating transition back to the hover, and finally making a VL by hooking back on.

Another VTOL pioneer was Bell. On 16 November 1954 it made the first free lift-off with a 'flat riser'. Known as the Model 65, it was assembled mainly from parts of other aircraft, and had two Fairchild J44 turbojets pivoted to the fuselage to point downwards or to the rear. A Turbomeca gas turbine above the fuselage supplied air to control jets at the wingtips and tail. This was possibly the first practical aeroplane to have 90° thrust vectoring. Hooked on jet lift, Bell went on to produce a much better aircraft, the X-14. Funded by the USAF but operated mainly by NASA, it lifted off on 19 February 1957 on the power of two Armstrong Siddeley Vipers with full jet deflection. Later it was re-engined with GE J85s. Subsequently Bell built the Lunar-Landing Research Vehicle, for training astronauts, and then switched to a totally different idea in the X-22A, described later.

In France, SNECMA had by 1952 become interested in the *Coléoptère* patented by the Bureau Technique Zborowski, a group of mainly former German engineers. The idea was for a minimum fuselage to be mounted upright and surrounded by an annular wing forming a duct. Inside this was to be a high-power propeller, or the body could contain a turbojet. The giant wing/duct could provide lift in forwards flight, and might even form a large ramjet.

In 1952 tests began with a Vampire with a jet reverser which (Chapter 5) was modified to deflect the jet by blowing compressed air into it, to divert it into surrounding cascades. In 1953 this was linked to the flight controls to govern the aircraft trajectory in flight. In 1954 tests began on a small Ecrevisse pulsejet in a vertical rig with remote joystick control, followed by an Atar DV turbojet with a special oil system adapted for running in the vertical attitude (not needed with the British 'Bedstead'). In 1955 an Atar DV was mounted on gimbals and remotely controlled while gyrostabilized.

On 22 September 1956 the Atar CP.400-P1 made the first of 250 'flights' under remote pilot control, with bleed air fed under gyrostabilizer control to four segments of jets surrounding the nozzle for pitch/yaw control, with two long bleed pipes for roll. On 30 March 1957 Auguste Morel made the first tethered flight in the CP.400-P2, free flights following on 14 May. In 1958 the P3 began testing, with Morel in a tilting ejection seat and with the engine inlets on each side of a fighter-type nose. This led to the CP.450 *Coléoptère*, flown tethered on 17 April 1959 and free from 6 May. This was intended to lead to operational VTOL fighters powered by a TF104 (SNECMA-modified Pratt & Whitney TF30) plus ramjet burning in the annular duct, but on one of the first free flights Morel was

The prolonged Flying Atar story was intended to lead to a Coléoptère fighter: 1, radar; 2, turbojet inlet; 3, cockpit; 4, accessories; 5, aux tank; 6, ramjet inlet; 7, main tank; 8, turbojet; 9, ramjet; 10, vectorable nozzle; 11, jet outlets; 12, controls; 13, legs.

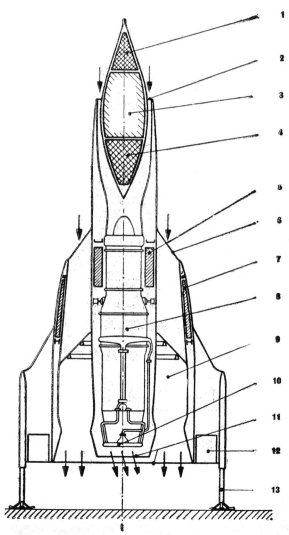

forced to eject, and the programme was abandoned.

Back in 1950 the US Navy had switched to turboprops. Seeking a way to operate air-defence fighters from the small available deck areas on warships or merchant ships, a turboprop seemed the only possible way to reconcile the demands of aircraft weight, thrust, and mission endurance. Over 40 years ago no practical jet fighter could lift off vertically, even with a mere minute's worth of fuel. So the answer could only be the Allison XT40 turboprop. Convair built two XFY-1s, and Lockheed two XFV-1s. They were very similar 'minimum aircraft', carrying little but the engine and a pivoting ejection seat for the pilot. The main difference was that Convair picked a tailless delta and Lockheed a tapered mid-wing and odd swept cruciform tail.

Both were designed to stand upright on four small wheels, two of those on the Convair being on the wingtips and the others on giant upper and lower fins. Lockheed used the tips of the four identical tail surfaces, but elected to begin by flying conventionally, with a temporary landing gear. Herman 'Fish' Salmon made the first XFV-1 take-off in March 1954; later he managed a transition to the hover, but he was never able to lift off vertically. Convair's James F. 'Skeets' Coleman had to lift off vertically if he was to fly at all, and achieved this on 1 August 1954. Later he demonstrated complete transitions, but it should have been obvious that the idea was a non-starter.

Although tail-standing jets took longer to be abandoned, it was clear from 1954 that practical machines ought to be flat-risers. Many were fitted with propulsive devices able to pivot through at least 90°. Apart from the Bell Model 65 already mentioned, one of the first was the same manufacturer's XV-3, funded by the US Army and first hovered on 23 August 1955. A 450 hp Wasp piston engine drove 23ft 'prop-rotors' pivoted to the ends of a fixed wing, so that it could convert into an aeroplane (first achieved on 18 December 1958).

Of course, this was a low-speed aircraft. Moreover, the 23ft prop-rotors posed a problem, in that if they jammed in the cruise (aeroplane) configuration, the machine could not land. Some designers therefore turned to ducted fans, with more blades running at higher speed so that diameter for a given thrust could be greatly reduced. An early example was the Doak VZ-4, funded by the US Army. First flown on 25 February 1958, this had a T53 geared to ducted fans pivoted on the ends of a fixed wing. This led to the Bell X-22A, with a short front wing and a larger wing at the tail, on which was mounted four 1,250 hp GE T58 engines. These were geared to four large ducted prop-rotors pivoted to the wings. First flown in March 1966, it was one of three Tri-Service transports. The XC-142A is described later, and the third, the Curtiss-Wright X-19A, was broadly similar to the X-22A but had 'radial lift force' propellers without ducts, driven by two 2,650 hp Lycoming T55s.

There have been several more recent tilt-rotor projects. Some, such as the Nord-Aviation Type 500 of 1967, used ducted fans. Others, especially those intended to lead to VTOL transports, used turboprops mounted on a tilting wing. The Hiller X-18 (two 5,850 hp T40s plus a J34 feeding tail-control nozzles) never achieved full transitions. Better results were achieved by five XC-142 transports built by LTV (Vought) helped by Hiller and Ryan. This design had four GE T64 turboprops on a short pivoting wing. In VTOL yaw control was effected by the ailerons, roll by differential prop-rotor pitch, and pitch by a small vertical-axis propeller at the tail. Testing from September 1964 reached a speed of 430 mph.

The XC-142 was one of many aircraft which did what it was meant to do and was then abandoned. Another, slightly later, tilt-wing was the Canadair CL-84. But occasionally a project appears to achieve long-term success, and one of these rarities was launched in May 1973 by the funding by NASA and the US Army of two Bell Aerospace XV-15s. It is not easy to see why the XV-15 did so much better than the CL-84, though it was a more lively performer. Weighing only slightly more than the CL-84, it had a similar T53 engine driving each prop-rotor (instead of one driving both). Another difference is that the tip nacelles pivoted on a fixed wing. First flown (hovering) on 3 May 1957, the XV-15s reached 382 mph on the level, showed excellent combat agility and precise hover control, and were so impressive that on 26 April 1983 Bell, jointly with Boeing, received a Navy design contract for what became the V-22 Osprey.

This is a much bigger version of the XV-15, with a transport fuselage. The tilting tip nacelles house 6,150 hp Allison T406 engines driving 38ft three-blade prop-rotors with cyclic as well as collective control. At one time it looked as though almost 1,000 V-22s would be produced for all four US armed forces, but opposition on political grounds has delayed and damaged the programme. Flight testing began on 19 March

The first V-22 Osprey in hovering flight.

1989, and in demanding tests 402 mph has been reached and simulated missions flown in medevac, assault, ASW, air-refuelling, fleet logistics, and search/rescue roles.

Returning to jets, the idea of tilting engines explored by the Bell Model 65 was continued in the abandoned Bell D-188A. This was intended to become the XF-109, but even the USAF looked askance at eight J85 engines (two vertical in the forward fuselage, two with afterburner in the tail and two pivoting on each wingtip). Germany likewise gave up the VJ.101 series, which again had tilting tip nacelles; these were to lead to the US/FRG fighter described later.

In 1961 the US Army awarded Lockheed-Georgia a contract for what became two XV-4A Hummingbirds to explore the concept of ejector lift. Earlier we saw how, in an ejector-type supersonic nozzle, the high-speed jet entrains a large airflow around it. The chief problem with lift jets is that, because the jet velocity is so high, they are extremely inefficient in hovering flight. (They are the opposite of the helicopter, which is inefficient in cruising flight.) Just as the high-BPR turbofan 'gears down' the turbojet to increase its propulsive efficiency, so can various forms of ejector augmentor 'gear down' the fast jet to make it a more efficient source of lift. To entrain as much extra air as possible, the jet is discharged through a large number of small nozzles. In the XV-4A the two 3,000 lb JT12A turbojets fed 20 transverse rows of multiple nozzles arranged along the two lift ducts,

to give a theoretical lift of 8,400 lb. The ejector chambers were angled back 12° so that with the fuselage level the aircraft would accelerate forward. At 92 mph one engine would be switched to its conventional jetpipe, the second following at 145 mph. The first XV-4A flew in August 1962, but the idea was abandoned, and the second machine was rebuilt with six J85 engines in the direct-lift mode.

A more complex form of ejector-augmented lift was tried in the Rockwell XFV-12A, intended by the US Navy as the American answer to the supposedly primitive Harrier. Planned as a Mach 2 carrier-based V/STOL fighter, this had a single engine, a Pratt & Whitney F401 augmented turbofan (as planned in 1968 for the F-14B). Upstream of the afterburner was a giant valve which could divert the whole gas flow along pipes running from root to tip along a canard and a main wing, blasting down from hundreds of nozzles in ejector ducts formed by opening doors in the upper and lower surfaces. Suffice to say the XFV-12A was a failure, and never made a free flight.

An alternative way of gearing down the jet is to feed it to drive turbine blades round the edge of a large lifting fan. This was tested from 25 May 1964 in two Ryan XV-5A Vertifans funded by the Army. Two J85s each rated at 2,658 lb could be switched from their normal jetpipes to drive large 36-blade fans inside each wing, plus a smaller fan in the nose for pitch control. Total trimmed lift force was 13,946 lb. Louvres pro-

No longer the Army XV-5A, the surviving Ryan Vertifan with wing fan doors open.

vided control, and on accelerating transition the thrust was switched at 140 mph to the normal jetpipes, acceleration then continuing to 547 mph and the lift fans all being covered by doors.

Like the XV-4, this led to a rebuild for hovering only, with fixed landing gear and no louvres.

In Britain, Griffith's scheme for separate lift jets was first tested by two Short SC.1s, small

Three-view of the Balzac, showing the bleed piping from the eight lift engines to the control jets.

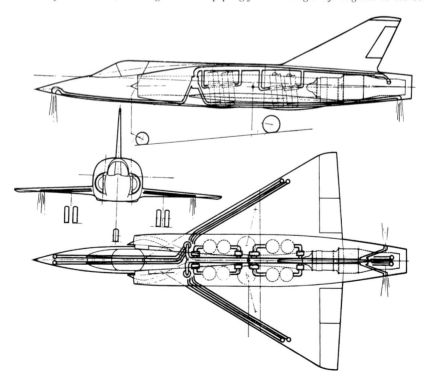

deltas with fixed landing gear and a fuselage occupied by four RB.108 turbojets for lift and a fifth for propulsion. The RB.108 was the first production lift jet, giving 2,030 lb plus 11 per cent bleed for flight control for a weight of 269 lb. The SC.1 triggered Martin-Baker's development of ejection seats usable at zero height and zero airspeed. Although first flown without lift jets on 2 April 1957, the SC.1s became complex, with gimballed lift jets, puffer-pipe control, and triplex full-authority autostabilization.

It was left to France to apply the idea, in the Dassault Balzac, a rebuilt Mirage III, followed by the bigger Mirage III-V fighters. The Balzac, lifted by eight RB.108s and with a Bristol Orpheus for propulsion, had the unique distinction of killing two pilots in two crashes, but the III-V became the first VTOL to exceed Mach 2, reaching 2.04 on 12 September 1966. It was lifted by eight RB.162 lift jets and propelled by a SNECMA TF106.

The RB.162 was the first aero engine to incorporate a large proportion of composite materials, mainly GRP. Rated at 4,400 lb (4,718 for emergency short periods), it weighed only 270 lb. Next came the Rolls-Royce/Allison XJ99, an engine of enormous potential representing a breakthrough in compact design. Apart from having the shortest combustor known (11.5 in), its chief advance was elimination of turbine stators, the HP turbine driving the four-stage HP spool in one direction and the LP turbine driving the two-stage LP in the other. Turbine gas temperature was typically 1,360°C, and the XJ99 gave 9,000 lb thrust for a weight around 440 lb. It was the key to the US/FRG advanced vertical strike aircraft, which was to have two afterburning main engines with deflectors and two pairs of XJ99s which in cruising flight would be retracted inside the forward fuselage.

Cancelled in 1968, the complex US/FRG was the antithesis of the simple British Harrier. This began life as a 1956 idea by French aircraft designer Michel Wibault for a VTOL engine of curious form. He proposed a turboprop geared to four centrifugal blowers delivering through four nozzles which could be vectored for thrust or lift. Bristol Aero-Engines simplified the idea into a turbofan with four nozzles all vectored together, two for the fan air and two for the jet. The engine first ran at 9,000 lb as the BE.53 in 1959, and in September 1960 at a rating of 11,000 lb lifted the first Hawker P.1127 off the ground.

The four nozzles were linked by chain drives from an air-turbine motor, rotating through an

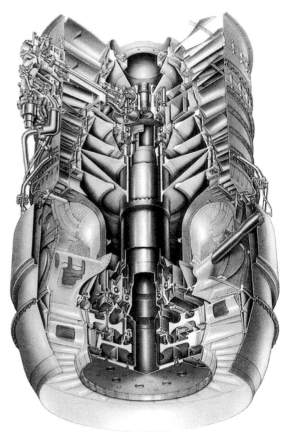

Cutaway of the Rolls-Royce Allison XJ99 lift jet.

angle which settled at 97.5° to give thrust, lift, or braking. They have always been part of the airframe, two on each side, disposed to even out the thrust on each side of the CG. Much research was needed to explore aerodynamic effects on the wing and tailplane. Bleed air for the reaction-control valves (RCVs) was originally taken from the LP spool, but to save bulk and weight the production system switched to eighth-stage HP. Nozzle power has to be considerable, over 150 hp, and could kill anyone getting in the way.

The Harrier entered RAF service in 1969 with a Pegasus rated at 19,200 lb. Over the years the engine has been uprated to 23,600 lb, though development has concentrated on extending life and reducing cost. Flight time exceeds 1.2 million hours, much of it in harsh front-line conditions, and apart from freeing the Harrier from the deadly danger of being parked on known airfields, the Pegasus also makes possible

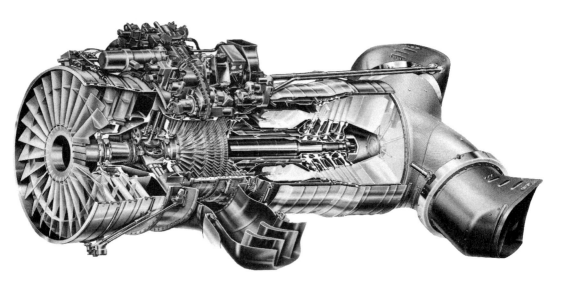

Cutaway of an early production Pegasus, showing the four vectoring nozzles, three-stage fan, eight-stage compressor, annular combustor with T-shaped vaporisers, and two-stage HP and LP turbines.

unnatural combat manoeuvres, including violent braking and 'square turns', using vectoring in forward flight (VIFF).

What are the drawbacks to this unique engine? Critics have suggested that it presents a juicy target to infra-red (IR) homing missiles, but this is arguable. The temperatures are lower than in other fighter engines, and much lower than afterburners, and the actual nozzles are even cooler and prevent the missile's seeker having a view of the hot turbine. For a subsonic aircraft, the engine could be criticised as being 'too powerful' except at take-off and landing, though no fighter pilot would agree. In the German VAK.191 and Soviet Yak-38 a smaller lift/cruise engine was augmented by two vertical lift jets.

A more valid criticism is that a high-BPR engine with four nozzles is unsuitable for supersonic flight. This is mainly because of the 'four-poster' layout; a 'three-poster' is better, with two fan nozzles and a single nozzle for the core jet. A single nozzle can be vectored even if it has an afterburner upstream, though the problems are severe.

Rolls-Royce pioneered a vectored afterburning engine in the 1960s with the RB.153, intended for the German VJ.101D. The nozzle was attached by a section of jetpipe able to revolve, with diagonal joints. Thus, as the linking section rotated, the nozzle first moved side-

ways and finally ended up pointing 90° down. The 101D was never built, but the same idea was adopted in the Soviet Union in the Soyuz (Kobchyenko) R-79, an augmented turbofan with dry and reheat thrusts of 24,200 lb and 34,170 lb. One of these forms the main power-plant of the Yak-141 STOVL fighter, the only Mach-2 jet-lift aircraft flying today. This aircraft also has two 9,040 lb Novikov RD-41 lift jets, the three engines being linked by a triplex electronic control system which automatically adjusts fuel flows and bleed air to the control jets.

An obvious problem in a jet-lift aircraft is getting the thrust in line with the CG. In the Harrier it has meant putting the engine in a place where, before you can remove it, you must first take off the wing. In the Yak-141 it meant stopping the 'fuselage' in the centre and carrying the tail on long girders (this is only a development of the structures seen in such fighters as the F-15, MiG-29, and Su-27). Another way of tackling the problem was seen in the unbuilt British Aerospace P.1214, a project by the Harrier team at Kingston. This simple and attractive machine matched the CG with the lift from the engine by stopping the body in the centre, and with the lift from the wings by making them forward-swept.

Thirty years ago the same team designed the Hawker P.1154, a stretched supersonic Harrier.

This would have had plenum-chamber burning (PCB), which in effect adds an afterburner to each front nozzle. The RAF and Royal Navy did all they could to damage the programme, and it was finally killed off in 1965. After wasting ten years, Britain resumed PCB development in 1975, and by 1985 had demonstrated a thrust boost from the front nozzles of 100 per cent, equivalent to about 50 per cent augmentation overall. If Britain was still a major power there would by now be a Super Harrier with a PCB-equipped engine in the 40,000 lb class.

As it is, Britain has virtually handed development of the Harrier to the USA, and its battle-proven vectored-thrust technology is the only type of propulsion already rejected for tomorrow's STOVL Strike Fighter (SSF). Administered by the US Advanced Research Projects Agency (ARPA), SSF is to fly from carriers and replace the Harrier from 2010. Details remain classified, but the aircraft would obviously have to be operable from small ski-ramp decks. Requirements include STO with a heavy load of offensive and defensive weapons, VL on return, supersonic speed, outstanding combat agility, and stealth design.

At first the assessors had an open mind, studying vectored thrust with PCB, lift/lift-cruise, hybrid fans, and tilting engines. Ejector-augmented lift looked promising, de Havilland Canada and General Dynamics getting as far as tunnel tests with the E-7, an ejector-lift tailless delta based on the F-16. The large ejector ducts along each side of the fuselage could be eliminated in high-speed flight by folding the outer walls against the fuselage and wing, but eventually the idea was abandoned.

By 1994 the number of possible propulsion systems had been whittled down to two, both comprising one or two main engines plus a remote lifting fan towards the nose to balance the thrust of the vectored main engine(s). To the author this seems a strange idea, less desirable than getting the main jet in the right place. Lockheed is working on an aircraft with a shaft-driven fan, the main engine being based on the F-22's P&W F119 in a team which includes Allison, Rolls-Royce, and Hercules. McDonnell Douglas and British Aerospace are using a fan driven by gas bleed from an engine based on the General Electric F120. Studies suggest that the good and bad points of the shaft-drive and gas-

How modern ejector lift might be applied to a fighter. The far side of the fuselage is in the wingborne cruise/combat mode.

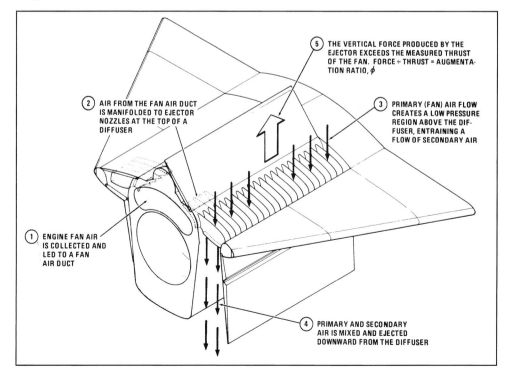

5 THE VERTICAL FORCE PRODUCED BY THE EJECTOR EXCEEDS THE MEASURED THRUST OF THE FAN. FORCE ÷ THRUST = AUGMENTATION RATIO, ϕ

2 AIR FROM THE FAN AIR DUCT IS MANIFOLDED TO EJECTOR NOZZLES AT THE TOP OF A DIFFUSER

3 PRIMARY (FAN) AIR FLOW CREATES A LOW PRESSURE REGION ABOVE THE DIFFUSER, ENTRAINING A FLOW OF SECONDARY AIR

1 ENGINE FAN AIR IS COLLECTED AND LED TO A FAN AIR DUCT

4 PRIMARY AND SECONDARY AIR IS MIXED AND EJECTED DOWNWARD FROM THE DIFFUSER

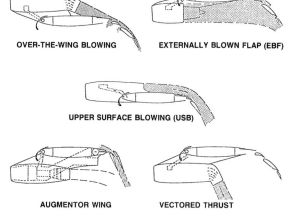

OVER-THE-WING BLOWING   EXTERNALLY BLOWN FLAP (EBF)

UPPER SURFACE BLOWING (USB)

AUGMENTOR WING   VECTORED THRUST

Five ways of using jet power to enhance the lift of a STOL transport (Boeing).

drive come out roughly even, so a complete non-flying example of each airframe will be tested on pylons and in tunnels at NASA Ames to see which performs better.

To conclude this chapter, a diagram shows some of the ways of using engine power to improve STOL performance of transports. Over-the-wing blowing is almost the same as upper-surface blowing (USB), the difference being that the latter mounts the powerful turbofan engine directly on the wing and squeezes the core jet and fan air out through a fishtail nozzle to blow over the depressed flaps. The Coanda effect, which keeps a fluid flow 'attached' to a smooth surface which curves away (try it with a horizontal bottle held under a tap), prevents the powerful jet from separating from the flap upper surface, so that lift coefficients up to something like 12 are theoretically possible. In other words, one can get about four times as much lift from a given size of wing as in normal aircraft.

The greatest pioneer USB aircraft was the Boeing YC-14, which from 1976 flew very impressively. Antonov adopted a similar layout in the An-72 and 74 without attempting to equal the YC-14's lift coefficients, but the US quiet short-haul research aircraft (QSRA) and Japanese Asuka took the idea further with four engines spread across the wing. The EBF was tested in the McDonnell Douglas YC-15 and actually adopted in a mild form on today's C-17 Globemaster III. Obviously, the powerful flaps have to be of heat-resistant titanium.

In the augmentor wing (AW), a large airflow is blown between the upper and lower elements of a biplane flap. The germ of the idea was the Jet Flap, proposed in 1955 by I. M. Davidson at the National Gas Turbine Establishment at Farnborough. This was tested from 1963 in the Hunting H.126, powered by a Bristol Siddeley Orpheus, the entire jet of which was collected in a large 'dustbin' and then piped to eight fishtail nozzles along the wing from tip to tip, blowing over the flaps and ailerons. There were obvious practical difficulties with so much near-red-hot ducting, and the severe stalling quality of the wing demanded such a speed margin as to nullify the STOL advantages.

Avro Aircraft in Toronto proposed blowing through spanwise slots, and de Havilland Canada then refined the idea to use cooler air from the fan or LP compressor, and to blow aft in cruising flight so that the sheet of air expelled would give propulsive thrust, making up for duct losses. STOL performance is achieved by the upper and lower flap elements vectoring the flow downwards, not only giving direct lift but also greatly augmenting circulation round the wing. The flaps also shield the blowing nozzles, reducing noise.

From 1972 the AW was tested by a modified DH Canada Buffalo with large biplane flaps blown by modified Rolls-Royce Spey (ex-Nimrod) engines which were also fitted with core-jet vectoring. The pilot could control both the flap blowing and the jet vectoring, and soon very precise landings were being made at 60 kt from a 7.5° approach, stopping in 300 ft. Rolls-Royce then designed the definitive RB.419-03 for an AW aircraft, much quieter and more fuel efficient than the Spey. It was assembled from the Dowty-Rotol variable-pitch geared fan of the M45SD (Chapter 16), the first three stages of a TF41 fan, and the core of a Spey 202 (Phantom). The results could be brilliant, if someone wanted a quiet transport able to land in a back yard.

15 Military jet engines

By 1960 the piston engine was virtually extinct as far as combat aircraft were concerned, and the main choice confronting an aircraft designer was 'what kind of turbojet do you want?' Centrifugal compressors were fast disappearing, and the main choice was whether the axial compressor would be a two-spool or a variable-stator; and for a supersonic aircraft an afterburner was almost mandatory. Not many new things seemed to be in prospect, though – as ever – there was inexorable pressure to wring more power from a given size of engine.

In fact, while today's engines show fantastic advances in the reduced bulk and weight of engine needed for a given propulsive task, enormous progress has also been made in other ways. Turbojets have in their turn become almost extinct, though one might still be the best answer if anyone wanted to fly at Mach 4. In their place the aircraft designer is asked what bypass ratio he needs for his turbofan, and whether it should have a plain jetpipe or an afterburner or full augmentation in both the core gas and the pure air from the fan duct.

A perhaps even greater advance, certainly appreciated by the pilots, is so-called 'carefree handling'; today's fighter engines can be slammed up or down the scale from flight idle to maximum afterburner, whilst being presented with incredible off-design airflows as the aircraft does anything from sliding sideways or performing a 'Cobra' manoeuvre to stopping in the sky and sliding tail-first. Another remarkable area of progress is engine reliability, time between overhauls (TBO), and total parts-life. The USA led in demanding numerical values for such things as maintenance man-hours per flight hour (MMH/FH) and shop visit rate (SVR), and the result is an improvement not only in engine life but also in reliability which, in some cases over the period 1954–94, is 1,000-fold!

Over this period some things have been abandoned. A notable example is the future strategic bomber which the USAF sought from 1954, thinking the choice lay between a CPB using 'Zip' high-energy chemical fuel or an NPB using the heat from a nuclear reactor. So far as the author knows, no laboratory today is working on either source of heat for aircraft propulsion. Even more fundamental is the failure of either the rocket or the ramjet to find a place in the propulsion of any aircraft flying in the 1990s, unless one includes the Shuttle Orbiter and its Russian *Buran* counterpart, which are not aircraft but spacecraft which can be guided back through the atmosphere to land on a runway.

In the 1950s the dominant engines in the USA were the Pratt & Whitney J57, and its bigger derivative the J75, and the General Electric J79. Both have been described in Chapter 13, though it was not then emphasized how, starting later, General Electric was able with the J79 to use multiple variable stators to achieve a similar p.r. in a single axial spool. This helped GE to rival the J57 in thrust whilst knocking roughly 1,000 lb off the weight (from 4,200 lb to 3,200 lb in non-afterburning versions) and reducing diameter from 40 in to barely 30. But Pratt & Whitney tried harder, and despite excellent GE studies, prototypes, and demonstrators, the Connecticut firm won consistently in 1960–80.

First, in November 1962, a political storm ensued when the Secretary for Defense announced that the potentially huge TFX programme had been won by General Dynamics using Pratt & Whitney engines. Intended to replace almost all of the fighter and attack aircraft, this programme was at first little short of a disaster. Even the carefully designed TF30 engine proved to have compressors which were aerodynamically much too close to the stall line, so that taken in conjunction with the short curving inlet ducts the stall tendency was unacceptable. Even the basic specification was unimpressive, with a maximum afterburning thrust of 18,500 lb, a dry weight of 4,125 lb and a dry sfc of about 0.689. Gradually the engineers, who in the lifetime of the TF30 were moved from Connecticut to Florida, transformed this important engine and by 1974 came up with the TF30-100, giving 25,100 lb thrust for a weight 50 lb less, and at last without the pilot having to be afraid to touch the power levers.

By 1965 it was clear that the F-111 was not going to be any kind of 'fighter' – which most people take to mean an *air-combat* aircraft – so the USAF requested funds for a new fighter, called FX and later the F-15. In December 1969 they picked the McDonnell submission, to be powered by two Pratt & Whitney F100 engines. This time the designers, led by R. T. Baseler and later by Richard J. Coar, realised they had to try harder. With its design based on a 1965 demo engine, the JTF16, the F100 entered service in the F-15 in 1974. It was soon evident that this time Pratt & Whitney had lived up to their slogan of 'dependable engines', though some of the credit goes to the superior installation, which most unusually has the entire inlet pivoted to behave better at large angles of attack (AOA).

The F100 began life at a maximum afterburner rating of 23,810 lb. Features included a 3-stage fan and 10-stage compressor, giving an overall p.r. of 24.8, an unprecedented value for a fighter engine. This was necessary to reduce sfc, because the USAF was very interested in economy for long range; but in fact all the sfc figures were higher than for the TF30, which is strange (a typical figure was 0.72). Whereas in the TF30 the HP/LP turbines had 1/3 stages, this time the work was divided 2/2. Another difference was that the relative size of the afterburner was greater, ending in an efficient balance-beam nozzle. Not least of the advances was that, despite its greater power than all TF30s except the Dash-100, the new engine was a full 1,000 lb lighter.

By the 1970s Pratt & Whitney had profound respect for GE, and strove to improve the already good F100. Compared with almost all previous engines the F100 was wonderful, though it did still demand tender loving care by both pilot and line personnel. In particular, it suffered from the stall/stagnation problem outlined in Chapter 5. An immediate palliative was to provide a unified fuel control (UFC) incorporating a power-lever angle prime (PLAP) which adjusted the engine fuel flow irrespective of pilot demand.

When, to the astonishment of many, the LightWeight Fighter programme led to the adoption of the F-16 by the USAF, Pratt & Whitney gained what became a giant new market for the F100. The choice of a single F-15 engine had played probably the key role in the selection of the F-16 over the F-17, with economies in reduced F100 price through increased production and enhanced mission radius compared with the twin-J101 aircraft. The result was the F100-200, with features required by a single-engine installation such as a back-up fuel control. Notably, despite unprecedented agility in combat, the F-16 fea-

The F100-229 is a near-total redesign, retaining the multi-strut inlet frame but increasing both fan airflow and pressure ratio. Dry weight is 3,705 lb (Pratt & Whitney).

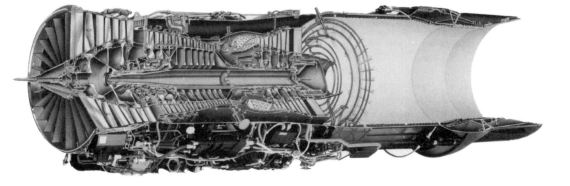

tured a fairly short duct from a plain fixed inlet.

Pressure for extended life, reduced maintenance costs and improved safety and reliability led by 1985 to the F100-220, which apart from digital control was notable for numerous changes in materials and cooling. But pressure from GE never eased, forcing the development of the F100-229. This was a near-total redesign, and the result is an outstanding engine which, as well as increasing maximum thrust to 29,100 lb, has been cleared to unprecedented installed life. For example, the most critical component, the HP turbine, has a time between inspections (TBI) of 4,300 mission cycles, typically equivalent to seven to nine years.

An offshoot was a turbojet derivative, the PW1120. This was lighter and simpler, and reduced overall diameter from 46.5 in (at the afterburner) to the 40.2 in at the inlet, maximum airflow being reduced from 228 lb/s to 178. Maximum thrust was 20,620 lb with the simple afterburner in operation. Israel selected it for the Lavi, but this was soon cancelled. Turning a turbofan into a turbojet is by no means common.

Another offshoot was the F401, essentially an F100 matched to the greater weight and salt-laden environment of the F-14 Tomcat. The first F401-powered F-14B flew in June 1973, but to save a modest sum the Navy cancelled the F401, leaving the F-14 as the A version, powered by the TF30. If anything, the TF30 caused even more problems in the F-14 than in the F-111, and certainly led to more aircraft losses. Navy Secretaries and Admirals called this 'the worst engine/airplane mismatch in history'. It led via the TF30-412 to the -414 and ultimately to the 414A, in which in a desperate measure to contain fan blades the casing was surrounded by steel, increasing weight from 3,992 to 4,201 lb.

During this 30-year effort GE never let up. In the 1960s the Ohio engineers planned the GE1, GE9, and GE15 as military demonstrator engines. In June 1970 the GE9 became the F101 to power the four-engined B-1 bomber. It was an augmented turbofan with a two-stage fan with variable inlet vanes handling an airflow of 350 lb/s, a nine-stage HP spool giving an overall p.r. of 27, and an advanced afterburner increasing take-off thrust from about 17,000 lb to 'the 30,000 lb class'. The original B-1A was later replaced by the B-1B, but rematching the engine to subsonic flight at low level did not change it very much. GE delivered the last of 469 F101-102 engines in December 1987.

Meanwhile, the little GE15 led to the J101, already mentioned as the engine of the Northrop YF-17, which lost out to the F-16 in the USAF LWF competition. This was not the end but a new beginning. McDonnell Douglas stepped in, seeing in the YF-17 a way to win a new competition for a Navy Air Combat Fighter (NACF). In June 1975 this succeeded, the F/A-18 Hornet being the winner with a pair of F404 engines. This kept the J101 core but had a fan (LP compressor) of 1 in greater diameter, raising BPR from 0.25 to 0.27, just sufficient to lift the engine from the J (turbojet) to the F (turbofan) category. About 100 lb heavier, at a little over 2,000 lb, the F404 had maximum thrust raised from 15,000 to 16,000 lb, to suit the Hornet's greater weight, but remained a straightforward two-spool engine with a diameter of only 34.8 in.

Seldom has an engine done so well. For a start, it quickly established every kind of record as the most troublefree and easily maintained engine the Navy had ever had. It followed by finding more applications than GE had dared to hope, not only winning the evaluations conducted by India, Sweden, Jugoslavia, and Singapore against the RB.199, but also being picked by Lockheed to power the F-117A 'stealth' bomber. In this application the F404-F1D2 has the afterburner replaced by a 'fishtail' jetpipe which spreads the jet out into a wide but shallow sheet expelled from a 2D nozzle forming the inboard trailing edge of the wing.

For the Swedish JAS39 Gripen the RM12 version was developed, with a thrust of 18,100 lb, improved birdstrike resistance, and special single-engine control. From 1992 the original F404-400 was replaced in production as standard Hornet engine by the -402 version, with changes as shown in a diagram (and visibly identified by the replacement of the titanium bypass duct chem-milled with criss-cross stiffening webs by a smooth black duct of PMR-15 graphite composite). Next came the F412, for the GD/McDonnell Douglas A-12 Avenger II 'stealth' attack aircraft. This had fly-by-wire FADEC control, and other features as depicted, and gave a take-off rating of 14,000 lb despite having no afterburner. The A-12 was cancelled, to be replaced by something yet to be decided. In any case, the F412's core was used in the F414 to power the F/A-18E/F Hornets, which are among the three biggest US aircraft programmes in the 1994–99 defence budgets. The F414 is again the subject of a diagram, showing how thrust has been increased to 22,000 lb.

Moreover, GE determined that it should recover its lost position in large fighter engines. Under Ed Woll the F101 was adapted with F404 features to create the F101X, first run in

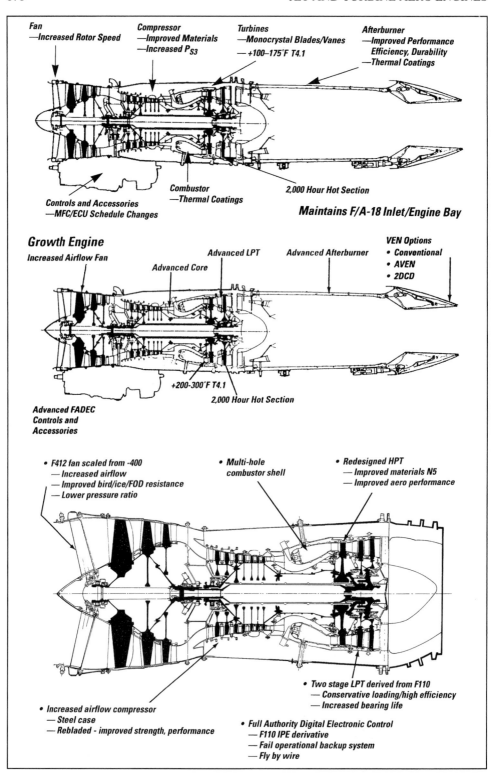

Fan
—Increased Rotor Speed

Compressor
—Improved Materials
—Increased P_{S3}

Turbines
—Monocrystal Blades/Vanes
— +100–175°F T4.1

Afterburner
—Improved Performance
   Efficiency, Durability
—Thermal Coatings

Controls and Accessories
—MFC/ECU Schedule Changes

Combustor
—Thermal Coatings

2,000 Hour Hot Section

Maintains F/A-18 Inlet/Engine Bay

Growth Engine

Increased Airflow Fan

Advanced Core

Advanced LPT

Advanced Afterburner

VEN Options
• Conventional
• AVEN
• 2DCD

+200–300°F T4.1

2,000 Hour Hot Section

**Advanced FADEC
Controls and
Accessories**

• F412 fan scaled from -400
   — Increased airflow
   — Improved bird/ice/FOD resistance
   — Lower pressure ratio

• Multi-hole
   combustor shell

• Redesigned HPT
   — Improved materials N5
   — Improved aero performance

• Two stage LPT derived from F110
   — Conservative loading/high efficiency
   — Increased bearing life

• Increased airflow compressor
   — Steel case
   — Rebladed - improved strength, performance

• Full Authority Digital Electronic Control
   — F110 IPE derivative
   — Fail operational backup system
   — Fly by wire

Left *Development of a fighter engine: top, the upgraded F404-402 (not mentioned, the PMR-15 bypass duct); centre, the F414 growth version for the F-18E/F; bottom, the largely redesigned F412 'stealth' engine* (General Electric).

December 1979. This was then renamed the F101DFE (derivative fighter engine), and a year later flew the F-16/101. It was just what the Navy wanted for the F-14, and a prototype called Super Tomcat flew in July 1981. The effort paid off, and in 1983 GE received an initial development contract funded by both the Navy and USAF, the engine becoming the F110.

Compared with the F101, the F110 is much slimmer, diameter being reduced from 55 in to 46.5, airflow being reduced to 255 lb/s, though p.r. is increased to 31. The afterburner and variable con/di nozzle are entirely different. The first engines delivered, in mid-1986, powered F-16C/D fighters, and after heated argument in Congress the USAF have subsequently bought F110-100 engines for the F-16 in competitive bidding against the F100. In 1987 the Navy began receiving the F-14A(Plus) followed by the F-14D, in all of which the TF30 is at last replaced, by the F110-400.

The new standard Tomcat engine has dry and afterburning thrusts of 16,000 and 27,000 lb. This dramatically improves flight performance, and also brings safety and reliability, as well as a new ability to do anything the pilot wishes with the power levers, no matter what the altitude, airspeed, AOA, or yaw. A remarkable range of angles of attack and yaw were tested with the original F101DFE engines, some of them beyond the normal F-14 manoeuvring envelope. Moreover, it was found that, with the F110 installed, afterburner was no longer needed for catapult launch, and while time-to-climb was reduced by 61 per cent, mission range was extended by 62 per cent! Seldom has a fighter been so improved.

In 1988 an F-16C flew with an F110-129 IPE (improved performance engine), which with a new FADEC and advanced features and materials increases airflow to 270 lb/s and dry/max thrusts to 17,000 and 29,000 lb. This is fast becoming the preferred engine for advanced F-16 versions. Looking beyond it, in 1991 GE ran the F110X at a dry thrust of 22,000 lb and maximum of 36,500 lb. This remarkable engine matches the -129 core with the fan of the F118 (see later), an advanced augmentor, and 'smart' adaptable FADEC control. Its thrust/weight ratio is 10, which 30 years ago was the goal for simple short-life lift jets.

The F118-100 is the engine of the B-2A 'stealth' bomber. Although it naturally builds on F101 experience, it is a dry (unaugmented) engine more akin to the F110, and having the same 46.5 in diameter. A major difference is a new fan with three blisk stages; as explained earlier, a blisk means that the blades and disc are fabricated as a single piece of material. Moreover, the blades of this fan have exceptionally large chord, handling higher airflow (280 lb/s) with increased efficiency. Take-off thrust is 19,000 lb, and special measures are taken to cool the jet and prevent any hot parts being seen by hostile IR seekers.

Despite the USAF's fixation with the F-15, by 1981 requests were issued to industry for ideas on a replacement. One of the demands was 'supercruise': sustained supersonic flight. In 1986 the choice narrowed to the Lockheed YF-22 and Northrop YF-23, two of each being ordered to evaluate the rival Pratt & Whitney F119 and GE F120 engines. GE's Brian Brimelow and P&W's Walt Bylciw (pro-

Many engineers and pilots would say the F110-129 was the greatest fighter engine in the world (General Electric).

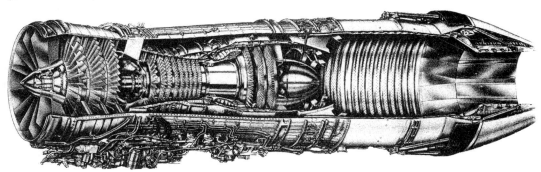

This F119 is fitted with the type of two-dimensional vectoring nozzle used on the YF-22 prototypes (Pratt & Whitney).

nounced Bilsher) each succeeded in convincing the author that their engine was clearly the best! Of course, both were impressive augmented turbofans in the 35,000 lb class, with contrarotating spools to eliminate turbine stators.

The big difference was that, while the F119 was rather the engine one might have expected, the F120 boldly adopted variable geometry. It had huge fan blades and a compressor giving an overall p.r. higher than in any previous fighter engine. This gave high thrust for take-off and climb and minimum fuel burn for loiter. But for supercruise the main bypass doors were shut, turning it into a turbojet with just a small bypass flow mainly for cooling and for extra oxygen for afterburning. Both engines were provided with hydraulically driven 2D nozzles able to vector ±20°, and both did well in flight test. In the event, the USAF picked what seemed the lower-risk F119.

These US fighter and bomber engines have been discussed at length because their histories parallel those of military engines in other countries. Since the withdrawal of the Spey 202/203 in British Phantoms, Britain has no military jet engine, apart from the Pegasus (Chapter 14). Rolls-Royce does share in important multinational programmes. The first was the Rolls-Royce Turbomeca Adour, launched in 1966 to power the Anglo-French Jaguar. A small and simple augmented turbofan, the Adour was almost Russian in having only seven compressor stages (two LP, five HP), overall p.r. being 11. Early versions gave 4,620 lb dry and 6,930 lb with afterburner, but later marks are rated up to 8,400 lb. Hawk trainer/attack aircraft have unaugmented versions rated at 5,200–5,900 lb.

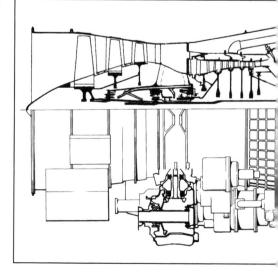

Next, in 1969, came the formation by Britain, West Germany, and Italy of Turbo-Union, to produce the RB.199 engine to power the Tornado. This again is a remarkably compact engine, the overall length with afterburner of 128 in comparing with 159 in for the rival F404 and 242 in of the TF30. One reason for the compactness is that the RB.199 is a three-spool engine, with 3/3/6 stages giving an overall p.r. 'greater than 23'. Early versions were rated at 9,100 lb dry and 16,000 lb with afterburner, but the Tornado ECR has the Mk 105 of 9,656/16,800 lb. A noteworthy feature of the RB.199 is its integral twin-bucket thrust reverser.

It is not easy to find reasons why none of the F404 market was won by the RB.199. Apart from the Tornado, its only application has been the EAP and the first two prototypes of the Eurofighter. Production Eurofighters will have the EJ.200, developed by a four-nation team (as before, plus Spain) on the basis of the Rolls-Royce XG40 demonstrator. Although it is more powerful than the RB.199, in the 20,000-lb class, the EJ.200 is rather smaller and much simpler; for example, it has less than 60 per cent as many aerofoils. To show the amazing development in axial-compressor design in the past 20 years, it has only eight stages in all (3+5) yet achieves a p.r. greater than 25. Another advantage of having fewer but larger (especially, wider-chord) blades is better resistance to impact damage. A novel feature is that the oil tank is a rapidly spinning drum to give positive gravity even under sustained negative-g manoeuvres.

The closest rival to the EJ.200 is the French SNECMA M88. SNECMA followed the Atar with the M53, for the Mirage 2000. This engine was most unusual because it had a single two-stage turbine driving both a fan and HP compressor. Usually it is poor design to make a fan and HP compressor rotate at the same speed, but the M53 offers the essential 'carefree handling' throughout the flight envelope. It shows its age in that, having the same 3+5-stage configuration as the EJ.200, it achieves a p.r. of only 9.8. In contrast, the M88 has two shafts, and its more modern design enables weight to be reduced from the M53's 3,380 lb to only 1,970, though maximum thrust is only slightly less.

SNECMA, like Rolls-Royce, has explored the turbo-rocket engine. This is essentially a turbo-ramjet (advanced turbojet/afterburner for supersonic use) which carries its oxygen with it. This makes it suitable for use at extreme altitudes, even above the sensible atmosphere. Turbo-rockets can be arranged in different ways, but most have a multi-stage core turbine driving a fan (LP compressor) via a reduction gear. Invariably the propellants are lox and kerosene, and these burn stoichiometrically at about 3,500°C, beyond what any turbine can stand. Accordingly, a large excess of fuel is pumped in to reduce the temperature, and the unburned surplus fuel is then mixed with air in the afterburner. Such an engine would be smaller and lighter than a conventional jet engine, but

A consortium called Eurojet Turbo GmbH was formed to produce the EJ.200.

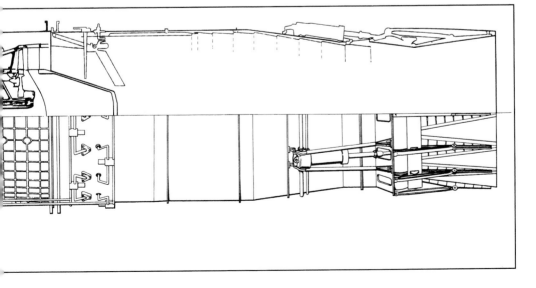

fuel consumption would be greater, so the application would be fast-climbing interceptors.

In the Soviet Union the principles of robust simplicity laid down by Tumansky were refined by later designers in the same bureau (Gavrilov, Metskhvarichvili, and Khachaturov), in a series of two-spool afterburning turbojets made in very large numbers. Tumansky himself managed the design of the R-15, a large afterburning turbojet for the Mach-3 MiG-25, and thus a single-spool steel engine with only five compressor stages. Today the bureau is called Soyuz (Alliance), and the General Constructor, V. K. Kobchyenko, produced the vectored R-79 engine mentioned in the preceding chapter.

The most immediate rival bureau, that of A. M. Lyul'ka, left the 1950s with the big AL-7 in full production, with a single nine-stage compressor and ratings typically around 14,500 lb dry and 20,000 with full afterburner. From this was developed the AL-21 family, with significant advances headed by a 14-stage compressor with variable stators ahead of the last five stages. With afterburner, the AL-21F-3 has ratings of 17,200 lb dry and 24,800 maximum. Today this bureau is called Saturn, and its main product is the two-spool AL-31F for the Su-27 family. Designer Dr Viktor M. Chyepkin is proud of this fast-responding engine, which for a weight of 3,373 lb has dry and maximum ratings of 17,857 and 27,560 lb, the AL-35 version being cleared to 29,900 lb. This is the only engine demonstrated in public in a manoeuvre (the 'Cobra') in which the inlets are pulled backwards at an AOA of up to 135°. Today Chyepkin is testing the next-generation SAT-41, an engine rather like a double-size EJ.200 with a very high turbine gas temperature, few parts, and '20 tonnes thrust for the

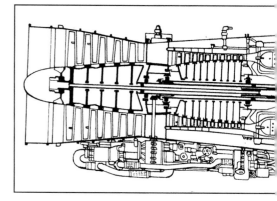

weight of a 10-tonne engine'.

At St Petersburg, the Klimov Corporation wishes there were fresh markets for the RD-33, designed by A. A. Sarkisov for the MiG-29. This is a two-spool (4+9-stage) turbofan of 0.4 BPR, rated in its original form at 11,110 lb dry and 18,300 lb with maximum afterburner, and later uprated to 19,335 lb, for a weight of 2,326 lb. Like the AL-31F, most RD-33 installations have an inlet screen to prevent FOD, automatically extended across the duct whenever the engine is running on the ground.

At Samara/Kuibyshyev the bureau headed by N. D. Kuznetsov produced a succession of large two-shaft afterburning engines for bombers which began with the NK-6 of 1960, rated at 44,090 lb, and led to the NK-25 of 1975, rated at 55,115 lb and used in the Tu-22M-3 swing-wing bomber. This family culminated in an outstanding three-shaft engine developed for the four-engined Tu-160 heavy bomber, the most powerful military aircraft of all time. Designated NK-321, this augmented turbofan has a three-stage fan (BPR 1.4) handling

The usual configuration for turbo-rocket studies, as described in the text (Rolls-Royce).

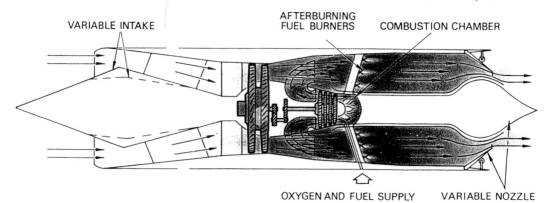

VARIABLE INTAKE   AFTERBURNING FUEL BURNERS   COMBUSTION CHAMBER

OXYGEN AND FUEL SUPPLY   VARIABLE NOZZLE

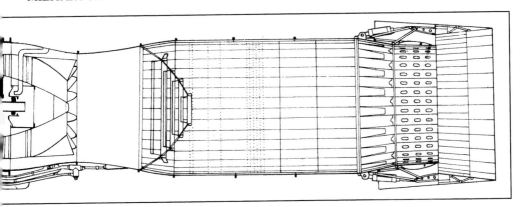

Longitudinal section of the D-30F6, the mighty engine of the MiG-31.

805 lb/s, a five-stage IP compressor, and seven-stage HP, turbine stages being 1/1/2. Diameter is 57.5 in, dry weight 7,496 lb and dry and maximum ratings are 30,843 and 55,077 lb.

If the NK-321 is the most powerful military engine, it is run close by the D-30F6. This was developed by Yuri Reshetnikov of the Perm-based 'Aviadvigatel' to power the MiG-31. Although it was originally based on the civil D-30 turbofan of 14,990 lb thrust, the F6 is a very different engine. A zero-stage on the fan increases airflow to 331 lb/s, and an enormous afterburner increases length to 277 in, and dry weight to 5,326 lb. Apart from the ability to cruise at Mach 2.83, the D-30F6 was required to have outstanding fuel economy, and p.r. at 21.15 is unprecedented for a Russian fighter engine. Dry and maximum ratings are 20,944 lb and 41,843. With almost 84,000 lb available, no wonder a MiG-31 take-off is so noticeable!

16 Civil jet engines

This chapter covers the engines of business and commuter jets, and even baby engines in sporting and training aircraft. At the other end of the scale are the long and slim engines for SSTs, and the big fans of up to more than 10 ft diameter that move people and cargo around the world. Conspicuously absent are large subsonic turbojets.

In Chapter 10 we sympathised with Whittle when, searching in 1936 for a way to 'gear down' the high-velocity turbojet to improve its propulsive efficiency for 500 mph aircraft, he invented what we today call the turbofan. He suggested using extra turbine shaft power to drive a bigger compressor which could accelerate a much greater airflow at lower velocity. He proposed several ways of doing this, some involving extra free-running turbine/fan stages downstream of the original turbojet. Nobody showed much interest, except that four years later Griffith began urging Rolls-Royce to build incredibly complicated engines that even today would be difficult to build.

Today the turbofan has virtually swept the turbojet from the skies. Even in supersonic fighters it is the preferred choice, partly because it provides so much oxygen for a downstream afterburner, though for Mach 2 the ideal BPR is less than 1. For the global fleets of civil jets the BPR has never ceased to rise, reaching 8.4-plus in the GE90, and upwards of 15 in unducted fans. It is therefore puzzling that from 1936 the turbofan was almost ignored as a kind of oddity, until 20 years later it began to creep in under the title 'bypass turbojet' with BPRs so trivial as to be hardly worthwhile.

We have already seen how Whittle's Power Jets company was, for political reasons, prohibited from making any engines just as the LR.1 was nearing completion. This turbofan was intended to halve the sfc of contemporary turbojets. It was designed in 1943–44 for long-range aircraft, the immediate need being bombers to attack Japan. When Power Jets was nationalized and told it must never build another engine, the loss of the LR.1 was (so the author was told in 1954) '. . . of no consequence. When Japan was defeated there was no longer any need for long-range bombers, so the engine was cancelled.'

A few weeks later specification B.35/46 was issued for a new long-range bomber, the ongoing need for which had been obvious to the RAF since 1918. Apart from the Short SA.4 Sperrin, an undervalued aircraft, the resulting 'V-bombers' built to this specification, and also the later B.9/48, all had engines buried in the wing roots. Thus, when the Mk 2 version of the Victor later went ahead with Rolls-Royce Conway engines, these were not true turbofans but merely 'bypass jets' able to fit inside the bomber's wing. BPR accordingly began at only 0.3, which did little except cool the outer casing.

It is strange that, when de Havilland were designing the Comet, Maj Halford did nothing to develop an efficient engine for long-range flight. Such engines were left to Rolls-Royce, Bristol, and Pratt & Whitney. By 1955 it was obvious that the 707 and DC-8 were really going to happen, and increasingly obvious that Britain's rival Vickers VC.7 would be cancelled (even in those days Britain's 'world's favourite airline' preferred Boeing to anything British). Accordingly, Rolls-Royce did all it could to get the Conway on the two US jetliners, and succeeded, but only to the extent of 69 aircraft out of 1,519. At this time the turbofan was still covered by Whittle's original patent. This would have expired in 1952, but in the High Court of Chancery Mr Justice Lloyd said that it had 'exceptional merit' and that 'unfortunate circumstances' (such as cancellation of the LR.1) had prevented its exploitation, so he gave it the maximum extension of a further 10 years, to expire in 1962. True to form, Britain had the idea and gave it away.

In 1955 GE's Pete Kappus asked the author; 'When are you going to do something with ducted-fan engines?' He had carried on roughly where Metrovick had left off, in particular studying free-running fans with double-deck blades added behind an existing turbojet. As he said; 'You can do it so the engine doesn't know what's happening downstream.' It made obvious sense, because for a small expenditure and low risk a noisy fuel-guzzling turbojet could be turned into a much quieter and more efficient turbofan. GE tested an engine called the X220, and on 26 December 1957 ran the CJ-805-23, an aft-fan version of the J79.

It was really aimed at the airlines, but because it was adopted only for the Convair 990 it made little impact on the civil market. But back in 1956 GE had convinced the USAF that this engine could transform the B-52. This made Pratt & Whitney wake up, and in a matter of months the J57 had been converted into the TF33 for the B-52H. The rest is history. Except for the small JT12 and Mach-3 J58, P&W ceased building turbojets. For the world's airlines they replaced the JT4A by the JT3D (civil TF33), an engine more powerful, 1,000 lb lighter, and with sfc reduced from 0.77 to 0.505! In passing, this instantly stopped Rolls-Royce's penetration of the market by the Conway, and it all happened precisely as Whittle's patent finally ran out.

This, in turn, made Rolls-Royce wake up, and it began urgent studies of new civil engines. By 1960 it looked as if the suddenly created JT3D was going to take the rest of the 'big jet' market, but Rolls still had good business with the smaller Comet and Caravelle. Quickly an aft fan was designed and fitted behind an airline-type Avon, achieving a world-record low sfc, but it got nowhere. Such an engine obviously could not fit in the Comet, and the Caravelle, after toying with GE's aft fan, switched to the JT8D.

The JT8D front-fan engine was derived by Pratt & Whitney from the Navy J52 turbojet to power the Boeing 727. It was probably the first jet engine to be developed without a military market. First tested in April 1961, it began life with a BPR of 1.1 and a rating of 14,000 lb. It was the right engine at the right time, and its adoption for the 727, 737, DC-9, and Caravelle made it the world's No 1 airline engine, almost

12,000 having logged about 430 million hours at the time of writing. It did what the airlines wanted, but did it noisily and with clouds of black smoke, and during the 1960s people began to think about something called the environment. In fact, fine particles of carbon do little harm, but by 1970 P&W was marketing various reduced-smoke conversion kits as well as noise-reduction options, because of growing public awareness of pollution.

Obviously, one could do better, and in October 1977 airlines launched the DC-9-80 powered by the JT8D-209. This was the world's first 'refan', a derivative engine produced by putting a bigger fan on an existing design. Whilst keeping changes to a minimum, P&W recognised that a BPR around 1 was too low, and a good compromise was to replace the JT8D's two-stage fan of 42 in diameter by a more efficient single stage with a diameter of 49.5 in. This increased mass flow from 319 lb/s to 469, and BPR to 1.78. The 6-stage LP spool and 3-stage turbine were all improved, and the bigger bypass flow was mixed with the core jet by a new 12-lobe mixer. The critical HP spool was left unchanged. Thrust jumped from 14,000 lb to 19,250, with plenty of further stretch in prospect. Today, despite failing to get on the 737, the single application of the JT8D-200 family on the DC-9-80 (now called the MD-80 family) has added 2,600 to the previous 12,000+ engines.

The obvious question is; 'Why didn't they do that in the first place?' There are several reasons, including the airlines' abhorrence of risk, and fear of moving too quickly. When the JT8D was designed, fuel cost $1.50 a barrel. By 1975 the price had jumped to $30. Growing environmental pressure was being reflected in legisla-

A plot of pressures and temperatures through a JT8D at sea-level take-off power (Pratt & Whitney).

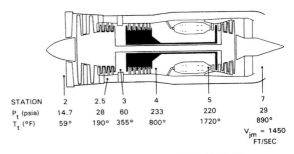

STATION	2	2.5	3	4	5	7
P_t (psia)	14.7	28	60	233	220	29
T_t (°F)	59°	190°	355°	800°	1720°	890°

V_{jm} = 1450 FT/SEC

AT SEA LEVEL STATIC TAKEOFF THRUST OF 14,000 LBS,
W_{af} = 165 LBS/SEC, W_{ap} = 150 LBS/SEC

tion which threatened to prevent noisy aircraft from flying. In Rolls-Royce's case, while Griffith rightly kept thinking of BPR around 20, a study was made to find optimum BPR for an engine in an external pod and, because of incorrect assumptions of nacelle drag, came up with the answer that it should not exceed 1.0.

This error made the competition from Rolls-Royce less effective. The Conway had been developed with military funding, and the changes to fit the engine to the 707 and DC-8 were mainly concerned with external dressing (though for the Boeing aircraft Rolls developed the complete pod, with reverser and noise-suppressing nozzle). Having killed off the rival Vickers VC.7, the British flag airline then took the amazing step of asking for a British rival to the US 'big jets' to be developed, specially matched to shorter runways, even though these runways were all being extended to match the US aircraft. The result was the VC.10, for which Rolls had to develop a largely new Conway with a bigger LP spool increasing airflow to 367 (later 375) lb/s and BPR to 0.6. With hindsight, of course, even this doubled BPR was a mere fraction of the optimum.

At this time, in 1957, the other British flag-carrier, BEA, was eagerly awaiting the turbo-prop Vanguard (Chapter 17). Within weeks, and for no evident reason, it became fashionable to regard anything with propellers as obsolete. BEA led the panic, and rushed to get government funds for a new short-haul jet and its engines. Such an aircraft already existed in the Caravelle, and had BEA bought it the French would have stayed with British engines in the form of the aft-fan Avon. As it was, BEA first bought the uneconomic Comet 4B and then ordered the DH.121, to be powered by three RB.141 Medway engines grouped at the tail. In one of the worst decisions ever taken by an airline, BEA then ordered the DH.121 to be cut back in size, with range reduced from 2,073 miles to only 921. So the DH.121 became the Trident. The Medway, which by this time had several other applications, was abandoned and replaced by the smaller RB.163 Spey.

This repeated the company's belief that BPR should be less than 1. First run in 1960, it had an airflow of 203 lb/s, and began life at 9,850 lb. From that moment on it was obvious that the Trident's smaller capacity, much shorter range, and longer field length would make it difficult to win against the 727, and everything possible was done to restore what had been thrown away, even to the extent of adding a booster turbojet used only for take-off! Even as the Spey was

going into production, in 1964, Rolls-Royce recognised that its BPR was too low, and studied ways of refanning it to reduce fuel burn and noise.

Almost unbelievably, despite a host of projects – and an outstanding three-shaft engine, the RB.203 Trent, which was actually run in December 1967 – no such engine appeared until the Tay was launched in 1983. By this time there was no British jetliner in production to take such an engine, and the launch customers were Gulfstream and Fokker. First run in August 1984, the Tay is a minimum-change engine, notable for an LP spool with a single-stage fan with snubberless but solid titanium blades handling 390 lb/s, driven by an LP turbine with an added third stage. From the outset both the bypass duct and nacelle for all applications have been made from CFRP, and a 12-lobe mixer leads to a single nozzle with target-type reverser. Development has been cautious, starting at a flat-rating of 13,850 lb and then moving to 15,100 lb by increasing fan diameter from 44 to 44.8 in. BPR clearly should have been at least 3, but was fixed at a mere 1.75 to suit customer installations and to boost high-altitude cruise thrust.

An obvious next move was the Tay 670, with a 48 in fan, rated at 18,000 lb to suit the Fokker 130 and the colossal global market for re-engining 727-200s, DC-9s, and 737s. Instead of (as was planned) running this engine in 1989, Rolls-Royce in 1990 concluded an agreement to form a German company called BMW Rolls-Royce, with the British firm supplying the knowledge and experience but only 49.5 per cent of the shareholding. The resulting BR 710 has the 48 in fan intended for the Tay 670, on a shaft by itself driven by a two-stage LP turbine. The HP spool is new, with an advanced 10-stage compressor with the first five stators variable. This engine starts off with a much higher BPR than the Tay, and has quickly gained applications from Gulfstream and Canadair at ratings just under 15,000 lb. Unless foolish decisions are taken, the 710, and planned derivatives with bigger fans and extra stages, will go into thousands of aircraft, including many at present fitted with the JT8D.

In 1994 Germany's MTU announced a turbo-fan project, the 'Model 17/20', said to be a joint effort with P&W, GE, SNECMA, and others, to fight for exactly the same market. Japan likewise announced a similar engine for the 70/80-seat YSX jetliner project. There will not be room for them all, and BMW Rolls have a head start.

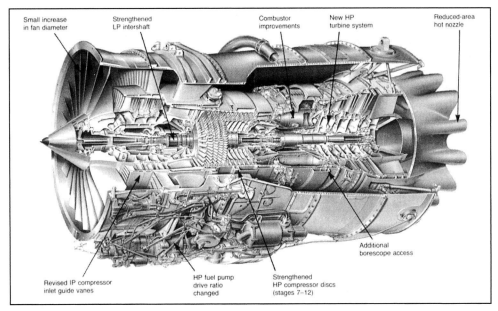

Features of the Tay 650 which, mainly from a 0.8-in increase in fan diameter, raises flat-rated thrust to 15,100 lb (Rolls-Royce).

Apart from a handful of sporting machines powered by Turbomeca turbojets, the small-jet civil market took off in the late 1950s when large corporations began replacing converted piston-engined bombers by the JetStar and Sabre (later Sabreliner), the first business jets. These had been designed for the USAF, using respectively four and two Pratt & Whitney J60 (JT12A) turbojets. In 1963 prototypes flew of the DH.125, Learjet, and Dassault Falcon, all designed for the civil market and all powered by turbojets designed for military applications. Fuel was cheap, and few people bothered about noise at airports, but the need for better small jet engines was self-evident.

First to respond was GE, whose J85 turbojet had in 1960 been offered in civil guise as the CJ610. Using this as the gas generator, a free-running fan was added downstream to produce the CF700. With a BPR of 1.6, mass flow was increased to 88 lb/s, and take-off thrust was set at 4,200 lb. Sfc was reduced from around 1.0 to only 0.645, and the CF700 replaced the JT12 in the Falcon 20. But with the market growing, rivals appeared.

Next to move was Pratt & Whitney Canada, whose JT15D first ran on 23 September 1967. This is a direct-drive front fan, with a single-stage HP turbine driving a centrifugal compressor, a two-stage LP driving the fan, and a folded annular combustor. BPR was set at 3.3, with air-flow of 75 lb/s to give a rating of 2,200 lb. Its success was ensured by selection for Cessna's Citation, and the Dash-4 and -5 versions with an axial booster added raise thrust beyond 3,000 lb. By 1994 more than 4,700 had been delivered.

Rolls-Royce Bristol, whose Viper had powered early 125 jets, ran prototypes of the excellent RB.401, but failed to get permission to launch it. Instead, the 125 switched to the American Garrett TFE731. First run in 1969, this followed the JT15D in having a single-stage HP turbine driving a centrifugal compressor and an LP turbine (with three stages) driving the fan and a four-stage LP compressor. The combustor was also of the folded annular type, but a major difference is that the fan is driven via a planetary reduction gear. BPR was 2.66 in the 731-2, rated at 3,500 lb, but climbed with successive models until in the latest Dash-5 engines it is 3.65, with thrust raised to 4,750 lb. The TFE731 has won the biggest share of the bizjet market, with deliveries exceeding 7,300.

In 1976 Dassault picked another Garrett engine, the ATF3, for the Falcon 200. This engine has three shafts and a unique reverse-flow layout. Rated at 5,440 lb in the Falcon, it has a high p.r. and good sfc, but its complexity has frightened off other customers. In contrast, in 1982 Garrett used the core of the mass-produced TPE331 turboprop to produce the F109 to power the T-46A trainer. In the event the USAF

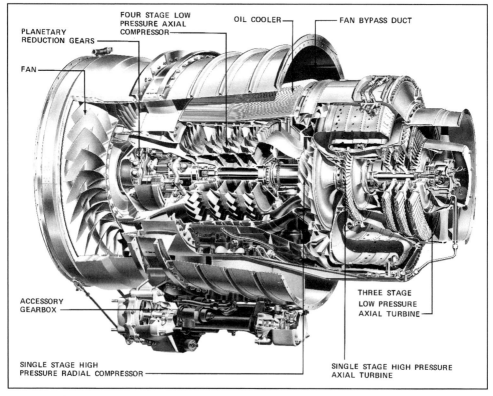

FOUR STAGE LOW
PRESSURE AXIAL
COMPRESSOR
PLANETARY
REDUCTION GEARS
OIL COOLER
FAN BYPASS DUCT
FAN
ACCESSORY
GEARBOX
THREE STAGE
LOW PRESSURE
AXIAL TURBINE
SINGLE STAGE HIGH
PRESSURE RADIAL COMPRESSOR
SINGLE STAGE HIGH PRESSURE
AXIAL TURBINE

Features of the very successful Garrett TFE731 geared turbofan.

cancelled this, but the engine lives on as the TFE 109, rated at around 1,600 lb thrust for smaller aircraft. Two-stage HP and LP turbines respectively drive tandem HP centrifugal compressors and a direct-drive front fan, and sfc is outstanding at 0.392.

Back in 1962 Turbomeca produced the Aubisque to power the Swedish Sk 60 trainer. This had a geared front fan followed by single axial and centrifugal compressors all driven by a two-stage turbine. Rating was 1,543 lb, and the ratio of thrust/weight was poor at 2.83. Seven years later came the Astafan, using an Astazou core – like other Turbomeca engines, running at constant speed – driving a reduction gear to a variable-pitch fan with a BPR of 7. Thrust was 1,870 or 2,710 lb in different versions, but despite sfc being as low as 0.31 there were no takers. The pitch variation was not used to confer reverse-thrust capability.

The author is astonished that so little has been done to develop fans with variable pitch. The principles are outlined in the next chapter. The chief pioneer was Britain's Dowty Rotol, which from 1966 worked on fans of low pressure ratio

(say, 1.05) and high BPR, geared down from the core. Following extensive testing with Astazou and Gnome units in the 1,000 hp class, testing began in 1975 of the Rolls-Royce M45SD-02. This used as drive unit an M45H, a two-shaft turbofan originally rated at 7,760 lb and produced in small numbers for the odd over-wing-pod VFW 614 German transport. This was given an extra LP turbine stage and a 2.38 reduction gear to a fan of 68 in diameter with 14 blades controlled by a ring bevel gear. With a BPR of 8.73, the SD-02 was rated at 10,000 lb with sfc of 0.32. It was to lead to the RB.410, to give 14,370 lb with sfc of 0.295, with low noise and qualities matched to future STOL aircraft. Astonishingly, we are still waiting for it.

The nearest approach, but with fixed blades and half the thrust, was launched in 1963 when Lycoming added a geared front fan to its T55 helicopter engine. The resulting PLF1 led by 1969 to the refined ALF502, with a 41.7 in fan giving a BPR close to 0.6 and a thrust in the 7,000 lb class. Of the 14 projected applications, two actually happened and made it worthwhile: the BAe 146, which became the Avro RJ, and

Canadair Challenger. Today Lycoming (now, like Garrett, an AlliedSignal company) is developing a family of LF500 engines with a common core used for various fan and shaft drives.

In a much smaller category, Microturbo was established in Toulouse in 1960 to make gas-turbine starters. Later it developed the TRS 18 turbojet to power many sport, research, and training aircraft, rated at 225–337 lb. In Michigan, Sam Williams founded a company to produce simple turbojets for drones, RPVs, and missiles. In the 1960s small turbofans appeared for manned seats, belts, and platforms. In the 1980s Williams developed the FJ44, a turbofan of 23 in overall diameter with a single-stage fan, axial and centrifugal compressors, an annular radial combustor, and 1+2 turbines. With a BPR of 3.28, this neat engine weighs 445 lb and is rated at 1,900 lb. In partnership with Rolls-Royce it suddenly found a wealth of applications, some of the more important being the CitationJet, Swearingen SJ30, and Promavia Squalus, suggesting that such an engine was overdue.

While the twin-FJ44 aircraft are the smallest practical bizjets, new engines are becoming available for larger and longer-ranged aircraft. In 1987 GE and Garrett formed the CFE compa-

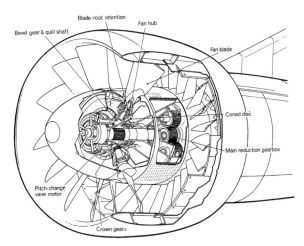

The Dowty variable-pitch (feathering and reversing) fan of the M45SD-02.

ny to market the CFE738, using the GE27 core (five axial plus a centrifugal driven by a two-stage turbine) and a Garrett fan handling 240 lb/s (BPR 5.3) driven by a three-stage turbine. Fan p.r. is 1.7 and overall p.r. no less than 35, so at take-off power of 6,000 lb the sfc is 0.372. This engine powers the Falcon 2000.

Cutaway of the Lycoming ALF502, showing the geared front fan and reverse-flow combustor.

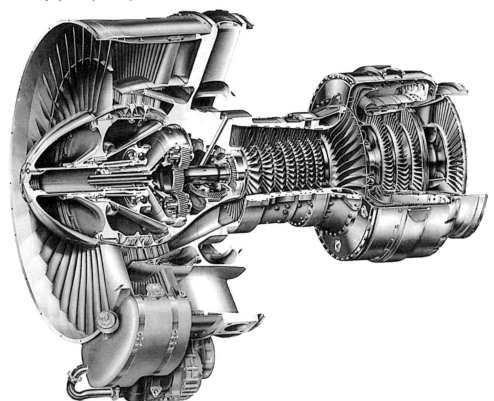

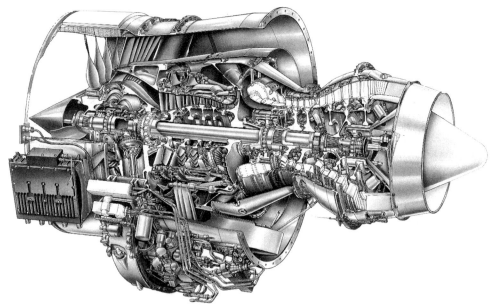

Cutaway of the PW305, with a direct-drive front fan; the large box beside the fan case is the FADEC (Pratt & Whitney Canada).

The immediate rival is the P&W Canada PW305, certificated three years earlier in 1990. Here the fan is driven by a three-stage turbine provided by partner MTU; the HP spool being four axial plus a centrifugal driven by a two-stage turbine. Lighter than the CFE engine (993 against 1,315 lb), the PW305 is rated at 5,225 lb and powers such aircraft as the Hawker 1000 and Learjet 60.

To fill the gap between the PW305 and the JT15D, the Montreal company ran the first PW500 on 29 October 1993. This high-BPR engine is aimed at the 3,000–4,000-lb bracket, with sfc 15 per cent lower than the JT15D. In 1994 one was being tested in a pod on the right side of the forward fuselage of the P&WC Boeing 720.

For the next size up, Allison had long wished to use the core of the T406 as the basis for a turbofan, and they ran the resulting GMA (now AE) 3007 in July 1991. The core, derived from the T56, has 14 stages with six variable stators and a two-stage turbine. The single-stage 38.5 in fan, driven by a three-stage LP turbine, gives a BPR of 5. The 3007 weighs about 1,580 lb and is flat-rated at 7,200 lb. The first applications are the Citation X bizjet and EMBRAER 145 regional airliner.

Moving up the size scale, in the late 1960s GE and France's SNECMA identified a need for a modern turbofan of high BPR in the ten-ton

class, meaning about 22,000 lb. SNECMA designed an engine called M56, while GE studied the GE13. In the immediate post-war years US companies had taken licences to make British jet engines, but now a US company was set to develop an engine with a foreign partner. What happened was that SNECMA, which had close ties with both P&W and Rolls, came to the USA to play off Pratt against GE, who were relative strangers. In December 1971 SNECMA chose GE.

The M56 fan was matched to the core of the F101 to form an engine called CFM56, and a joint company called CFM International was created. Then, because the F101 powered the B-1, the US State Department refused an export licence. This held things up until September 1973, when a licence was granted. The first CFM56 went on test at Evendale on 20 June 1974, exceeding target thrusts and sfc on the same day.

This time there was no pussyfooting; BPR was set at 6. Thus the fan diameter was initially 68.3 in, handling 780 lb/s. A unique feature was that its 44 slender titanium blades had tip shrouds forming a continuous peripheral ring. It rotated with a three-stage LP core booster, the slightly modified F101 HP spool giving overall p.r. around 25. Turbine stages were 1+4, and total dry weight about 4,600 lb. Take-off rating was set at 22,000 lb and cruise sfc about 0.65.

The CFM56 first flew in the No 2 position on a YC-15 STOL prototype, soon followed by a refined installation with full-length fan ducts on the SNECMA Caravelle. Then Boeing offered the engine as a retrofit on 707s, and completely re-engined a 707 demonstrator. All seemed set fair to start signing customers, especially as the new engine was aimed at 'derivative or re-engined versions of the One-Eleven, Caravelle, Mercure, 737, DC-9, 727, four-engined A300, 707, DC-8 and many projects including the DC-11 and Boeing-Aeritalia 751'. Today, with CFM56 sales approaching 9,000, it is hard to believe that CFM went from 1971 to 1979 without a single order!

It is typical of the airline business that, whoever the people behind it may be, or however good the product, a 'new name' finds it almost impossible to get started. In the event the breakthrough came when, in March and April 1979, United, Delta, and Flying Tiger ordered the DC-8-70, re-engined with the CFM56-2 by Cammacorp; later in that year Capitol, Cargolux, and Spantax followed suit. A major reason was that the new engine would enable stretched DC-8s – previously in most respects the noisiest subsonic airliners – to fly cargo at night. Altogether, 110 DC-8s were re-engined, and their dispatch reliability of 99.77 per cent made it certain that the CFM56 would go into other aircraft.

Strangely, the re-engined 707 never happened, but other versions of Dash-2 engine eventually went into a variety of military variants, as well as into some 350 KC-135 tankers. The latter had been notable for their noise and smoke on take-off, which also needed water injection, but what influenced the USAF and Armée de l'Air was that the new engine transformed the mission radius and fuel uplift. Had the CFM56 gone no further it still would have been judged a success. But in March 1982 CFM ran the first Dash-3 engine with a new fan, with only 38 blades, with mid-span shrouds, and with diameter reduced to 60.0 in to handle 688 lb/s. Boeing picked this to power the 737-300, -400, and -500. CFM offered the 56-3C1 as a common engine for all 737 versions at ratings from 18,500–23,500 lb. At the time of writing over 6,000 of these engines had been ordered, a record for a single application for a civil airline engine.

In September 1984 the CFM56-5A1 was launched to power the A320. This used an improved 68.3 in fan, with 36 larger blades handling 852 lb/s, plus other advanced features and FADEC control, and proved so reliable, flat-rated at 25,000 lb, that it was soon cleared for 120 min ETOPS operations. Not mentioned previously, ETOPS (Extended-range Twin OPerationS) permits a twin-engined civil transport to fly across oceans and deserts provided it is never more than either 120 or 180 min away from a suitable airfield. From the 5A CFM then developed a family of Dash-5C engines with a 72.3 in fan handling 1,027 lb/s, a four-stage LP booster, active-clearance on the HP spool and new turbine section with five LP stages, an integrated mixer nozzle, and FADEC control. Rated at 31,200 lb, it was the launch engine for the A340. Today CFM has a range of Dash-5 engines rated at up to 34,000 lb for the A319, 320, 321, and 340.

Thus, CFM has succeeded beyond its wildest dreams, and it seems churlish to suggest that anyone might come from behind and produce a better engine. But, while in 1972 Rolls-Royce was in no position to do anything (see later), Pratt & Whitney, in partnership with MTU of West Germany and Fiat of Italy, launched the JT10D in the 23,000 lb class. It fell into the trap of repeatedly aiming higher to meet customer demand, and eventually had redesigned the engine so often that it called it the PW2037 in the 37,000 lb class, as described later. But P&W still wanted to compete with CFM.

Meanwhile, by 1980 Rolls-Royce was once again able to play a positive role. Its Bristol team linked with Japanese Aero Engines (a group formed by IHI, KHI, and MHI) to produce an all-new engine called RJ.500, with a

Cutaway of a CFM56-3, the small-fan version used in all current types of 737 (CFM International).

59 in fan and a rating of 20,000 lb. This too was abandoned. The upshot was that on 11 March 1983 Pratt & Whitney, MTU, Fiat, and Japanese Aero Engines met at Derby with Rolls-Royce to form IAE, International Aero Engines. The objective: to develop an advanced engine for future 150-seat airliners. Called V2500, it was to run in 1986 at 25,000 lb.

With such a wealth of talent it should have been plain sailing. A two-shaft engine, the V2500 had a modern big-blade 63 in fan rotating with a single booster stage, all driven by a five-stage LP turbine, and a 10-stage HP compressor with controlled-diffusion end-bend blading driven by a two-stage turbine with single-crystal blades in powder-metallurgy discs. Airflow was 789 lb/s, BPR of 5.8 and design p.r. 36. Then trouble struck. First, the heavily loaded HP compressor, a Rolls responsibility, failed to reach design figures, and the engine had to be redesigned with the fan given recambered blades and moved forward to make room for a three-stage booster. The new V2500 ran in November 1987.

Unfortunately, in 1986 IAE had sketched various derivative ultra-high-bypass (UHB) engines. In December 1986 it was decided to market the SuperFan, in which the V2500 core plus extra LP stages drove through a 3:1 reduction gearbox to a 110 in fan with variable-pitch blades. It was exactly what long-haul airlines wanted, and when Airbus discovered that the projected A340 could accept an engine of this diameter without a change to the landing gear,

IAE offered the SuperFan in early 1987, and it was on this basis that the A340 went ahead in April 1987. On almost the same day the IAE board got cold feet!

There was absolutely nothing wrong with the design of the SuperFan, almost all of which, including the gearbox, was based on known technology. The sole problem was the commercial risk of failing to meet the in-service date of spring 1992, with consequent severe penalties. IAE was at pains to stress its long-term commitment to such engines, but to the media it was a headline story of 'cancellation', leaving egg on many faces. The A340 continued, and so far this superb aircraft is available only with various sub-types of CFM56-5C.

This was an unnecessary blow to IAE, whose V2500 quickly showed itself to be simpler, tougher, more reliable, quieter, remarkably more fuel-efficient, and to produce less pollution than its rival. Such qualities eventually translate into dominance in the marketplace, and the V2500 has now been produced at ratings from 18,500–35,000 lb for all single-aisle Airbus aircraft and the MD-90. Current production is centred on A5 and D5 engines with a fourth booster stage and p.r. of 30-33. It would be logical for similar engines to get on the A340 and many re-engining programmes.

Very special engines are needed for supersonic transports (SSTs), and only one is in service. To power the Anglo-French Concorde Bristol, later Bristol Siddeley and from 1966 Rolls-Royce, developed the subsonic Olympus into an

Cutaway of the initial production V2500; later models have a fourth stage on the core booster (International Aero Engines).

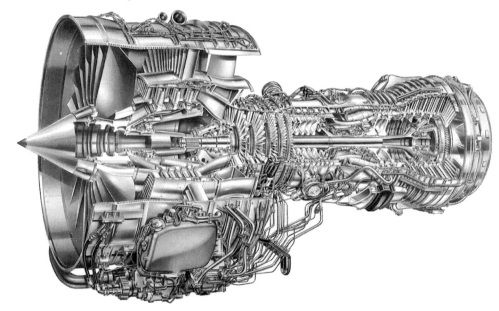

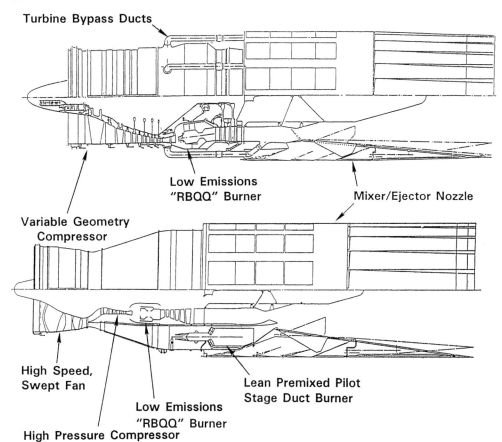

Turbine Bypass Ducts

Low Emissions "RBQQ" Burner

Mixer/Ejector Nozzle

Variable Geometry Compressor

High Speed, Swept Fan

Lean Premixed Pilot Stage Duct Burner

Low Emissions "RBQQ" Burner

High Pressure Compressor

Several groups of design teams in at least five countries are studying propulsion of the next SST. Most are agreed that some form of variable geometry is essential, such as a valved aft fan or tandem front fans; here the variable-stream engine (upper) and turbine bypass engine (lower) are matched to very high Mach numbers.

uprated afterburning version for the Mach 2 TSR.2 bomber, and used this in turn to develop the Olympus 593. The number '593' became part of the basic designation, while the engine grew with ever-greater airflow from the '593 Mk 601' to the '593 Mk 621'. The latter has single-stage turbines driving a seven-stage LP and seven-stage HP spool, handling 410 lb/s (which makes the core at least as powerful as the giant subsonic turbofans today giving over 100,000 lb thrust). With modest afterburner the rating is 39,940 lb, for a weight of 5,793 lb, including a variable primary nozzle with retractable 'spades' to break up the periphery and reduce noise, and upper/lower secondary nozzle eyelids which adjust the jet according to flight regime and on landing rotate to form a reverser.

In 1963 President Kennedy announced that the USA would develop a bigger and faster SST.

On the last day of 1966 the winning combination was the Boeing 2707, powered by the GE4, an afterburning turbojet selected over P&W's duct-burning turbofan. Predictably, it began big and grew bigger, ending as the GE4/J5P with a nine-stage compressor with the first two and last five stators variable, handling the record core airflow of 646 lb/s. Take-off rating was 69,900 lb, using the biggest-ever afterburner, but on 29 March 1971 Congress voted 49:48 to cancel the programme.

The remainder of this chapter is devoted to the engines that since the 1960s have transformed air transport. This revolution was the introduction of engines that were not only of much higher BPR, but which were also of unprecedented size, so that the core could drive a giant fan providing almost all the propulsive thrust. At a stroke this enabled 'wide-body' aircraft to be designed which more than doubled the number of passengers whilst carrying them in greater comfort, more than doubled the tonnage of cargo whilst removing previous limits on the bulk of items, dramatically reduced sfc (which both increased aircraft range and improved airline profitability or made possible cheaper travel), and eliminated noise in new aircraft as a factor threatening to restrict airline operations.

It would have happened, of course, but it happened sooner because of the US Air Force. Unlike Britain, where lone voices calling for long-range aerospace policy have been consistently ignored, several groups in the United States make it their business to look far ahead so that decisions are taken correctly 10 to 20 years before their effects are felt. In one of its periodical 'view from the bridge' studies, the USAF in 1962 undertook Project Forecast to determine its future programmes. It was headed by Gen Bernard A. Schriever, who had impressed the author with his management of USAF ICBM and space programmes.

He collected truckloads of every kind of numerical data, written background, and oral evidence. One early result was the appreciation that the existing vehicles of the Military Air Transport Service, even the C-135 and forthcoming C-141, were inadequate. A Military Airlift Command was planned, and a requirement was issued for a CX-HLS, an airlift transport much larger than anything previously seen. Unlike the cancelled XC-132 of 1957, it would be a jet. Schriever was interested to see that both P&W & GE were already thinking of big turbofans in which BPR would jump from near unity to perhaps 6 or 8.

Suddenly engines were likely to be required in numbers at such prices that a single ship-set (the engines for one aircraft) could cost $5–$10 million. A serious game had become even more important, and anyone with imagination could see that such engines were bound to become the mainline propulsion of the world's airlines. Pratt & Whitney was so determined to win that it demonstrated hardware. GE was driven so hard by Gerhard Neumann that it demonstrated a complete engine. Although the GE1/6 was a half-scale of the engine for the CX, it demonstrated things that had not been done before: a p.r. exceeding 25, a turbine entry temperature of 1,377°C, and a BPR of 8. Thanks to the CJ805, GE could offer a proven form of reverser. The Assistant Secretary of the Air Force went to Evendale and saw the 1/6 running at 15,830 lb with the unprecedented sfc of 0.336.

In August 1965 GE was awarded a record engine contract for $459,055,600 for the TF39 to power a C-5 which, five months later, was ordered – against the unanimous vote of the Air Force – from Lockheed. Looking back, the TF39 was a strange engine. The core was like a slim tube in comparison with the enormous case surrounding the 8 ft fan. Moreover, the fan was of a '1½-stage' type. The spinner rotated with a first fan stage with short blades inside a ring held by fixed inlet guide vanes inside the outer case. The second stage had blades of full height (36 in), with mid-span snubbers in line with the ring surrounding the first stage. Today it is universal practice to put any core booster downstream of the fan, but the TF39 '½-stage' had a possible advantage in protecting the main fan against FOD. The long drive shaft transmitted torque from a six-stage LP turbine. Around it was the 16-stage HP spool, with 7 variable stators, driven by a 2-stage HP turbine with air-cooled blades. The reverser, with a translating cowl and blocker doors, acted on the fan duct only, which provided 86 per cent of the thrust. Fan p.r. was 1.55 and airflow 1,549 lb/s, HP p.r. 16.8 and airflow 172 lb/s, and overall p.r. 26. The engine weighed 7,026 lb and was rated at 41,100 lb. In 1968–71 GE delivered 464 TF39-1 engines, followed in 1986–88 by 200 TF39-1Cs rated at 43,000 lb and with long-life and reliability features.

This gave GE the chance to get back into the airlines. Its failure with the CV-880 and -990 had not been because of the engines, and the airlines had voted for GE on the SST. But the immediate opening was a giant new civil jet, much faster than the C-5, to be built by Boeing for PanAm, and though GE offered a CTF39 this could not quite provide the required cruise thrust. GE decided it could not afford the diversion of effort to produce a suitable engine, and left this programme to P&W, whilst seeing how it could use TF39 technology in a new civil turbofan. This was an odd decision, because the company intended to produce this new engine anyway, which it was to develop to thrusts greater than Boeing initially needed.

Thus, though Pratt & Whitney lost out on the C-5, it won an initial monopoly position on an aircraft destined to be built in far greater numbers, the Boeing 747. Its higher cruising speed was reflected not only in greater wing sweep, but also in an engine BPR of 5 rather than 8. By 1965 the JTF14 design had been completed, and in 1966, when PanAm placed its historic order for 25 aircraft, the definitive engine had been run as the JT9D. The single-stage fan of 91 in diameter handled 1,484 lb/s at a p.r. of 1.6 at 3,650 rpm, with 46 slender (aspect ratio 4.66) forged titanium blades with two rings of part-span shrouds at 70 and 90 per cent radius. This fan was overhung ahead of the front bearing, and was preceded by a huge blunt rotating spinner to open out the core inlet where there were three booster stages turning with the fan on a massive shaft supported only in a bearing at each end and driven by a four-stage turbine. The HP spool had 11 stages, with the first four stators variable, driven by a 2-stage turbine overhung beyond the rear bearing. Overall p.r. was 21.5. The annular combustion chamber was claimed to give higher temperature-rise in a shorter length than ever before.

The JT9D ran in December 1966 and, like the TF39 before it, first flew in the right inner position of a B-52E in June 1968. The 747 first flew on 9 February 1969, and airline service began on 22 January 1970. The engine was the JT9D-3, rated at 43,500 lb and weighing about 8,800 lb. Even at the time it was regarded as a typically massive and conservative design – for example, it used two rings of fan-blade snubbers – and to enable the rating to be increased to 45,000 lb water injection was added in the Dash-3A engine, which today no airline would countenance.

Despite all this, and unlike the JT8D, engine problems were many and serious. The first to make a major impact was that, as power was applied on take-off, the thrust was transmitted via the rear mount, the front fan-case mount being floating; the engine bent nose-down, the casings distorted from circular to oval and, while gaps appeared in one plane, blade rubs occurred at right-angles. Vortices from the inboard pylons caused flutter problems, and then by late 1969 problems were encountered when starting. In any crosswind the variation in the pitot pressure sensed by the barometric fuel control caused over-rich mixture and turbine overheat, demanding shutdown. When the aircraft was parked tail to wind, the fan could be blown round 'the wrong way', reducing core airflow during start-up and again giving a turbine-overheat warning. By spring 1970, when Boeing's new Everett plant was full of engineless 747s, aircraft in service began to be stranded in many countries with first HP turbines damaged by failure of the rivets securing the blade-retaining plates. These and similar troubles took a long time to resolve.

At this time Pratt & Whitney had trouble almost everywhere except with the JT8D. Their engineers were extended trying to restore their 'dependable engines' reputation, but gradually better JT9Ds appeared, and thanks to hours-building faster than with any previous engine – deliveries were 653 by the end of 1970 and 1,132 by late 1972 – the problems were forgotten. Thanks to Japan Air Lines the JT9D got on the DC-10, and the resulting engine (via the Dash-15/25/20) became the Dash-59A, rated at 53,000 lb without water. This version was restressed with a stiffer frame, which put weight up to 9,140 lb but enabled a Y-frame added to prevent ovalization to be eliminated, and a common 747/DC-10 nacelle to be developed with the help of Rohr. From this engine onwards the fan was 1in larger, and a 4th core booster stage was added.

Subsequent engineering changes were for product-improvement. In 1974 a family of JT9D-7R4 models was offered, notable for a better fan with 40 wider blades with only a single ring of snubbers, a zero-stage on the core booster, a larger LP turbine, and many other changes. From these an almost entirely redesigned engine, called PW4000, went on test in April 1984. The total number of parts was said to have been reduced by almost half. The fan was again redesigned with 38 blades with aft part-span shrouds, handling 1,705 lb/s with a p.r. of 1.7, BPR falling to 4.85. The combustor was machined from a forged ring, and the HP turbine incorporated single-crystal blades in the first stage and DS blades in the second. Active-clearance control aerofoils and FADEC control were among other improvements. To meet market needs the engine was offered in three sizes, the 4052 (Boeing) and 4152 (A310) having a 93.6 in fan and thrust around 52,000 lb, the 4168 (A330) a 99.8 in fan and rating of 68,000 lb, and the 4084 (Boeing 777) a totally new 112 in fan with hollow titanium wide-chord blades at last eliminating snubbers, rated at 84,000 lb.

Back in the mid-1960s, Rolls-Royce at last recognised that BPR had to jump from 0.6 to something around 5, and began studying three-shaft engines in which the LP turbine drives the fan, an IP (intermediate pressure) turbine drives the IP compressor or core booster, and the small HP spool runs in the centre. As explained in Chapter 2, this can actually simplify engines by reducing the number of stages, make them shorter and more rigid, improve performance retention over long periods, and in theory make them quieter, especially on the approach. In July 1966 a full-scale two-shaft engine, the RB.178-16, was run to prove principles, and the three-shaft RB.178-51 was proposed at a rating of 45,000 lb for the planned European Airbus,

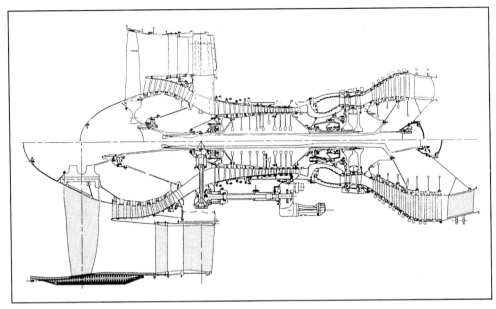

Comparative longitudinal sections through the PW4000 (upper half) and the PW4084 demanded by the 777; a seven-stage LP turbine is required to drive the big fan and six-stage LP spool (Pratt & Whitney).

which became the A300. Political insistence on British engines was to bedevil the Airbus programme and cause Britain's foolish withdrawal.

In 1965 Bristol Siddeley completed project design of a rival two-shaft engine, the BS.123, but when this was costed out at £450,000 each it seemed better to join with SNECMA (partner on Concorde and financially tied to Pratt & Whitney) and offer a licence-built JT9D. Rolls saw prominent danger signals, and in October 1966 nipped this scheme in the bud by purchasing Bristol Siddeley and, in June 1967, getting the JT9D replaced by the newly invented RB.207, a big three-shaft engine rated at 47,500 lb and upwards. But by this time US carriers, notably American Airlines, were calling for a wide-body smaller than the 747 matched to US trunk routes. They were thinking of three engines in the 30,000 lb class, and Sir Denning Pearson of Rolls felt that it was far more important to get on the US trijet than on the non-existent Airbus.

By 1967 there were two rival trijets, the DC-10 and Lockheed 1011 TriStar. In February 1968 American picked the former, but left the choice of engine open. Pratt & Whitney's JT9D was too big, but GE offered the CF6 tailor-made for the job. Rolls rushed to offer a smaller RB.207, the RB.211. This was not only a three-shaft engine, but it promised a fan of unrivalled

efficiency, with wide blades needing no part-span shrouds and made in a thin 'lenticular' profile from a CFRP composite called Hyfil. Rolls promised to deliver certificated RB.211s at 42,000 lb at a very keen fixed price from September 1971. It seemed a superior choice, and in March 1968 Lockheed launched the TriStar on the basis of the RB.211.

This was a bigger challenge than the Derby engineers had ever faced. It was so big that it broke the company and came within an ace of breaking Lockheed. First, the Hyfil fan blades failed bird-impact and rain/hail tests, and had to be replaced by a larger number of narrower titanium blades with snubbers, wiping out the RB.211's advantages in efficiency and weight. The TET of 1,260–1,290°C was too much for the forged HP turbine blades. Not least, the engine was miles from giving the guaranteed performance. With millions going out each week and no RB.211 being delivered, the famous firm rushed towards a crisis.

At the end of 1970 Dr Stanley Hooker was recalled from retirement to look at the ailing RB.211. He was stunned to find the shaft speeds totally mismatched, and an HP turbine efficiency 65 per cent instead of 85. Batch 2 engines at TET of 1,202°C gave only 34,200 lb, and an installed ship-set weighed 38,441 lb, against the promised 34,566. Morale was non-existent. It seems beyond belief that one man could transform this giant programme, but after one week the shops had been set to produce new NGVs for all three turbines and new HP rotor blades made by casting and in a better material. Thus one particular engine, which with difficulty had been coaxed to 37,000 lb, ran at 41,500 lb with the corrected NGVs and at over 43,500 with cast blades, still at 1,227°C.

If Hooker had been called a year earlier the crisis would have been averted, but just as the improved engine went on test Rolls-Royce was declared bankrupt. The government stepped in and formed Rolls-Royce (1971) Ltd to continue military programmes. Eventually it was persuaded also to take the RB.211 on board, funded on a cheeseparing daily basis. But Hooker singlehandedly so transformed the engineering team, while government appointee Sir Kenneth Keith so transformed the company, that morale soared and the RB.211 improved almost by the hour.

The RB.211-22C was certificated in February 1972, in some respects even better than the original guarantees. Meanwhile, Hooker could see there would be a need for a version in the 50,000 lb class, and quickly designed the 211-524 with a fan of unchanged diameter (85.5 in) handling 1,450 lb/s instead of 1,220 and a new IP spool giving increased core airflow. This was

the basis for a series of fine engines rated at up to 60,600 lb for TriStars, 747s, and 767s, adding sales of over 900 to the 676 Dash-22s. 'World reliability records' may be rather childish, but it so happens that the title is currently held by an RB.211-524B4 of Delta, which in mid-1994 was still in the centre position of an L-1011 after six years, 27,000 h, and over 4,200 flights.

In August 1978 Boeing took the unprecedented step of launching an aircraft with a foreign engine. The 757 went ahead with a lightened RB.211 with a 73.2 in fan and an IP compressor with six instead of seven stages. Called the 535C, this was rated at 37,400 lb, and entered service on the first day of 1983. Miffed, Pratt & Whitney produced a rival, the PW2037 (PW2000 rated at 37,000 lb), which had little in common with the JT9D. The 78.5 -in fan, driven by a 5-stage LP turbine, has 36 snubbered blades handling 1,340 lb/s with BPR 6; overall p.r. is 26.9 and dry weight 7,300 lb. The 2037 was designed above all for low fuel burn, and it was picked by Delta, and later by Northwest, United, and United Parcel Service, whose orders were so big that at one time P&W could claim to have outsold the 535. But Rolls responded with the 535E4, with a 74.5 in fan with only 22 wide-chord blades made by diffusion bonding titanium skins on a honeycomb core, enabling snubbers to be eliminated. With

This section through a Trent 700 nacelle shows such features as noise-suppressing linings, accessories in the pod cowl and the translating cowl reverser (Rolls-Royce).

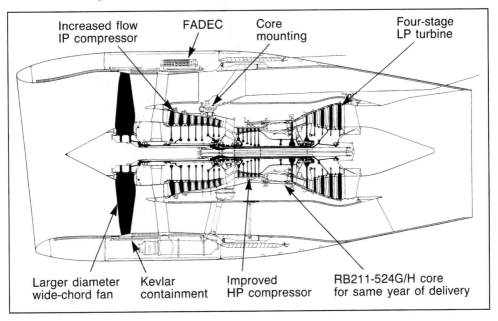

other improvements, and a thrust of up to 43,100 lb, this kept the British engine competitive, but it is its reliability that at the time of writing enables Rolls to claim over 80 per cent of the 757 customers.

In 1988 it was clear that the A330 and a projected rival from Boeing (which became the 777) would need engines of even greater power. Work began on the RB.211-524L, with a 26-blade fan scaled up from the 535E4, an 8-stage IP spool and improved 6-stage HP, a low-emissions combustor and new-technology turbines, the 4-stage LP driving the fan despite having to turn at reduced speed to match the bigger fan. In 1989 this engine was named the Trent, and it is now in production in two forms. The Trent 700 has a 97.4 in fan and is offered for the A330 at 64,000, 67,500, 71,100, or 75,150 lb. On 31 January 1994 an A330 became the first Airbus aircraft to fly with Rolls engines, and certification was due later that year. The Trent 800 has a 110 in fan and is offered for the 777 at 77,900, 80,080, 86,500, or 91,300 lb. Like the 700, the 800 passed the severe fan-blade off (FBO) test, in which a giant blade is severed while the engine is running at red-line overspeed, with flying colours. The Trent blade opens a new chapter, with superplastic forming and diffusion bonding titanium to leave canted spars from root to tip. Although customers choose engines mainly for political/financial reasons, on technical grounds the Trent is hard to beat. It offers more than adequate thrust, unsurpassed fuel economy, outstanding ease of maintenance, and the lowest weight of all the big fan engines.

When, in March 1968, Lockheed picked Rolls-Royce, it was a black day for GE. Using TF39 technology it had offered for the new trijets the CF6, with a slightly modified core and a five-stage LP turbine driving a fan cut back to 86.4 in rotating with a single-stage core booster. Originally planned for 32,000 lb, the CF6-6 grew to 36,000 and then 40,000 lb, with 1,307 lb/s airflow and a BPR of 5.9, to match growth of the projected aircraft. All was not lost: a month after launch of the RB.211 two giant US domestics, American and United, picked the CF6 to power the DC-10. Then came National, followed by a request from the KSSU group – KLM, Swissair, SAS, and UTA – for a DC-10 with longer range. The result was the DC-10-30, for which GE offered the CF6-50 with a three-stage booster and higher temperatures, to start life at 49,000 lb; airflow began at 1,439 lb/s, with BPR reduced to 4.4.

The Dash-50 engine put GE in a strong position. A firm programme, ahead of either rival in thrust, it had a big weight advantage over any JT9D, and Rolls was in no position even to think about competing. The complete nacelle already designed by Rohr for the DC-10-30 was the obvious engine for the Airbus A300, and this went ahead in late 1969. GE never looked back. By 1 January 1980 nearly 50 airlines had signed for DC-10s, while in the longer term the A300/310 gained 90. Moreover, in 1972, six years after turning its back on the 747, GE was asked by the USAF to power the E-4A Airborne Command Post version, and since then various CF6 engines have competed strongly on 747s. The 2,000th CF6-50 was shipped in mid-1989, by which time the CF6-80C2 had been running seven years with fewer parts, lower emissions, and a 93 in fan and four-stage booster giving up to 61,500 lb with sfc 13.5 per cent lower than any Dash-50. It surpassed Dash-50 sales on the

Cutaway CF6-6 pod, as designed for the DC-10 in 1968. The amount of space between the core and the bulging inner cowl is unusual (General Electric).

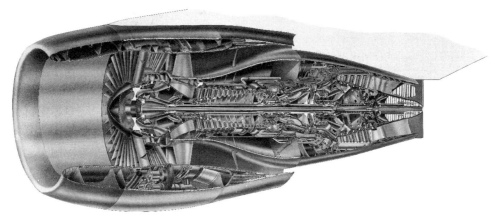

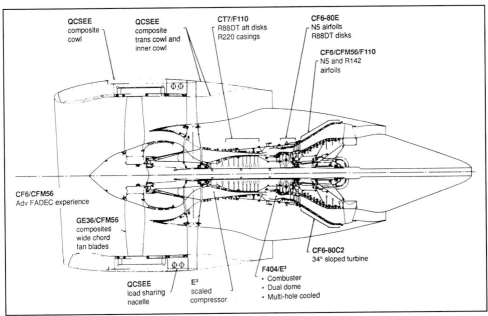

Already featured in detail in Chapters 2–6, the monster GE90 was based on experience gained with numerous previous engines, some being listed here. The E^3 was the Energy-Efficient Engine, and the QCSEE the Quiet Clean Energy-Efficient, both subsidised by NASA (General Electric).

A300/310, 747/767, and MD-11.

Also in 1989, GE decided to begin working on a completely new engine rated at over 75,000 lb to be ready for larger twin-engined aircraft from the mid-1990s. A feature would be a giant fan with wide-chord blades needing no mid-span shrouds (which Rolls had introduced in 1984). This engine, the GE90, first ran in 1993. Most of its features have been described in earlier chapters. Not least of its achievements has been to make the Boeing 777s ordered for British Airways not even British-engined. It should be an outstanding engine, though compared with the Trent 800 it is similar in power but considerably larger and heavier.

In the Soviet Union the design bureau of N. D. Kuznetsov produced a profusion of mainly military turbojets and turbofans, including very powerful augmented engines for supersonic transports and bombers. The first to enter passenger service, on the Il-62 in 1967, was the NK-8, a two-shaft engine with BPR around unity and thrusts of 20,950–23,150 lb. Versions have been qualified on liquefied natural gas (LNG), LH_2 (liquid hydrogen), and other cryogenic (refrigerated) liquid fuels. Another variant, the 28,660 lb NK-86, powers the Il-86. The usual engine of the Tu-144 SST was the NK-144 augmented turbofan, of which many versions were produced with BPR close to unity and a thrust of 44,100 lb.

The first turbofan in Soviet service, in the Tu-124 in 1962, was the D-20 designed by the bureau of P. A. Solovyev. A two-shaft engine of unity BPR handling 249 lb/s, it was rated at 11,905 lb. Ten years later the D-30 was developed for the Tu-134, with mass flow increased to 265 lb/s and thrust to 14,990 lb. A little later the designation D-30K was used for a totally different engine with BPR raised to 2.3 and an airflow of 600 lb/s, and this found several applications (in the Il-62M and Tu-154 replacing the NK-8) at ratings from 22,050 to 26,850 lb. In 1984 Solovyev began testing the PS-90, a modern transport engine of BPR 4.8, with a 74.8 in fan, overall p.r. of 35 and take-off rating of 35,275 lb. Although its designation honours the designer, Solovyev was succeeded in 1989 by Yuri E. Reshetnikov, who was eager to tell the author that the PS-90 is replaced by the RR535 in the Tu-204 and by the PW2337 in the Il-96M only in aircraft for export. Today the team is called 'Aviadvigatel' (aircraft engine), and its many impressive projects simply need a stable political environment.

The same can be said of ZMKB Progress at Zaporozhye in Ukraine. This group added to its shaft engines the AI-25 small turbofan in the

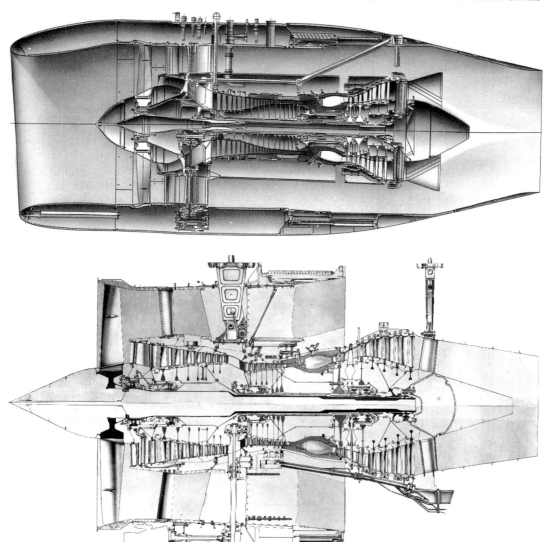

Top *Cross-section of the Aviadvigatel PS-90A12 nacelle, as chosen for the Yak-242. Notable features include the front mount carrying the weight and the rear mount reacting the thrust, the cascade-type reverser, and the hot/cold mixer nozzle.*

Above *Longitudinal section of a D-18T, complete with fan case and pylon mount struts (ZMKB Progress).*

mid-1960s, with BPR of 2.2 and ratings around 3,300 lb. In 1971 testing began of the D-36, a conservatively designed three-shaft turbofan of BPR 6.3, handling 562 lb/s and rated at 14,330 lb. The first 'wide-body' engine, the D-18T, first ran in 1980. Again a conservatively designed three-shaft engine, this has a BPR of 5.6, a 91.73 in fan handling 1,687 lb/s, and a take-off rating of 51,660 lb for a weight of about 9,100 lb.

When crude oil became expensive in the early 1970s, designers in several countries began looking carefully at propfans. Today such engines are unfashionable, except in Russia and Ukraine, where people believe numerical values rather than fashion. These are covered in the next chapter.

17 Turboprops and propfans

Except for Guillaume and Whittle, every proposal prior to the Second World War for using a gas turbine to propel aircraft thought of the engine as a means for driving a propeller. For the next half-century, turbine aircraft were either turboprops or jets. What few people then envisaged was that the two types of engine would become linked by a spectrum of engines, so that by the 21st century many engines would be an intermediate species. If almost all of the thrust in a turbofan comes from a single giant fan, does that make it a turboprop? Alternatively, if the fan is surrounded by a duct, or even by a short fixed shroud, does that make it a jet engine?

Another way of emphasizing the transformation that has taken place is to look at the turbines. One of the earliest turboprops, the Rolls-Royce Clyde, had an axial compressor upstream of a centrifugal on a different shaft. A single-stage LP turbine drove both the axial compressor and, via a gearbox, the contrarotating propeller. Today, the GE90 turbofan needs a turbine with *seven* stages just to drive the fan! Each stage adds to the length and weight, and above all to the cost. So many stages are needed to provide the required torque with high efficiency; which is not quite the same as providing the required power at the low rotational speed of a giant fan.

The spectrum of engines was completed by the progressive increase in BPR. The first turbofans, or bypass jets, were little different from turbojets. As BPR increased, eventually designers found it difficult to match the speed of small turbines to that of large fans, and it became necessary to insert a reduction gear. When the fan had 40 or more slender blades the engine was

Whittle's patent for 'No 3 Augmentor' ducted the gas to outboard turbine blades mounted outboard of the fan blades. This gave good matching of blade velocities, but at the expense of inelegant ducting.

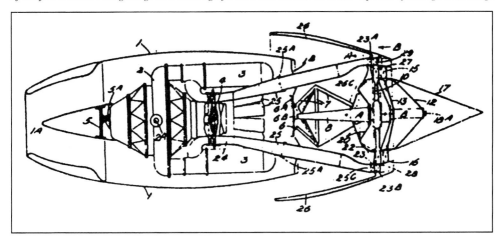

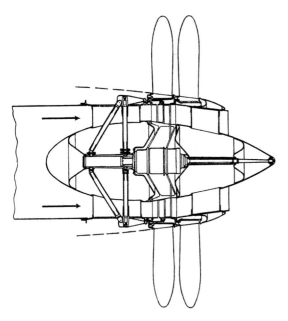

Metrovick also tested the F.5, an open-fan augmentor. The entire free-running unit was mounted downstream of an F.2/4 turbojet.

clearly a turbofan, but the perfection of wide-chord blades needing no part-span shroud was the final clue to completing the spectrum, so that today the designer has infinite choice in assembling the core engine, extra turbine, fan (with or without variable-pitch blades), and optional gearbox in any arrangement, tractor or pusher.

All this could have happened more than half a century sooner, had it not been for the way decisions are dictated by fashion. At the end of the Second World War, Metrovick actually built and tested both ducted and open fans, as explained earlier. Amazingly, it abandoned this work, because nobody was very interested. The author's predecessor as Technical Editor of *Flight*, C. B. Bailey-Watson, wrote in May 1951; 'It is entirely possible that the last has not been seen of the open-fan type augmented power unit for aircraft, in that there are many indications that such units will conveniently bridge the transition stage from the operational zone of the turboprop to that of the turbojet.' Few had such foresight.

It is remarkable that, of the plethora of pioneer turboprop projects, the first to be run should have been designed by György Jendrassik, an engineer at Ganz Wagon Works, Budapest. His Cs-1 was a remarkably sound and compact engine in the 1,000 hp class, with a 15-stage compressor, a folded reverse-flow can-annular combustor, and an 11-stage turbine. It ran in August 1940, and was to have powered the RMI-1 twin-engined fighter, but the Germans provided the 1,475 hp DB605 piston engine. Subsequent turboprops by BMW, Junkers, General Electric, Northrop, Wright, SNECMA, and SOCEMA have been mentioned earlier.

Aviation (and the airlines in particular) tends to be a world where nothing new is acceptable until many people have accepted it. If anyone can break through this impasse – and appearing to offer less risk is far more important than bril-

The SOCEMA TGA.1 was designed from 1941, when France was occupied. The only big problem was combustion, though nobody today would machine the large-diameter compressor/turbine rotor from welded steel forgings.

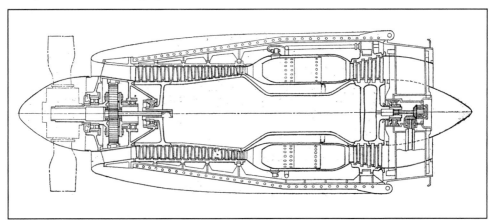

liant technical performance – they cry all the way to the bank for anything up to 40 years. In the case of the Rolls-Royce Dart it is longer, because the first Dart was sold in 1946 and sales of Dart spares will go on at least until 2005. This success was due to the Dart being, in the words of a former Rolls managing director, 'agricultural machinery'.

Rolls began learning about turboprops with the first Trent, a simple lash-up of a Derwent II turbojet with a cropped compressor driving a quill shaft to a reduction gear carrying a small five-blade Rotol propeller. Very quickly this showed the need for a proper control system linking the engine and propeller, with various safety interlocks. This knowledge made possible the impressive Clyde, which reached 4,200 hp plus 830 lb residual jet thrust, equivalent to 4,543 ehp, but the company could not see a worthwhile market. Instead it got Lionel Haworth to form a tiny group at Derby to design a simple engine in the 1,000 hp class.

This was the first gas turbine produced at Derby, and it flew in the nose of a Lancaster in October 1947 at 750 hp. Later development was done with Wellingtons and Pionairs (BEA DC-3 freighters), one of which overtook an amazed USAF B-29 pilot at 33,000 ft. Features included

a two-stage turbine driving through a spherical coupling to tandem centrifugal compressors, and a planetary/epicyclic reduction gear to a 10 ft four-blade propeller. The two-stage compressors followed the aerodynamics established with high-altitude Griffons and Eagles, p.r. being about 5.5.

The Dart was up against the supposedly more advanced axial Armstrong Siddeley Mamba, Napier Naiad, and Fedden Cotswold. The RAF decided not to buy a three-seat turboprop trainer, and BEA bought the Ambassador instead of the Viscount. Instead of a total loss of interest, G. R. (later Sir George) Edwards picked the Rolls engine for the Viscount, and BEA even carried fare-paying passengers in the prototype in July–August 1950. Rolls has a habit of developing engines to give much greater power, and the key to success was raising the Dart's power from 990 to 1,400 shp, allowing Edwards to stretch the Viscount from 32 to 53 seats.

This was just the beginning. With a larger propeller the Dart powered the F.27 Friendship, in 1956 it got a three-stage turbine, and power climbed past the 2,000 hp level to peak in the 1960s at 3,245 ehp. Throughout its life the Dart suffered from supposed experts who knew that centrifugal compressors were obsolete, and by

Cutaway of a later Dart with three-stage turbine. Above the air inlet is the ducted oil cooler, the hot air from this being discharged upwards. The drawing suggests that the tandem compressors and propeller reduction gear are driven through separate shafts, but this is not the case, there being only a single drive shaft.

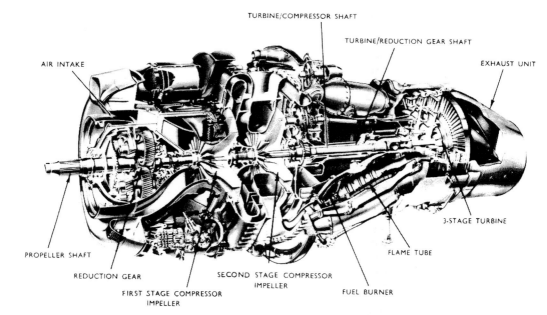

the late 1950s it was also common knowledge that turboprops were obsolete. Despite this, 7,100 Darts were sold, flying over 130 million hours, far more than any other British aero engine. It is only sad that, when Rolls should have been producing a successor, which could have sold a further 7,100 and (we can now see) could have had tandem centrifugal compressors, it was diverted by the problems of the RB.211.

The only other Rolls turboprop has been the Tyne. The author 'assisted' (as the French say) at the first run, in April 1955, when as the RB.109 it was aimed at an eventual 4,000 hp. It was a two-shaft engine with a 15,250 rpm HP spool running round an LP shaft linking a three-stage turbine to a six-stage axial compressor and 0.064 double-epicyclic reduction gear driving a four-blade reversing propeller. The objective was power with low fuel burn, and sfc came out around 0.52, compared with about 0.7 for mid-1950s Darts. The Tyne was slim, but it could have been of twice the diameter without affecting the nacelles in which it was installed, and the same could be said of most other turboprops. Sadly, the Tyne hit the market just as the 707 made turboprops seem passé. This was especially tough on the Vanguard, the capacious, fast, and economical 'Viscount successor', which sold just 43 to two airlines – less than one-tenth of the Viscount total. A trickle of 6,100 hp Tynes are still being delivered, mainly from plants in France, Belgium, and Germany, for the Atlantique; Tynes also found a market in ships and electricity generation.

Armstrong Siddeley's Python, run in April 1945, was the sole outcome of the 20 years of axial research at the RAE. Rated at 4,110 ehp, it powered the Wyvern carrier strike fighter, a demanding application which did not need its reverse-flow layout. To rival the Dart, ASM next produced the Mamba, a neat single-shaft axial engine which began at 1,000 hp. It soon faded, though the Double Mamba did have a 25-year innings in the Gannet, in which both power sections gave 3,875 ehp for take-off, one half then being shut down for cruise. One Gannet AEW.3 is retained by US propeller maker Hamilton Standard (HamStan) for propfan noise research.

Napier's Naiad, an axial in the 1,500 hp class, and Double Naiad got nowhere, and the tiny penetration of the market by the painstakingly-developed single-shaft 3,500 hp Eland was killed off when Napier was taken over by Rolls. Even Bristol's colossal effort made only a pin-prick on US dominance of the world scene. Bristol was so late getting started that it had to find a market previously ignored, and it picked on long-range turboprops of the highest efficiency possible, so that complexity and weight of the engine would be more than countered by fuel savings.

Details of the first effort, called Theseus, were released in 1945. Frank Owner aimed to beat the most economical piston engine at 300 mph at 20,000 ft, but scaled down the target power from 4,000 to 2,000 hp. He used a nine-stage axial compressor followed by a centrifugal (together achieving a p.r. of only 5), delivering through eight pipes to a matrix-type heat exchanger with 1,700 stainless-steel tubes. From here the hot air reached the eight combus-

Cutaway of a Tyne, showing the reduction gear centred in the annular inlet. The projecting accessories were no problem; there was tons of room in the nacelle (Rolls-Royce).

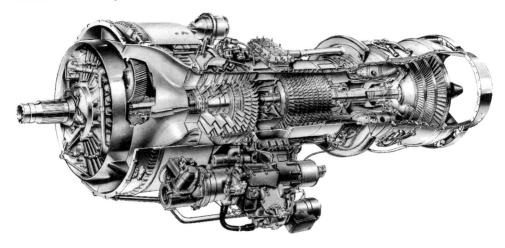

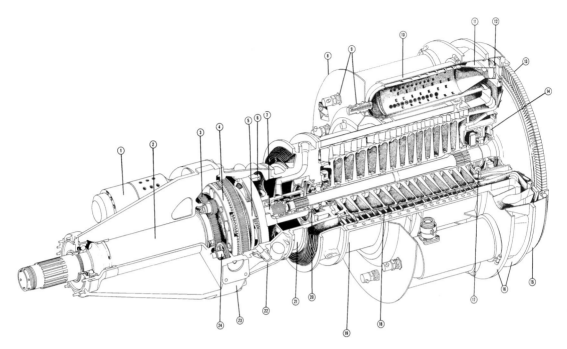

*Though it never went into production, the XT31 deserves a good look because it first ran on 15 May 1943, possibly the second to run in the world: **1**, starter; **2**, prop shaft; **3**, accessory drive; **4**, low-speed planet cage; **5**, high-speed planet cage; **6**, torque arm; **7**, input pinion; **8**, firewall; **9**, fuel nozzles; **10**, combustion chamber; **11**, transition liner; **12**, turbine nozzle; **13**, turbine rotor; **14**, turbine bearing; **15**, nozzle case; **16**, main frame; **17**, oil pump; **18**, compressor; **19**, stator case; **20**, oil pump; **21**, inlet case; **22**, intermediate case; **23**, forward case; **24**, fuel regulator drive (General Electric).*

tion chambers interleaved with the air pipes before escaping via a two-stage compressor turbine, a single-stage turbine driving the propeller reduction gear, a heat exchanger, and a jetpipe. In December 1946 the Theseus became the world's first certificated turboprop, but its heat exchanger was so troublesome that it was omitted, leaving an engine weighing 2,300 lb and giving less than 2,300 hp.

Owner followed this with the Proteus, again with two reversals of flow, but for a different reason. The air entered near the back, flowed forward through a 12-stage axial compressor followed by two centrifugals, and then flowed back through eight long and slim chambers to escape via a 2-stage compressor turbine and single-stage power turbine. The result was a nightmare. Even after eliminating the first centrifugal stage, Owner had to admit; 'We set out to achieve the lowest sfc in the world, regardless of weight and bulk; so far we've achieved the weight and bulk'. New chief engineer Stanley Hooker completely redesigned the Proteus, and

by 1954 had increased power from 2,500 to 3,800 shp whilst cutting 1,000 lb off the weight.

This eliminated almost all of the problems, but it left two. In February 1954 a propeller input pinion stripped its teeth, freeing the turbine from load. In a split second it had oversped so violently that it burst into fragments, one of which passed through the oil tank and started a fire which left the second Britannia a wreck and probably lost a sale to KLM. A year later Britannias of BOAC began to suffer flameouts in icing conditions. There was no danger whatsoever, but the airline did all it could to turn this molehill into a mountain, delayed entry to service by two years, and nearly broke the Bristol company.

Subsequently the Proteus flew millions of hours, and still drives the biggest hovercraft across the Channel. It was followed by the BE.25 Orion, an excellent turboprop with straight-through flow through two axial spools, the three-stage LP turbine driving the LP spool as well as the gearbox to prevent dangerous

Comparison between the T57, with HamStan B48 propeller, and a J57 turbojet (Pratt & Whitney).

overspeed. A unique feature is that it was flat-rated at 5,150 hp to 19,000 ft, and thus still gave nearly 3,500 hp at 36,000 ft, with economy rivalling the best diesels, but by 1957 everyone 'knew' that turboprops were obsolete, and the Orion was cancelled. Since then, of roughly 60,000 turboprops sold around the world, not one new type of engine has been British.

GE's pioneer XT31 got nowhere, neither did turboprop versions of the T58 and British Gnome version. GE did achieve modest sales of the T64 in the DHC-5 Buffalo and Aeritalia G222, though this again is mainly a helo engine. On the other hand, the T700 helo engine has a useful application as the 1,735 hp CT7 turboprop in the Saab 340. Former engine giant Wright soon gave up the T35, but then spent much effort producing the T47 and T49, turboprop versions of the Olympus and Sapphire, and in 1955 even flew two T49s in the XB-47D.

Pratt & Whitney's T34 in the 6,000 hp class found only a tiny market in the Douglas C-133, and the monster 15,000 hp T57 (for the Douglas XC-132) was cancelled. It was left to Allison to push the turboprop in the USA, and in 1947 this division of General Motors at Indianapolis first ran the XT38. This was a single-shaft engine with a slim (20 in constant diameter) axial compressor and four-stage turbine. The most novel feature was that the reduction gear was carried far ahead of the power section so that the air inlet duct could curve in behind it. Despite the long and heavy drive shaft, the first T38s weighed only 1,225 lb, for a power of 2,750 hp. After testing in the nose of a B-17, two powered the Convair 240 Turbo Liner on 29 December 1950. Small numbers of twinned XT40 engines powered the Convair XP5Y and R3Y Tradewind, Douglas XA2D Skyshark, and NAA XA2J, with different lengths of extension shaft. Others powered VTOLs (Chapter 14).

The T38 had been funded by the US Navy, but its refined successor, the T56, was funded by the USAF, despite having an even number. The compressor was cut back from 19 to 14 stages whilst increasing the p.r. to 9.25, and the eight combustion cans were replaced by a can-annular chamber. By 1955 the T56-1 was in production at 3,750 ehp for the Lockheed C-130A Hercules, driving 15 ft 1 in Aeroproducts three-blade propellers on gearboxes mounted above the line of the compressor inlet. Two years later, virtually the same engine, but with a 13 ft 6 in four-blade propeller carried on a gear-box mounted below the power section (so the inlet was above the spinner), went into produc-tion as the Allison 501 for the Lockheed Electra airliner. The T56 grew modestly in power, to 4,050, 4,910 and finally 5,250 ehp. Apart from the C-130, Electra, and P-3 Orion (using a T56 with low gearbox), this engine went into only the E-2 Hawkeye/Greyhound and a few CV-580 conversions, yet deliveries by 1994 exceeded 15,300. All versions run at 13,820 rpm, the pro-peller speed usually being 1,020.

Advanced fixed-wing and helicopter versions never happened, but in the 1980s one of these led to the 501-M80C, with a free-turbine putting 6,000 shp into the output shaft. Although it is classed as a turboshaft, it could equally be con-sidered a turboprop, because as the T406 it dri-ves the complex 38 ft prop-rotors of the V-22 Osprey. Although it has a 14-stage compressor similar in size to that of the T56, p.r. is increased to 16.7 and airflow from 32.4 to 35.5 lb/s, new features including variable stators and an annular chamber. Almost the same power section is used in the turboprop, which began life as the GMA 2100 and became the AE

2100 after a 1993 management buyout to form Allison Engines. Weighing 1,548 lb, the 2100 drives a Dowty six-blade propeller in the Saab 2000, in which it is flat-rated at 4,152 shp. In the C-130J it would be fully rated and give 31 per cent more thrust than previous C-130 engines.

In 1958 Allison received a contract for a small turboshaft engine, the T63 (next chapter), and in 1969 a civil turboprop version was cer-tificated designated 250-B15, rated at 317 shp. Although only three of the 20-odd Model 250 versions are turboprops, rated at 420 or 450 shp, they power a multitude of aircraft, showing the demand for such an engine. Various conversions are marketed by Soloy, including the Dual Pac, in which two 250-C30S power sections (each rated at 700 shp, and not yet sold in single ver-sion) drive through a free-wheeling gearbox to enable single-propeller aircraft to be certificated as twin-engined.

In 1947 Turbomeca of France began produc-ing small gas turbines. The first turboprops, the 400 hp Marcadau of 1953 and 650 hp Bastan of 1957, established the principle of running at constant speed, varying the power by control-ling fuel flow and propeller pitch. The only tur-boprop to sell in numbers so far has been the Astazou, with one axial stage upstream of the centrifugal compressor and usually rated at 800–1,100 shp. Today a handful of aircraft are powered by single examples of the Arrius 1D, a free-turbine engine of 420 shp.

A typical AE2100, with variable stators and a totally new gearbox but still resting on over 180 million T56 hours. In November 1994 Allison became a Rolls-Royce company (Allison Engines).

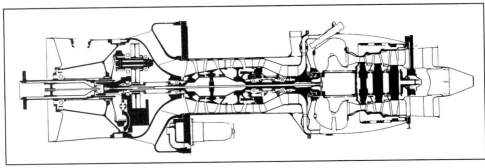

The Astazou XX (Astazou 20) differed from lower-powered versions in having three axial stages upstream of the centrifugal, but it was a typical Turbomeca engine. Reduction gear and prop shaft at left.

Pratt & Whitney Canada has found a bigger market for its PT6, designed in 1958 as a replacement for the 550 hp Wasp piston engine. It approached Montreal banks and the Quebec government, seeking a $10 million loan to get started. Asked the obvious question; 'How many engines have you sold?', the answer was; 'Six, but we can see about 34 more.' So the company did not get its loan, which the investors probably regret, because so far sales of the PT6A turboprop version exceed 31,000, to 5,490 customers. All have the inlet near the back to a reverse-flow three- or four-stage axial plus centrifugal compressor, an annular combustor, a single-stage compressor turbine, and a single-stage (high-power models, two-stage) power turbine, the jetpipe(s) emerging at the front. PT6A models are rated from 500 to 1,650 hp.

For years, P&WC has been wondering whether to throw out the axial compressor. In 1982 it began design of the PW100, to take over the market vacated by the Dart. It followed the Rolls engine in using two centrifugal compressors in series, but unlike the earlier engine the gas generator – P&WC call it the turbomachine – looks small in comparison with the inlet case, gearbox, and mass of auxiliaries. Already the list of variants is long, with 3,500 sold at ratings from 1,800 to 2,890 hp.

The only serious rival to the PT6A is the TPE331, designed in 1964 by AiResearch, produced for years by Garrett, and now a product of AlliedSignal. The design team never consid-

A total contrast with the previous figure, this is the arrangement adopted by P&WC for the PT6.

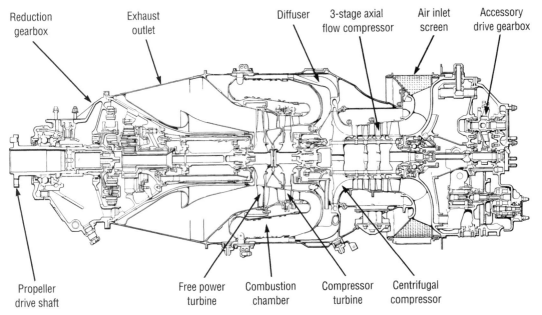

ered an axial compressor, using two modern titanium wheels which in early versions achieved a p.r. of 8.0, rising in later versions to 10.37. All versions are single-shaft, the three-stage turbine driving via the compressors to a helical spur gear and planetary output with ratio of 20.865 or 26.3. The military T76 had the inlet above the spinner, but commercial versions have a chin inlet. Sales exceed 13,300, at ratings from 575 to 1,960 hp.

In recent years aircraft designers have become more adventurous. The Embraer/FAMA CBA-123, for example, is powered by two Garrett TPF351 turboprops notable not only for being free-turbine engines (like most modern turboprops), but also for being designed as pushers. Air is rammed in at the front of the nacelle and goes straight through the two centrifugals, handling 14 lb/s at p.r. of 13.3, and into the folded combustor. It escapes through a two-stage HP turbine and three-stage power turbine, driving the epicyclic gearbox downstream of the twin jetpipes. A 2,100 hp engine, it is torque-limited to 1,450 shp, so this is available to high altitude for fast cruise.

Both giants of general-aviation piston engines, Continental and Lycoming, have produced turboprops. Teledyne Continental's TP-500 is a rather heavy (394 lb) single-shaft engine with p.r. 8 from a single centrifugal, giving sfc of about 0.7 cruising at 25,000ft. It was certificated at 480 hp in October 1988, but no sales have been announced. Textron Lycoming

developed a '101' family of prop and shaft engines with one axial and one centrifugal compressor and a free power turbine. A brochure weight for an LTP 101 turboprop is 335 lb, with versions rated at 620 or 700 hp. They have found enough applications to sustain viable production, but perhaps it was a mistake to have these engines for a time managed by the reciprocating-engine division at Williamsport. Now that they are included in the package bought by AlliedSignal they may really go places.

In the previous chapter it was noted how in the 1960s Lycoming used the T55 gas generator to drive a geared fan, resulting in the ALF502 airline/bizjet engines. No turboprop version was launched, but in 1994 the market is being hit by the LP512, one of the upgraded LF500 family. In the 7,500 hp class, the 512 is claimed to be '30 in shorter and several hundred pounds lighter than the competition . . . fitting a low-drag nacelle in 70/90-seat airplanes and with 30,000+ cycles life'. Certification is due in 1996.

Another piston-engine team which has achieved success is Walter, at Prague-Jinonice. Although the company became Motorlet in 1946, the engines retain the famous name. The main product has been various versions of M601, rated at 500–800 shp, with layout resembling a PT6 but with only two axial stages. More than 4,300 have been delivered, mainly for L-410s. The bigger L-610 required a new engine, the M602, first flown in 1988. Rated at

Though based on TPE331 experience, the TPF351 is a free-turbine engine, and is also arranged as a pusher. Note the two stages of planetary gears (AlliedSignal Garrett).

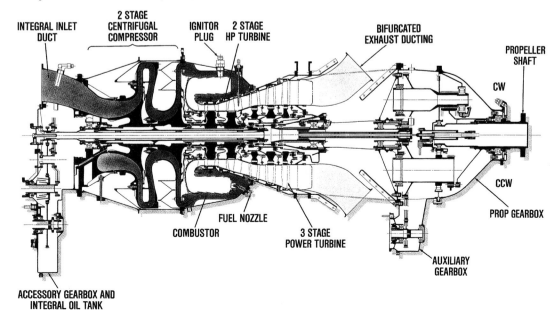

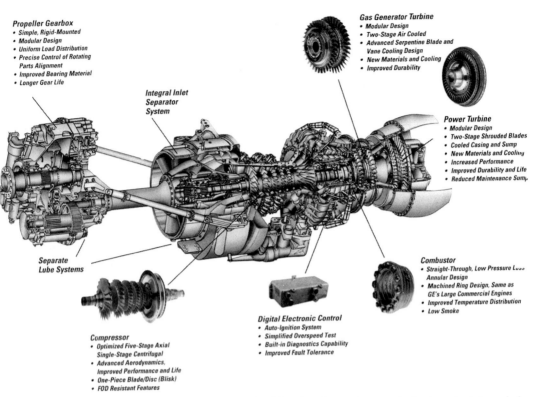

Propeller Gearbox
* Simple, Rigid-Mounted
* Modular Design
* Uniform Load Distribution
* Precise Control of Rotating Parts Alignment
* Improved Bearing Material
* Longer Gear Life

Integral Inlet Separator System

Gas Generator Turbine
* Modular Design
* Two-Stage Air Cooled
* Advanced Serpentine Blade and Vane Cooling Design
* New Materials and Cooling
* Improved Durability

Power Turbine
* Modular Design
* Two-Stage Shrouded Blades
* Cooled Casing and Sump
* New Materials and Cooling
* Increased Performance
* Improved Durability and Life
* Reduced Maintenance Sump

Separate Lube Systems

Combustor
* Straight-Through, Low Pressure Loss Annular Design
* Machined Ring Design, Same as GE's Large Commercial Engines
* Improved Temperature Distribution
* Low Smoke

Digital Electronic Control
* Auto-Ignition System
* Simplified Overspeed Test
* Built-in Diagnostics Capability
* Improved Fault Tolerance

Compressor
* Optimized Five-Stage Axial Single-Stage Centrifugal
* Advanced Aerodynamics, Improved Performance and Life
* One-Piece Blade/Disc (Blisk)
* FOD Resistant Features

A page from a typical sales brochure, emphasizing that the product can be both up-to-date and also very experienced. The subject is the CT7-9 turboprop, based on the core of the T700 helo engine; it powers the Saab 340, CN-235, and L-610G (General Electric).

1,840 shp, this again emulates Pratt & Whitney Canada in that, like the PW100, it has a 'straight-through' layout with a chin inlet to tandem centrifugal compressors. It differs in that each impeller is on its own shaft, a third being needed for the propeller. The rotational speeds are lower than in PW100s, reducing p.r. to 13 overall and raising sfc to a best of 0.559, compared with around 0.48; and the M602 weighs

about 1,279 lb compared with 861–990, so no decision has yet been taken on production.

Many companies have tested prototype turboprops but failed to find a market. They have included de Havilland, Rover, BMW, Daimler-Benz, Fiat, Alfa Romeo, Teledyne CAE, and Boeing. To this list may be added several of the names invented since 1988 in the former Soviet Union.

In 1946, a team of mainly German engineers was set to work at Kuibyshyev under N. D. Kuznetsov. By far their greatest achievement was the NK-12, which aimed at unprecedented power by handling over 120 lb/s (a typical airflow in post-1956 versions is 143 lb). Easily the biggest turboprop core, it required a 65 hp gas-turbine starter. Initially designed for 12,000 shp, the NK-12 was upgraded by 1957 to 15,000 hp, a figure not even approached today by any other propeller engine.

Features include a 14-stage compressor, a can-annular combustor, and a five-stage turbine. At the front is a double reduction gear which

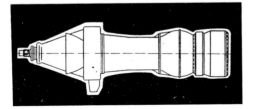

To the same scale, the Dobrynin VD-4K turbocharged piston engine (4,300 hp) of the Tu-85 and the NK-12M (15,000 hp) of the Tu-95.

splits the torque to a pair of four-blade propellers of the AV-68 family from 18ft 4in to 20ft
4in diameter. These propellers rotate at low rpm
in very coarse pitch to enable the Tu-95 'Bear'
and Tu-114 family to fly at up to 550 mph, an
unprecedented speed for propeller-driven aircraft. What nobody expected in the early 1950s
was that this basically conventional turboprop,
designed almost with wartime technology,
should still be an important engine at the end of
the century.

The only other widely used turboprops in the
former Soviet Union are single-shaft engines
designed under Aleksandr Ivchyenko at
Zaporozhye, Ukraine. The AI-20 has a 10-stage
compressor handling 45.6 lb/s and is rated at
4,000–5,042 shp. The AI-24 has a similar spool
scaled down to 31.7 lb/s and is rated at
2,550–2,781 hp. By modern standards both are
heavy and uneconomic, but have the benefit of
many millions of hours in harsh environments.

Today, three groups in Russia are producing
turboprops. In Leningrad Klimov Corporation,
headed by A. A. Sarkisov, which has a link with
P&W Canada, is in production with the TV7-
117S to power the Il-114. A two-stage turbine
with cooled blades drives the seven-ax/centrifu-
gal compressor with p.r. of 16. A two-stage LP
turbine drives the planetary reduction gear to
the Stupino SV-34 propeller with six CFRP
blades. The 117S was certificated at 2,466 shp,
and weighs 1,146 lb. Cruise sfc is competitive at
0.397.

At Omsk the Mars bureau, headed by V. G.
Kostogriz, has produced small numbers of the
TVD-100 with a two-ax/centrifugal compressor
(p.r., 9.2) to develop 720 shp. The related TVD-
20 has a second HP turbine stage to drive a
seven-ax/centrifugal to give 1,380 shp with
rather better specifics (0.506 instead of 0.646),
weight being 628 lb. At Rybinsk, under A. S.
Novikov, the TVD-1500 has a single-crystal
two-stage turbine (1,267C) driving a three-
ax/centrifugal compressor (p.r., 14.4) and two-
stage LP turbine driving the two-stage spur plus
planetary reduction gear, mounted ahead of and
below the gas generator, to put out 1,300 shp for
a weight of 529 lb and sfc of only 0.374.

The TVD-1500 is clearly a competitive
engine, which could be made in large numbers
for agricultural aircraft. But for faster applications the former Soviet designers have, unlike
their Western counterparts, never abandoned
development of propfans. It is still arguable how
such engines should be arranged, where they
become distinct from turbofans, and even
whether they should be built at all. In the

*Not least of the remarkable features of the NK-
12 family is that these 15,000-hp engines were
designed 45 years ago, and are so basically
conventional. This view shows the integral
planetary gearbox and inlet.*

author's opinion their introduction is merely a
matter of time. Humans owe it to themselves as
much as to the environment to burn as little non-
renewable fuel as possible.

The profan emerged in the early 1970s, just
as the price of fuel began to soar. For several
years designers had studied propellers with
more blades, but with reduced diameter to
increase the P/D (power divided by square of
the diameter). New materials made possible
blades with swept or scimitar-like profiles, thinner than before. With supersonic tips, such propellers can penetrate beyond the traditional limit
on aircraft speed of around Mach 0.68. By raising BPR to 25 or more, they enable jet speed to
be combined with propeller propulsive efficiency. Put another way, the turbofan, with propulsive efficiency of 60–65 per cent, could be
given a 'fan' offering 75–78 per cent, plus variable pitch to match airspeed or reverse thrust
after landing. Compared with advanced turbofans, single-rotation propfans promised to
reduce sfc by at least 25 per cent, and contrarotating types by nearly 40 per cent.

HamStan pioneered the idea, and NASA took
it up. Soon forecasts were showing how expensive fuel could be countered by the new technology, at the same time restoring an American
near-monopoly of the world jetliner market. So
in 1974 models and full-scale blade elements
from HamStan began to undergo aerodynamic,
structural and noise testing, mainly at NASA
Ames and Lewis laboratories.

Assumed parameters were cruising at 35,000 ft
at Mach 0.8 (800 ft/s) with an eight-blade

External view of the UDF; interior details are in previous chapters (General Electric).

propfan loaded at 37.5 shp/D. This compares with around 12 shp/D for existing turboprops, which of course could not get near Mach 0.8. Various blade shapes led to one called SR-3, broader and more curved than previous patterns, with a tip sweep of 45°, an acoustic planform designed to make noise from different parts of each blade almost cancel out, and a thickness/chord ratio outboard of 0.6 radius of only 2 per cent.

The programme was supported by Federal funds. A Fiscal 1980 ban on augmentations held back NASA three years, but in 1981 strenuous efforts were made to get back on schedule. Small-scale propfans were tested above a NASA JetStar. To test the most difficult installation HamStan developed a single-rotation propfan driven by a 6,000 shp Allison T701 mounted by Lockheed-Georgia ahead of the swept wing of a Gulfstream II. One of the many NASA contracts went to GE, which shared the cost of developing a totally different engine which GE registered as the UDF, UnDucted Fan. Kenneth O. Johnson argued that no gearbox would be needed; all they need do was extend GE's free-running aft-fan concept with contrarotating variable-pitch propfan blades giving a BPR of about 34. The challenging objective was a power loading ten times that of modern propellers: 65 shp/D. Thus, a 10ft UDF could transmit not 1,500 hp (as in, for example, a Saab 340) but 15,000 hp. The small diameter made direct drive feasible, though to take out the required power needed a 12-stage turbine with 6 stages turning one way intermeshing

with 6 turning in opposition. SNECMA took a 25 per cent share, and the UDF impressed visitors to the 1988 Farnborough airshow hung on the left side of an MD-80.

The main competition was a joint programme by PW/Allison, using a 6+6-blade HamStan 11 ft 7 in contrapropfan driven via a 13,000 hp gearbox. The 578-DX demonstrator was run at Indianapolis in pusher configuration, and then, after GE had completed UDF tests, flown on the same MD-80 from April 1989. By this time many companies had studied the prospects. Rolls-Royce favoured the ultra-high bypass (UHBP) turbofan, with a reduction gear, but also devoted a lot of effort to the RB.509 and RB.529. The former was an unshrouded engine like the PW/Allison but shorter and lighter, with the gearbox integrated into the hubs. Rolls even built a 15,000 hp gearbox test facility. The RB.529 Contrafan had no gearbox, but used free aft turbines to drive single-stage fans (in a duct) directly.

When, in 1987, the IAE SuperFan was shelved, MTU continued working on Crisp (contrarotating integrated shrouded propfan), but then decided to link with Pratt & Whitney on the advanced ducted propeller (ADP). This NASA-funded project led to a full-size engine being run at West Palm Beach in November 1992 and at NASA Ames from June 1993. It matches a PW2000 core with a 40,000 hp gearbox driving a variable-pitch fan of 118 in diameter. BPR is 15, and thrust 53,000 lb in the forward direction and nearly as much in reverse. For obvious geometrical reasons, the braking thrust of v-p fans is much greater than that of limited-angle jet reversers.

What the author cannot comprehend is why, having done all this, a fall in the price of fuel resulted in propfans in the Western world being put on hold, if not discarded. A typical quote is; 'When the price of fuel began to fall, airlines lost interest, and the engine makers are now improving their conventional turbofans.' Most householders would be only too happy to use something that would knock 40 per cent off their energy bills, quite apart from which, even if fuel was given away, a more efficient engine enables any aircraft either to fly further or carry more.

Russians and Ukrainians are less swayed by fashion, and their propfan programmes continue to be top priority. A profusion of projects by the former Kuznetsov bureau have led to the Samara enterprise testing the NK-93 from 1992, flying it in 1994, and offering it for such aircraft as the Il-96M and Tu-204 for service from 1997.

It is a three-shaft engine using a turbofan as core, with seven LP and eight HP stages, each driven by single-stage turbines. A three-stage turbine drives the 30,000 hp gearbox with seven satellite pinions taking the power to the 114.2 in fans. The front fan has 8 blades and handles 40 per cent of the power; the rear fan has 10 blades, the blades being swept 30° and having 110° of pitch movement. Overall p.r. is 37. The NK-93 weighs 8,047 lb and is rated at 39,683 lb, with cruise sfc of 0.49.

Of several other propfan developments in former Soviet plants, by far the most advanced in timing is the D-27. This has been developed at Zaporozhye, Ukraine, using D-36/236 experience applied to the '27' gas generator with high cycle parameters. A D-236 prototype engine with a 10,850 hp gearbox driving a 13 ft 10 in Stupino D-36 tractor propfan with eight front blades and six at the rear was tested on a Yak-42LL exhibited at the 1991 Paris airshow. The production D-27 has a 14,000 hp gearbox driving an SV-27 with wider but thinner blades with more pronounced scimitar leading-edge curvature. The tractor D-27 powers the An-70, which to the author is the obvious C-130 replacement. A pusher version is to

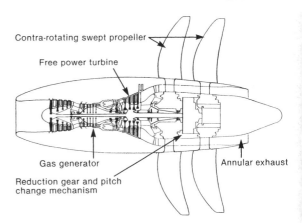

Rolls-Royce proposed this more compact form of geared contra-pusher, with the gearbox integrated between the hubs.

power the An-180, with a calculated cruise sfc of 0.286. This is roughly half the fuel burn of today's 757. It may be difficult for the Russians and Ukrainians to reap the deserved commercial reward, but at least in the next century they will be the ones without egg on their faces.

Rated at 13,880 to 16,250 hp, the ZMKB (Zaporozhye, Ukraine) D-27 is today's only production propfan.

18 Helicopters

In the mid-1950s the helicopter was transformed by replacing piston engines by turboshaft engines. The transformation was as great as that imparted to fighters by jet engines, but designers of rotary-wing machines were confused by the wealth of possibilities. Many sought to do away with shaft drive, replacing it by two concepts which made possible a single-rotor machine with no anti-torque tail rotor. One was gas drive, in which the lifting rotor is turned by compressed air or hot gas fed up

Graph from a 1955 paper by Maurice J. Brennan (then of Hawker Siddeley); it still applies today, and indicates that for flights longer than 40 min the turboshaft engine and its fuel is lighter than rival systems.

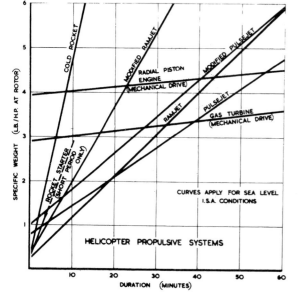

through the hub. The other was tip drive, in which the 'engines' are mounted on the tips of the blades. Today these schemes are rare, so they will be got out of the way first.

A pioneer of gas drive was A. G. Forsyth, who worked at Fairey Aviation. In 1942 he patented an efficient tip pressure-jet, but naturally (this was Britain) nobody was interested until, in 1945, it was discovered that an Austrian, Friedrich Doblhoff, had during the war built four small helicopters with a single rotor driven by tip combustion chambers fed with air from a piston-engined compressor. So in 1946 Doblhoff's patents were handed to Fairey, though Forsyth and Drs Bennett and Hislop used a more advanced scheme in modifying the second Gyrodyne into the Jet Gyrodyne. This was the research vehicle for a helicopter which in cruising flight could convert into an autogyro, propelled by propeller(s) and lifted by a wing and freewheeling rotor. The Jet Gyrodyne hovered in January 1954 and made its first transition on 1 March of that year. It remained piston-engined, but the principle was used in the Rotodyne described later.

In France, the SNCASO 1100 Ariel I of 1949 had a Mathis piston engine driving a compressor feeding similar rotor-tip burners, and the SO 1110 Ariel III, flown on 18 April 1951, was the first to use a gas-turbine (a Turbomeca Arrius) to pump the air. In the SO 1310 Farfadet, flown on 9 June 1953, the same rotor was combined with a Turbomeca Artouste turboprop in the nose and a fixed wing to increase speed to 150 mph.

SNCASO also produced a helicopter with cold tip jets, so-called because they expelled the compressed air without burning fuel. The resulting SO 1220 Djinn, first flown on 2 January 1953, led to the two-seat SO 1221. The Djinn's engine was a Turbomeca Palouste, and an identical scheme was used in the Fairey Ultra-Light, which was neater than the Djinn and cruised at 95 mph, compared with 60 mph. It says something about the two nations that SNCASO sold 150 Djinns, while Fairey built six Ultra-Lights and sold four.

Possibly the largest helicopter rotor blades ever built were a pair used to lift the XH-17 on 23 October 1952. This 'flying crane' was designed by Kellett in 1948 but passed to Hughes to build. It comprised a very stalky four-legged landing gear, able to straddle heavy loads, a small box-like fuselage housing a cockpit and fuel, two GE5500 engines (modified J35 turbojets) hung on each side, and large air ducts leading through the hub of the 130 ft rotor to a row of four jets at each blade tip. It hovered at 42,000 lb, but was soon abandoned. Much later, Hughes flew a gas-drive helicopter of its own design. From 1960 the Army funded 'hot-cycle' research which led to the XV-9A, with a 55 ft rotor driven by the entire hot gas flow from two GE T64 engines. While at take-off and landing the ducting and blades got nearly red hot, in cruising flight the rotor cooled in autorotation. In passing it is surely odd that, while the XH-17 used turbojets merely to provide compressed air, the XV-9A used shaft-drive engines merely to feed their exhaust.

Another gas-drive programme was launched by Percival Aircraft in Britain in 1951, with a Ministry order for the P.74. A contract went to Napier for the Oryx, comprising an axial compressor, five cans, and a two-stage turbine which also drove an auxiliary compressor. In the P.74 two of these 750 hp gas producers were mounted vertically inside the cabin, feeding the mixed core gas and air from the auxiliary compressor to jets on the tips of the 55 ft 4 in rotor. In 1956, at full power the P.74 stayed firmly on the ground.

The 'nearest miss' of the gas-drive machines was the Fairey Rotodyne. This began in 1947 as a projected 18,000 lb transport powered by two 1,000 hp Dart or Mamba turboprops which at take-off and landing would provide bleed air for the rotor. In cruising flight the Rotodyne was to be a twin-turboprop airliner, with normal wing and tail controls, the only unusual feature being the freewheeling (autogyro) rotor. At first flight, on 6 November 1957, the engines had become 3,500 hp Napier Elands, each driving a 14 ft 6 in four-blade propeller and an hydraulic clutch which in vertical flight drove an auxiliary nine-stage compressor feeding pressure (combustion) jets on the four tips of the 104 ft rotor. The

Rotodyne continued to grow, reaching 58,500 lb with Tyne turboprops before cancellation in 1962. The only thorny problem was noise, resembling a hardworking steam locomotive.

One of the last gas-drive machines was the Fiat 7002, first flown in January 1961. For this Fiat developed a special engine, the 4700, with a centrifugal compressor (11.2 lb/s, p.r. 4) and annular combustor; a free power turbine drove a separate centrifugal blower (7.7 lb/s, p.r. 3). The 4700 was designed to run in the vertical attitude to put the service compressor closer to the rotor. It weighed 335 lb and was rated at 610 gas hp.

In the United States in 1947–54 a profusion of small helicopters took to the air with single rotors driven by ramjets or pulsejets on the tips. A further 15 were built in other countries, but none achieved lasting success, one of the near misses being the Kolibrie, designed by SOBEH in the Netherlands and made in small numbers by NHI and then by Aviolanda. Its see-saw rotor was powered by ramjets. Despite the obvious attractions, the tip-drive helicopter is today almost extinct.

In contrast, the traditional shaft-drive machine has gone from strength to strength. Switching to turboshaft power at a stroke more than doubled the power that could be obtained from a given bulk and weight of engine, and also opened up levels of power far beyond anything available from practical piston engines. It dramatically reduced the incidence of engine failure, made possible use of less-volatile (supposedly safer) fuels, including the same as used by ships and army vehicles, greatly reduced vibration, and eliminated the previous problem of providing cooling in hovering flight. Of course, there remains the need to provide an oil radiator, and in large helicopters this has forced draught by a fan absorbing up to several hundred horsepower.

Ironically, the supplier of the first turboshaft engines to fly a helicopter was Boeing, which soon abandoned its engine business. The engine, an XT50-2 of 175 hp, with a centrifugal compressor, two cans, and two-stage turbine, was fitted under Navy contract to a Kaman XHOK-1 twin-rotor 'eggbeater' which began flying on 10 December 1951. On 26 March 1954 a bigger Kaman HTK-1 began flying with the 240 hp piston engine replaced by two 190 hp Boeing YT50 turboshafts.

Suitable turboshaft engines for small helicopters were slow in coming. One of the first to become available was the Turbomeca Artouste, almost a Palas turbojet with a second turbine stage. In July 1954 this powered two heli-

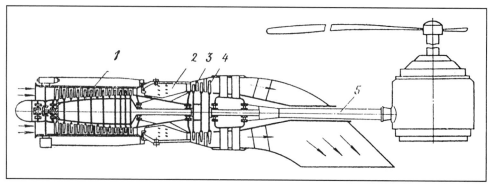

A Russian drawing showing the classic arrangement for a shaft-drive turbine-engined helicopter:
1, compressor; 2, combustor; 3, compressor turbine; 4, power turbine; 5, drive shaft to the main
gearbox and rotors. The jetpipe is usually turned to one side.

copters, neither of which went into production. Aerotechnica of Spain contracted Nord of France to build prototypes of the AC.13A three-seater, powered by an Artouste I rated at 260 hp, of which only 170 hp went to the rotor; yaw control in the hover was effected by running the jetpipe to the end of a tail boom and fitting it with left/right valves. Nord called the AC.13A the 1750 Norelfe. In Connecticut, Sikorsky converted an S-52 into the S-59 (Army XH-39) with a Continental XT51-3 mounted above the fuselage close behind the main gearbox. The XT51 was a 425 hp licence-built Artouste II derated to 320 hp to match the transmission. This four-seater set a speed record of 156 mph

on 29 August 1954.

The French Sud-Est group got into helicopters with the German wartime Fa 223, and among several prototypes was the SE.3120 Alouette of 1952. On 12 March 1955 the SE.3130 Alouette II flew on the 400 hp of an Artouste II, soon setting a record climb to 26,932ft. Nobody then knew that this would be the real start of the turbine-helicopter era. Sud and later Aérospatiale sold 1,300 Alouette IIs, followed by over 400 high-altitude Lamas and 1,382 Alouette IIIs (plus another 415 made under licence), making the team at Marseilles-Marignane No 1 in Europe. Today it is Eurocopter France.

A little-known 'first': the Boeing-engined Kaman K-600 (Navy XHOK-1 No 125477).

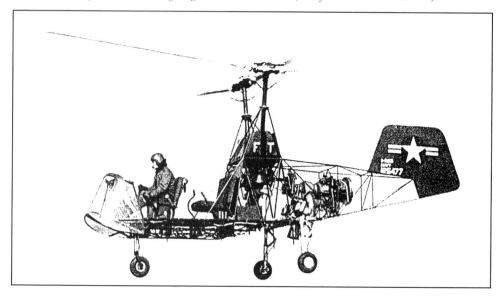

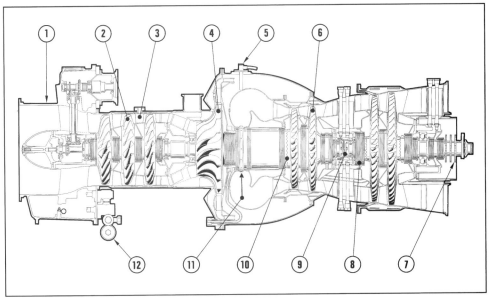

Largest of the Turbomeca helicopter engines, the Makila has three axial stages upstream of the centrifugal: 1, inlet; 2, titanium axial rotors; 3, cast steel stators; 4, steel centrifugal; 5, igniter; 6, gas-generator turbine with cooled blades; 7, three phonic wheels measuring power-turbine speed; 8, pressurized bearing labyrinths; 9, star-type bearing with pressure/vent air pipes; 10, curvic couplings (two); 11, annular combustor with fuel holes in main shaft; 12, fuel/oil heat exchanger.

Since then many types of helicopter have been made at Marignane in large numbers, including the three-engined SA.321 Super Frelon, twin-engined SA.330 Puma/332 Super Puma/532 Cougar, SA.341/342 Gazelle, SA.365/366 Dauphin/565 Panther, AS.350 Ecureuil/355 TwinStar, and Fennec. Throughout, the chief source of engines has been Turbomeca. Early types included the single-shaft Artouste III (around 390 lb, 590 hp, 2,525 delivered) and free-turbine Turmo (500–800 lb, 1,400–1,600 hp, 2,012 delivered). Single-shaft Astazous are still being made (2,480 so far), with the centrifugal impeller preceded by from one to three axials to give 500–1,000 hp. Today's engines all have free power turbines: the Arrius (unrelated to the 1951 engine, centrifugal, 480–640 hp), Arriel (one axial added, 640–750 hp), TM 333 (two axials added, 912–1,060 hp), and Makila (three axials added, 1,750–2,100 hp).

On 1 January 1953 General Electric at Lynn proposed a turboshaft engine for US Navy helicopters, to give 800 hp for a weight of 400 lb. On the basis of one $2.6 million order, the Small Aircraft Engine Department was formed in October 1953. The T58 was notable for having a 10-stage variable-stator compressor (12.4 lb/s, p.r. 8.3) and a free power turbine with rear drive. It actually gave 1,050 hp for a weight of 250 lb. It was the key to the Sikorsky S-62 and then the twin-engined S-61 Sea King. T58s were produced at Lynn at ratings up to 1,870 hp until 1984 for a host of helicopters and surface/marine applications, and in Britain (as the Gnome), Italy, and Japan.

In 1957 former Power Jets engineer Denis P. Edkins designed a bigger Navy engine, the GE T64. This had a 14-stage compressor and was aimed at 2,600 hp. Eventually turboprop as well as turboshaft versions went into production, and T64s are still being delivered for CH/MH-53Es at 4,750 shp for a weight of 755 lb.

In 1971 the Army picked GE's T700 to power the UH-60A Black Hawk. This time the emphasis was on robust reliability in harsh environments. The compressor, with five blisk axials and a centrifugal, handles 10 lb/s at p.r. around 15, to give a free-turbine output of 1,622 shp for a weight of 437 lb. Subsequent military, naval and civil CT7 versions are rated up to 2,000 hp, and licence deals with Britain, Italy, Japan, Australia, and South Korea are aimed at keeping out rivals such as the RTM 322.

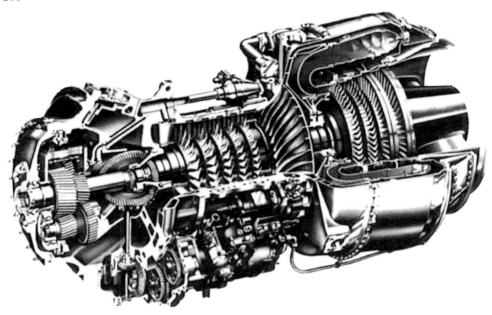

Cutaway of the T53, which in mid-1994 was nudging 20,000 sales (AlliedSignal Lycoming).

Under Operation Paperclip many German engineers were only too eager after the Second World War to work in the USA. Dr Anselm Franz, who in 1942–45 had headed the Jumo 004B team, went to Lycoming to become V-P Turbine Engineering to manage a 1952 USAF/Army contract for an engine in the 600 shp class. This ran in 1953 as the T53 (LTC1), with five axials ahead of a centrifugal (10.8 lb/s, p.r. 5), a folded annular combustor, and a free-turbine output of 825 shp. It flew in a Kaman

Most Allison 250 helicopter engines are installed the other way up, with jetpipes pointing diagonally upward, but the nameplates on this C20 (a model retaining the axial compressor) show which way is up.

HOK, but the breakthrough was selection in 1955 to power the Bell XH-40 utility helicopter for the Army. This became the HU-1, and finally the UH-1 and AH-1. Thousands served in Vietnam, boosting deliveries to over 19,000 T53 engines, rated at up to 1,550 shp, including a few made in Germany, Italy, Japan, and Taiwan, plus turboprops for the OV-1 Mohawk.

In 1954 Lycoming received its first contract for a 1,600 hp engine, the T55 (LTC4). This added two more axial stages to a larger compressor, and deliveries exceed 4,100 at ratings up to 4,867 shp in the MH-47E Chinook. The core was the basis of airline turbofans.

In 1972 Lycoming (then Avco Lycoming, later Textron Lycoming, and now AlliedSignal Lycoming) announced the LT101 family mentioned in the previous chapter. After prolonged redesigns the LTS101 turboshaft has found major applications at 550–684 shp, for a weight of 240–270 lb.

In June 1958 Allison received an Army contract for the T63, the first for a gas turbine in the 250 hp class. A notable feature was a six-stage axial in front of the centrifugal compressor (p.r. 6.2), delivering through a long can on each side to a two-stage gas-generator and power turbines. Over the years the T63 and commercial Model 250 have found many applications, including the JetRanger, at powers from 317–420 shp. In 1977 the C28 series replaced the axial stages by a single new centrifugal (p.r. 8.4) handling a greater airflow, increasing power to 550 hp and in the C30 family to 700 hp. Demand for the Model 250 is still brisk, with over 26,000 delivered.

Pratt & Whitney proposed several high-power engines, but achieved only limited sales of the T73. A J60 turbojet with two power-turbine stages added, it gave up to 4,800 shp for a weight of 935 lb. The application was the Sikorsky CH-54 Skycrane, which used a rotor system based on that of the piston-engined S-56. Back in 1958, Westland in Britain had at its own expense flown prototypes of the Westminster, with the same rotor system powered by two 3,500 hp Napier Elands. This could have been a world-beater, but the government said; 'Take over Fairey, drop the Westminster and work on the Rotodyne, which we'll cancel in due course.'

Pratt & Whitney Canada, however, achieved large sales of the PT6T Twin-Pac and military T400, both with two PT6 power sections driving a single gearbox. Most are rated at 1,800 shp and weigh about 650 lb. The Sikorsky S-76B is powered by two separate PT6B engines, each of 981 shp. In 1983 P&WC announced the PW200 family, with a single centrifugal compressor (p.r. 8) and single-stage core and power turbines. The PW206A powers the McDonnell Douglas Explorer at 638 shp and the 206B the Eurocopter 135 at 555 hp.

Discounting the Blackburn/Bristol Siddeley/Rolls-Royce Nimbus, a licence-built Artouste with two added axial stages to give up to 968 hp, Britain has produced just two helicopter engines. The first was the Napier Gazelle, first run on 3 December 1955. Designed to operate in any attitude, it had an 11-stage axial compressor (16.1 lb/s, p.r. 5.9) and free-turbine output. Bulky and heavy (typically 830 lb), it powered the Wessex and Belvedere, ratings being 1,260–1,650 shp. The second engine, designed in the mid-1960s as the Bristol Siddeley BS.360, became the Rolls-Royce Gem, which is complicated in comparison with rivals. The air passes through a four-stage axial, a centrifugal on a different shaft, a folded combustor, single-stage HP and LP turbines, and, finally, a two-stage power turbine. Versions weigh around 400 lb and have contingency ratings from 900–1,200 shp.

Rolls-Royce is a partner in two important collaborative programmes. For the EH 101/Merlin, the RTM 322 with Turbomeca is rated at 2,312 shp for a weight of 538 lb. This engine has performed brilliantly, which may enable it to beat GE in several important helicopters. For the

Cutaway Gem, showing the three-shaft layout, with drive at the far end (Rolls-Royce).

Eurocopter Tiger/Gerfaut Rolls is partnered by Turbomeca and MTU on the MTR 390, with tandem centrifugal stages (7.05 lb/s, p.r. 13) and an output of 1,556 shp for a weight of 373 lb, showing how far we have come since the Gazelle.

Tandem centrifugal compressors are also used in the most important new US helicopter engine, the outstanding T800. A product of LHTEC (Light Helicopter Turbine Engine Co, formed by Allison and AlliedSignal Garrett), it gives 1,350 hp for a weight of 315 lb, with sfc around 0.46. T800s are flying in many types of helicopter, but the main application is the RAH-66 Comanche.

This leaves just the Russians, who have a reputation for liking bigness. In 1953 Pavel A. Solovyev, newly appointed a General Constructor, was charged with producing engines for a helicopter of unprecedented size and power, the Mi-6. Its 114 ft 10 in five-blade main rotor required over 10,000 hp, and in 1957 the D-25V was qualified at 5,575 shp. Features include a nine-stage axial compressor (p.r. 5.6)

driven by a single-stage turbine, and a two-stage power turbine with rear drive. Typical weight is 2,921 lb, and the helicopter's R-7 gearbox weighs 7,054 lb. The D-25VF is a version rated at 6,500 shp; in one application, the Ka-22, the power turbine of each engine drove both a rotor and a propeller.

Most Soviet helicopters have flown with engines designed under Sergei P. Isotov. The

Undoubtedly a leader in the 1,300–1,500-hp class, the T800 has tandem centrifugal compressors and a large integral inlet particle separator (LHTEC).

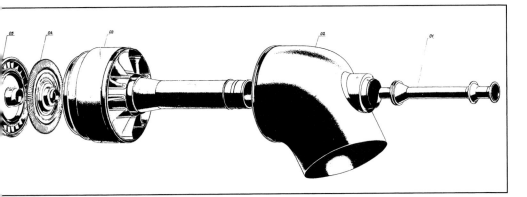

*All modern engines are modular, and the modules of the world's most powerful helicopter engine, the D-136, are: **01**, drive shaft; **02**, jetpipe; **03**, rear frame; **04**, power turbine; **05**, nozzle ring; **06**, gas-generator turbine; **07**, combustion chamber; **08**, HP compressor and accessory drives; **09**, LP compressor with three blow-off ducts* (Zaporozhye KMB).

GTD-350, used in the Mi-2, resembles the Allison 250 but has seven axials ahead of the centrifugal (4.83 lb/s, p.r. 6.05), to give 444 hp for a weight of 307 lb; this engine is made only in Poland. Taken together, Isotov's TV2-117 and TV3-117 have been made in larger numbers than any other helicopter engine (39,000), for such machines as the Ka-27/29/32 and Mi-8/14/17/24/25/28/35/40. Features include a 10-stage variable-stator compressor and 2-stage core and power turbines with rear drive. Ratings are 1,600–2,300 hp for weights of 620–740 lb. These are now products of the Klimov Corporation, which is developing the TV7-117 to power the Mi-38. Rated at 3,500 hp, it has the same core as the 117S turboprop described earlier.

The Ka-25 was powered by the TVD-10 (military designation GTD-3) designed by Glushenkov at Omsk but later handed to PZL-Rzeszw in Poland, which produces it for Swidnik helicopters as the PZL-10W. It has six axial and one centrifugal stage (10.14 lb/s, p.r. 7.4), and gives 1,134 hp for a weight of 310 lb. Kobchyenko, of the Soyuz bureau, designed the TV-O-100 to power the Ka-126. This has two axials ahead of the centrifugal (p.r. 9.2) and gives 720 shp for a weight with many accessories and a particle extractor of 353 lb.

In Ukraine, ZMKB 'Progress' used the five gas-generator modules of the D-36 turbofan to produce the D-136 to power the Mi-26. Thus the 136 has a 6-stage LP spool and 7-stage HP, and turbine stages 1/2/1 for rear drive. Its dry weight is 2,315 lb. The take-off rating is 11,400 shp, for an sfc of 0.456. No other helicopter engine, existing or planned, comes close to this power.

In 1950 Westland proposed a family of helicopters driven by turbojets hung on the blade tips. The biggest, with three AS Sapphires, was to have a 196 ft rotor lifting 206,000 lb, a typical load being 450 troops. The firm then discovered that such machines were impractical, but today a 40,000 shp helicopter engine would be no problem at all (though the gearbox might). An acceptably quiet city-centre Jumbo Jet is technically feasible, but the author doubts whether he will live to travel in it.

Jet and Turbine update

Gas-turbine engines for aircraft are the subject of intense competition. This means that the technology changes and develops almost by the hour, so in two or three years quite a lot can happen. In fact, much has happened since the first edition of this book was published in 1994, so the following has been added to continue the story at least to the end of the century, with a little crystal-ball gazing beyond that.

Back in the 1950s, apart from a very few research aircraft propelled by rockets or ramjets, jet aircraft were all powered by turbojets. The Civil Jet Age was launched with only slightly modified versions of the same turbojet engines that powered the latest fighters and bombers. For example, the Comet and Caravelle were powered by the Avon, which was also fitted to the Hunter, Lightning, Valiant, Scimitar and Sea Vixen. The Tu-104 was powered by the RD-3M turbojet, which was used in the Tu-16 and M-4 bombers. The 707 and DC-8 utilised the JT3 turbojet which, as the J57, had previously powered the B-52, F-100, F-101, F-102 and F8U, F4D and A3D.

Today, military and civil aircraft use engines which in most cases are unrelated. Combat aircraft need engines of small diameter giving the ultimate in performance; fuel economy is usually a secondary factor, and noise doesn't matter. In contrast, the design of engines for civil airliners is dominated by the need for minimal fuel burn and the least possible environmental impact,

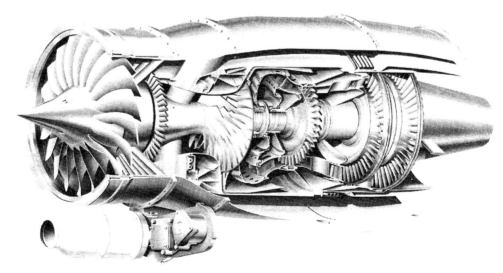

Cutaway drawing of the Williams Rolls FJ44 (Page 187). As the smallest turbofan on the market, this has already picked up a wide range of applications.

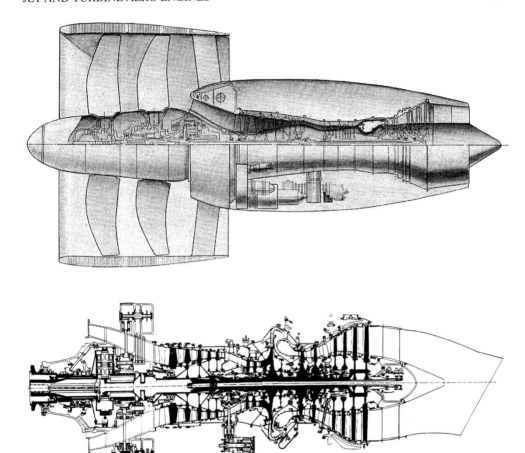

Top, Longitudinal section through Kuznetsov's NK-93 propfan (see page 210).
Above, Longitudinal section through ZMKB's D-27 propfan (see page 211).

while diameter hardly matters. Accordingly, the dramatic developments in recent years have been along two entirely different paths.

By far the most impressive development has been in the sheer size and power of engines for the largest airliners. Nothing remotely like this has ever happened before. Merely to write down the figures is impressive enough, but engineers think in pictures and it is when one plots what has happened that one sees the enormity of the progress made.

The diagram (overleaf) shows how the sustained growth in the power of airline jet engines has taken a fantastic leap in the past three years, from 60,000 to over 100,000 lb thrust! We may be sure that nothing like this – a sudden jump of some 70 per cent – will happen again for many years to come.

The reason for this giant leap is entirely because of one aircraft, the Boeing 777. When Boeing produces a new big jet airliner, Pratt & Whitney, General Electric and Rolls-Royce all have to be in on it. They cannot afford not to. General Electric bitterly regretted coming very late to the 747, and Rolls-Royce came even later and picked up a few crumbs. With the original Airbus, the positions were reversed. The A300B was launched with General Electric, Pratt & Whitney managed to pick up a gradually increasing part of the business, and Rolls-Royce foolishly thought this non-American aircraft was not worth bothering with, and have regretted it ever since.

Thanks to the 'Triple-7' all three of these engine suppliers can now offer turbofans in the '100-k' (100,000 lb thrust) class. These engines,

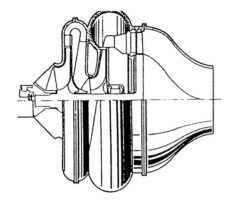

For historical interest, this is a longitudinal section through the first Soviet turbojet design, Arkhip Lyul'ka's RTD of 1937. Never run, it would have had two centrifugal compressors in series, driven by an axial-flow turbine, an excellent arrangement. When page 131 was written this engine was unknown.

has a completely new 112-in fan with more efficient hollow titanium blades (technology pioneered by Rolls-Royce) driven with a six-stage LP spool by an LP turbine with a record seven stages. The PW4090 has an improved HP spool and higher TET, while the PW4098 has a slightly larger fan rotating with a seven-stage LP spool. Thus, they effectively offer 3½ different engines, and so far these have outsold their rivals.

In contrast, GE just designed the GE90 to be the largest aircraft engine in the world. Its 123-in fan has snubberless blades of graphite/epoxy composite, and it rotates with an LP compressor with only three stages on a shaft driven by a six-stage turbine. This huge engine entered service at

in refined forms, are likely to suffice for the next 30 years at least. After all, the biggest passenger jet in prospect, the 530/850-seat Airbus A3XX, only needs engines in the '70-k' class. A four-engined jet with '100-k' engines would weigh about 1,700,000 lb and carry 1,200 passengers. In the author's opinion, the airlines are unlikely to be ready for such a vehicle until about 2025 at the earliest. But how wonderful it will be in that year to have suitable engines not just available but available with a background of millions of hours in the Triple-7!

On the other hand, all three engine giants, especially GE and P&W, have complained that if all three are going to continue chasing each major airframe there will be precious little profit. GE's Murphy has deliberately delayed development of a '102-k' version of the GE90, saying "Wait for a solid launch market, we must have a business case to go forward with a new engine". P&W President Krapek said "Three engine makers on a wing makes returns very difficult . . . The returns are a long time away, if ever. One manufacturer has said it will take 28 years, if at all . . . We can't keep doing this".

To get to 100 k the three Western engine companies have followed rather different routes. As explained previously, Pratt & Whitney developed their JT9D into a new baseline engine designated PW4000 with a 93.6-in fan turning on the same shaft as a four-stage LP compressor (called a core booster or, in the USA, a low spool) driven by a four-stage turbine. Using this as the basis, they then developed a succession of growth versions. The PW4168 for the A330 has a 99.8-in fan driven together with a five-stage LP spool by a five-stage LP turbine. The PW4084 for the 777

A plot of the thrust of airline jet engines shows the incredible leap in power that is being made at the end of the century.

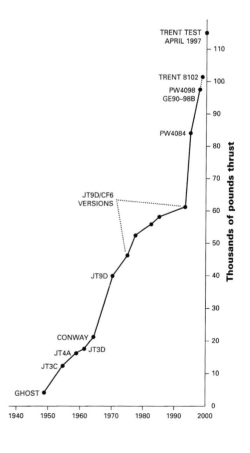

a takeoff rating of only 76,000 lb, but different ratings are available at up to 90,000 lb, with 98 k promised by 1999. Despite GE's technical excellence, and the fact that early versions were dramatically de-rated, the GE90 had a troubled early history due to a wide variety of problems, some of which (such as the reverser) were not GE's responsibility. Moreover, while they offer the GE90 for the 777, they are left with the older and unrelated CF6-80C2 or -80E for the smaller twin-jets.

Rolls-Royce started with an improved RB.211 which, as related on page 196, was developed into an engine called the Trent. Today the Trent has evolved into a family of engines which, partly because of the design flexibility inherent in the three-shaft layout, are uniquely amenable to a wide variety of propulsion requirements, with takeoff thrust from 55 k to 110 k. A 2:1 spread of power has never before been simultaneously available from a single basic engine.

The original Trent 700 and 800 are briefly described on p.196, and by 1997 the latter had been tested at thrusts up to a record 114,500 lb. The Trent 900 is essentially a Trent 800 with a core scaled down to a linear ratio of about 0.9, to give a BPR (bypass ratio) of 8.2. This results in a thrust band of 70 to 80 k. The Trent 900 was originally intended for the stretched 747-500X and -600X, which did not interest the market sufficiently and were abandoned by Boeing in early 1997. The Trent 900 would be ideal for the A3XX, and also for stretched versions of the 767 and A330. In early 1997 Rolls-Royce signed a Memo of Understanding for the A3XX to enter service powered by the Trent 900 in 2003. In the long term this could be extremely important to Rolls-Royce.

In 1994 Airbus studied ways of stretching the A340, and soon reached gross weights which put it beyond the reach of its existing CFM56 engines. After various political manoeuvres General Electric (50/50 partner in the CFM56) went ahead alone with a largely new engine in the 50-k bracket, but then failed to agree on how to share the engine development cost. At the same time, the stretched A340, now firmed up as the A340-500 and -600, kept growing until these aircraft needed an engine in the 56-62 k class. The unexpected upshot was that Rolls-Royce said, in effect, "With modern computer-assisted design it is relatively simple to scale a core up and down and still retain near-perfect aerodynamics and mechanical design. If we scale down the Trent 800 core a bit more, to about 0.81, and match it with the fan of the Trent 700, perhaps adding a stage to the LP turbine because of the reduced core gas flow, the result would be an engine with a BPR of about 8.5, with world-beating fuel economy and quietness." In June 1997 this was selected as the launch engine for the impressive new A340s. It is called the Trent 500.

Thus, Rolls-Royce alone can offer a family of what is basically the same hi-tech engine for the amazing range of thrusts from 56 to 102 k, and clearly with more to come. It remains to be seen what effect this will have on market shares.

From the sidelines, all these developments are fascinating. To produce engines for the 777 the three main rivals have spent a reported US$6 billion. GE has spent by far the most, and so far – not entirely because of the huge GE90's various problems – has had the smallest return. P&W's various sizes of PW4000 have sold almost as well as its two rivals combined, but Rolls has come from behind and, by sheer technical excellence, appears likely to become No 1 in the course of time. To some extent this is because of the unique flexibility of the three-shaft architecture, which enables the fan, LP and HP spools to be scaled up and down without the slightest loss in efficiency.

Incidentally, the unprecedented growth in the power of the largest airline engines has made it possible to produce a 747 Twin. Whether the market would welcome such an aircraft is problematical, but with minor wing redesign a 747 with two engines in the 110-k class could be a remarkably low-cost people-mover.

As predicted in earlier pages, BPR has risen inexorably year by year. Bearing in mind that the BPR for a turboprop is typically in the region of 50, it is difficult to understand why engine manufacturers have been so timid in producing turbofans with optimum BPR. As explained earlier, the first turbofans typically had a BPR of unity or less. In some cases this was because the engine's diameter was restricted by the need to fit inside the root of a wing, but even when the engine was installed in an external nacelle or pod the BPR was seldom higher than 1.0.

For example, the world's best-selling civil jet engine, the Pratt & Whitney JT8D was designed with a BPR of 1.0. This was far below the optimum, and made the engine less fuel-efficient and much noisier than it need have been, even though in every case it was mounted outside the airframe where there was no severe constraint on overall diameter (even on the 737). After 15 years Pratt & Whitney decided to refan the JT8D, resulting in the JT8D-200 series in which BPR is increased to about 1.75. This was still nothing like enough, and while later 737s have the CFM56 engine the latest DC-9 derivative, the MD-95, has BMW Rolls-Royce 715s.

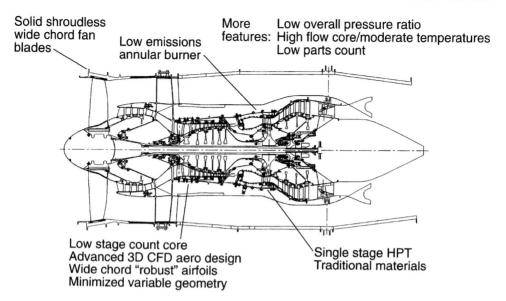

Solid shroudless wide chord fan blades

Low emissions annular burner

More features:
Low overall pressure ratio
High flow core/moderate temperatures
Low parts count

Low stage count core
Advanced 3D CFD aero design
Wide chord "robust" airfoils
Minimized variable geometry

Single stage HPT
Traditional materials

Longitudinal section through the PW6000. According to Pratt & Whitney's President, this small/ medium-sized engine is 'crucial to our future'.

Pratt & Whitney are kicking themselves for losing a gigantic market. Even if they never appreciated that the old 737 and DC-9 would be capable of development to remain competitive into the next century, they must have recognised that there would always be a need for twin-jets in approximately this size, and they should have had an engine ready for it. As it is, they have lost most of the market to the CFM56, and are now coming from behind with an all-new engine called the PW6000.

This is aimed at a size class rather below that of the latest 737s and MD-90/MD-95, namely a range of new and projected aircraft with 70 to 110 seats. The only aircraft at present in this category are the Fokker 70 and 100, whose manufacturer went bankrupt, and the RJ70 to RJ115 which are likely to be replaced by a new design with twin engines. In 1997 this category of jetliner was the most eagerly contested of all, with eight existing or projected designs involving 15 manufacturers in nine countries. This time Pratt & Whitney is determined to secure at least a significant share of the market, despite competition from GE with growth versions of CF34, BMW Rolls-Royce with the BR 715 and CFMI with the CFM56-9.

The new Pratt & Whitney engine, essentially a JT8D successor, is the PW6000, which will be produced initially either as the PW6020 (20-k class) with a 55-in fan or as the PW6021 with a 56-in fan. How can the Connecticut firm hope to start from scratch and succeed in so competitive a

field? The answer is partly because they are a big and respected company with a gigantic customer base in virtually every country of the world. Another factor is that, because the PW6000 is a totally new design, they hope to make it simpler than any rivals, with fewer parts, and thus make it commercially more attractive. BPR is still modest, at about 5.2; in the author's opinion, 8.5 would be better.

According to P&W President Krapek "The 6000 is crucial to our future. It's our way back into the narrow-body business we lost". Excluding the nacelle, Pratt & Whitney estimate development cost at $650 million. Thus, as this small engine will be sold for a fraction of the price of wide-body engines – target price is $3 million, only a fraction of which is profit – large numbers must be sold if the programme is ever to become profitable. We must wait and see if the PW6000 ever gets anywhere near the sales total of the JT8D, which was 14,550 in June 1997 with a few Dash-200 versions still being made!

Unlike Rolls with the Trent, CFMI has left its GE-supplied core alone, as far as possible, and has created engines with different thrust levels by fitting different sizes of fan. The bigger fans need extra LP turbine stages; thus, while most CFM56 engines have a four-stage LP turbine, the powerful CFM56-5 family have five while the low-thrust (18.5-23-k) CFM56-9 needs only three. This is welcome for the Dash-9, because it reduces the number of parts, which since 1980 has come to the fore as one of the basic factors determining not only the price of an engine but also the cost of maintaining it.

On the other hand, making a lower-thrust version of an existing engine, whilst sticking to an almost unchanged core, runs directly counter to the modern trend of progressively increasing BPR. The original CFM56-2 of the 1970s had a BPR of 6. Today CFMI would like to raise that to about 8 or 9, but the only way to produce a version in the 20-k class (the CFM56-9) with an unchanged core is to fit a smaller fan, so that the BPR of the new version is only just over 5.0. Beyond doubt, others will have to emulate Rolls-Royce, and scale their cores up and down.

The only example in the West of an all-new engine for widebodies is the GP7000, being developed by an unprecedented Engine Alliance (now called just Alliance) by GE and Pratt & Whitney. This was launched in 1996 to meet competition from the Trent 900 in the propulsion of the proposed Boeing 747-500X and -600X. The British engine was judged to be such a formidable competitor that a derivative engine based on the CF6-80, GE90 or PW4000 would not be good enough. The answer was a new engine perfectly matched to the job and, because of the enormous development bill, and the need to pare selling price of the eventual engine to the lowest possible level because of the competitive situation, the two engine giants agreed to work 50/50 in creating the new engine.

GE are responsible for the HP compressor, combustor and two-stage HP turbine, while P&W handle the fan and co-rotating LP compressor, LP turbine and assembly and test. This gives approximately the required work-split. BPR will

be in the region of 7.8. Cancellation left the original version, the GP7176 (the designation meant 76,000 lb thrust), without an application. Wisely, the Alliance partners decided to keep going, this time aimed at the A3XX and possible stretched 767s and a rapidly formulated 747 Growth. By the 1997 Paris airshow in June the core booster had changed from having two stages to having three and then four, while the LP turbine grew an extra stage, from five to six.

At last manufacturers are getting BPR about right, though the higher this is pitched, the greater some other problems become. For example, the higher the BPR, the lower the cruise thrust for any given takeoff rating. For example, a JT8D with a BPR of about 1 will develop 14,000 lb on takeoff but will still put out about 4,200 lb at cruise rpm at Mach 0.8 at 35,000 ft, or 30 per cent of the takeoff thrust. Tomorrow's engines in the 76-k class, with a BPR of about 8, will have a high-altitude cruise thrust of only about 9,000 lb, or a mere 11.8 per cent.

A second problem with high-BPR engines is matching the LP turbine to the fan. With the HP spool there is no problem, just as there was no problem with early turbofans of low BPR. Today we have huge fans which inevitably run at much lower rpm, yet they have to be driven by a small turbine downstream of the slim HP spool and diminutive combustor.

The engine designer has three choices. The simplest is to do his utmost to open out the slope of the multi-stage LP turbine to make each successive stage bigger. This can be seen in the drawing of the GP7176, where even the first stage of the LP turbine has a much greater diameter than the HP turbine. This helps to match the

Longitudinal section through the original scheme for the Alliance GP7176. GE and Pratt & Whitney have now added two stages to the LP compressor and one to the LP turbine.

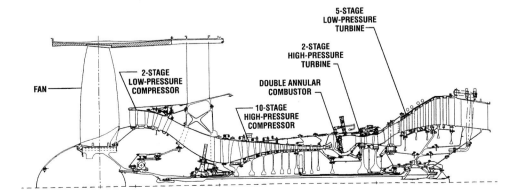

turbine to the fan shaft speed, but even so the limitation on speed means that the LP turbine must have many stages. In the GP7176 there were five, but the LP turbine of all the biggest PW4000 versions has seven. This means ponderous weight and high cost.

Alternatively, he can run the LP turbine at high speed and drive the fan through a reduction gearbox. This enables the LP turbine to be much smaller, with perhaps only two or three stages, saving a great deal of weight and cost. The down side is that a gearbox which may have to transmit 60,000 horsepower is a huge development challenge, besides being itself heavy, costly and an extra source of potential failure. However well it is designed, it will also waste a great deal of heat, which must be dumped overboard by a large oil cooler. So far nobody has had the nerve, or development funds, to produce a large geared turbofan, though small geared engines were tested in the 1960s (page 186), and the TFE731 and LF500 are modern examples.

The third alternative is to use the core as a gas generator to drive an independent free-running turbine downstream on which the propulsive fan blades are directly mounted (a drawing of one proposal is on page 69). This has a great deal to commend it. Both the turbine blades and fan can be perfectly matched, and there is no need for extra turbine stages, or for a gearbox or even for a long and heavy shaft.

Back in 1955 General Electric planned a single-rotation aft-fan engine, and actually managed to sell it to the airlines (page 183). This bold pioneer engine, the CJ-805-23, could well have started a fashion. As it was, it was fitted to only one type of aircraft which, for reasons unconnected with propulsion, found few customers and lost money for Convair and for GE. The aft-fan engine itself served with good fuel efficiency and reliability for over 30 years, but it proved to be an aberration. It didn't have to be like this.

Where major procurement decisions are concerned, the airline business is ruled at least as much by fashion as by the balance sheet, and today engine manufacturers are quite happy to draw pictures of unconventional engines – such as contra-rotating aft fans with a BPR of around 20 – but have no intention of sticking their necks out by funding a demonstrator which could actually lead to customers. Risks are quite high enough even with traditional direct-drive front fans!

Having criticised Pratt & Whitney for failing to produce a successor to the world's best-selling airline jet engine until others had taken over the market, a similar charge could be levelled at Rolls-Royce for doing all the hard work to establish the turboprop throughout the world and then leaving the entire market to rivals. This was partly because, as explained on page 202, the RB.211 rather monopolised management attention when a Dart successor was being planned. This was no excuse; the costs and risks of such an engine were minimal, and the market had already been established.

A relevant factor was the abysmal failure of the superb Vickers-Armstrongs Vanguard in the market-place. This was because it hit the market in 1960, which was precisely the wrong time. At this time the short-haul jet was all the rage, and turboprops were fast becoming unsaleable except to power military airlifters such as the C-130. Armstrong Whitworth's AW.650 Argosy civil cargo aircraft, powered by four Darts, would have sunk almost without trace had the RAF not placed an order for 56.

It was just at this time that turboshaft engines began to take over from piston engines to power the world's helicopters, but these were mainly quite small, and not many were also developed with gearboxes to drive propellers. In Britain de Havilland took a licence for the General Electric T58, calling it the Gnome, and for a while toyed with the idea of producing a turboprop version to power a replacement for the Dove, but abandoned the idea. Thus, from the early 1960s the turboprop market was so depressed that it had only one way to go, upwards! Remarkably, the country which pioneered the turboprop, Great Britain, has abysmally failed to share in what soon became explosive growth.

In the field of turboprops the principal development in the high-power field is the FLA (Future Large Aircraft). This is intended to be a capable airlifter to replace such aircraft as the C-130 and Transall. Its history shows, yet again, the amazing ability of European planners to sit round tables for ten years without daring to come to a decision. Thus, having through this procedure thrown away all the many chances of doing the job properly, it is now decided that European air forces not only need an FLA but need it urgently. Lack of an FLA has opened the floodgates to Lockheed Martin with the C-130J. Moreover, though they accept the need, European governments don't want to pay for development, but have told AMC, the specially formed Airbus Military Company, that if they develop an FLA then the air forces will probably buy some (but fewer than assumed during the wasted ten years).

When this book was written the FLA was still rather vague; the only thing confident about it was that it was to be a jet. In April 1994 it was decided that, for several sound reasons, such as combat radius, steeper approach, shorter field length and better manoeuvrability on the ground, the FLA ought to be powered by 10,000 hp turbo-props! Unfortunately, apart from the D-27 in Ukraine, no suitable engine exists.

Candidate engines are the AlliedSignal AS-812F, two Allison projects, the BMW Rolls-Royce BR700TP and the SNECMA M138. In the author's opinion all are at least six years away from operational service. Even though all utilise gas generators based on portions of existing engines, the pacing item is the propeller gearbox. The two most likely are team efforts. One comprises BMW Rolls-Royce (itself a team) plus Hispano-Suiza and ZF, using a core based on the BR710. The other comprises SNECMA 36.5 per cent, MTU 36.5 per cent and FiatAvio 27 per cent, and uses the HP spool of the M88 with LP parts related to the CFM56. Propellers would come from Dowty Aerospace or Ratier-Figeac. They would have eight scimitar-like blades, and a diameter of about 17 ft.

A major development at almost the other end of the scale is the Ayres Loadmaster. Ayres, based at Albany, Georgia, is developing this unusual aircraft to meet the demands of high-priority cargo, especially small packages. Other uses might include multisensor operations, medevac and even front-line troop repositioning. The Loadmaster looks rather like a small Cessna, but its span is well over 60 ft and it will carry over three tons or 19 passengers. The propeller is driven by the impressive LHTEC CTP800-4T, a 2,700-hp package which promises to sets new standards of economy and durability. It comprises two power sections derived from the original T800 helicopter engine (page 218) driving a common gearbox. One rather wishes that British Aerospace could have invented the Loadmaster (apart from the discontinued J41 and ATP, they haven't produced a new aeroplane for 23 years), because within a few months of announcing it Ayres had taken orders for 50 from FedEx, with options for an additional 200. In addition to this, to meet other orders, in June 1997 Ayres signed a Memo of Understanding with LHTEC for 100 shipsets of CTP800 engines, worth over US$100 million, with engine deliveries to begin in 1999.

A rather similar aircraft, but smaller and with precisely half the power, is the Cessna Dual Pac Grand Caravan. This is powered by a Soloy Dual Pac, a neat package comprising two 661-shp Pratt & Whitney Canada PT6A-114A power sections driving into a common gearbox. Like the Loadmaster, this aircraft is designed to fly at maximum weight with either power section inoperative.

Today production of the PT6 turboprop has exceeded 34,000, and it should soon surpass GE's J47 as the all-time best-selling aircraft gas-turbine engine. The Montreal manufacturer's more powerful PW100 has replaced the Dart in several types of aircraft, and with 4,100 engines already sold, and demand increasing, appears certain to pass the pioneer British engine's total of 7,100. Allison's Model 250 is fast approaching 30,000 sales, but most of these are turboshaft engines for helicopters. In contrast, AlliedSignal have sold nearly 14,000 TPE331 engines entirely to drive propellers. One application of the TPE331 is the Embraer Tucano, so the RAF's pilot trainer today – successor to the famed Tiger Moth and JP – is a Brazilian aircraft with an American engine.

This would have seemed inconceivable even as little as 30 years ago. At that time the British aircraft and engine industries were capable virtually across the board. In many fields they had pioneered new types of aircraft and led the way to meet the demands of new markets. For example, in 1962 de Havilland flew the first D.H.125, one of the first purpose-designed business jets, and this soon sold all over the world powered by Rolls-Royce Viper engines. It was increasingly apparent that these heavy, noisy and fuel-hungry turbojets needed to be replaced by a modern small turbofan. Rolls-Royce designed the RB.401, and actually built and tested successive prototype engines, but then walked away from the small/medium bizjet market!

Worse was to come. Having re-engined the 125 with American and Canadian turbofans, British Aerospace came to the incredible conclusion that building business jets was not 'core business' – presumably building concrete harbours and Rover cars was – and sold the entire 125 programme to Raytheon. Today this giant US firm produces the 125 in Kansas, and as I write has just taken a single order for 20. Moreover, it has developed the 125 much further, into the Hawker Horizon, with a 'stand-up' cabin and 6,500-lb P&W Canada PW308A engines. The author is not a businessman, and would like someone who is one to explain to him the sense in doing all the difficult high-risk work in establishing a production line of world-beating aircraft, with British engines, and subsequently first throwing away the engine market and then the entire project.

In sharpest contrast, Pratt & Whitney Canada thinks aviation a good business to be in. It had a desperate and disheartening struggle to get started (page 206), but gradually built up secure and growing business with the PT6A turboprop and JT15D turbofan. These are small engines, and you have to sell dozens to earn as much as the price of a single engine for a wide-body, but if your sales are numbered in thousands then the company can become quite ambitious. Thus, the PT6 began at 400-500 shp, and is today in production in an amazing variety of versions with ratings up to 1,700 shp. The bigger turboprop, the PW100, began at 1,800 shp, is now in production in many versions at up to 2,750 shp, and is being further developed into the PW150 rated at up to 7,500 shp. The PW150 replaces the first centrifugal compressor by a two-stage axial handling much greater airflow. This increase in power opens up large new markets to the Montreal company, starting with the stretched DHC-8-400.

Having sold substantial numbers of PT6B and PT6T (twin) engines for helicopters, P&WC are now in production with a rapidly growing family of turboshaft engines called PW200. These are based on the principle that you can make an engine simple and still achieve good performance. For example, they have only a single centrifugal compressor, but by being made of titanium alloy it can run fast enough to have a pressure ratio of 8, with higher values in prospect. Most of this family are in the 600-shp class, but the core has the potential to reach 1,000 shp, and we may be certain P&WC will make it get there.

When it comes to jet engines P&WC have been even more aggressive. Having developed the little JT15D from 2,200 lb to 3,045 lb, they then produced the PW305 (page 188) in the 5,000-lb class. This has now been developed into the PW306 family, with a larger fan and forced mixer nozzle, in the 6,000-lb class, and in 1997 was developing the PW308A with an initial rating of 6,500 lb, for the Raytheon Hawker Horizon. P&WC will supply the complete Horizon propulsion system, with nacelle and reverser. Likewise, the PW500 (page 188) has now been further developed to 4,450 lb for the fast-selling Cessna Citation Excel.

In the author's view it is an indictment of British management that the United Kingdom should not only have got rid of its globally established business jet but should also have walked away from the propulsion of such aircraft. Whether Rolls-Royce will make an attempt to get back into this business is problematical. At present it has a tiny toehold at the extreme ends of the spectrum via collaboration with Williams (page 187) and the purchase of Allison.

Longitudinal section through the SNECMA/P&W Canada SPW14. Note how the centrifugal compressor rotates with the axial; the LP turbine drives the fan only.

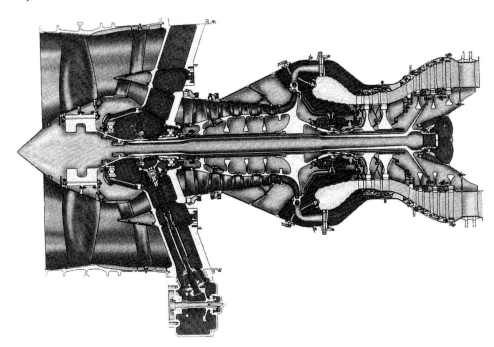

Meanwhile, P&WC is aggressively expanding upwards into the 12 to 16-k market, via a Joint Venture with SNECMA. The two French-speaking partners are developing the SPW family, of which the baseline member will be the SPW14, the figure denoting thousands of pounds of thrust. This will be an all-new engine with minimum parts-count, with a single-stage fan with modern wide-chord blades and a compressor with four axial stages and one centrifugal. In 1996 the SPW14 was named as one of two finalists in the AI(R) twin-jet competition, eliminating the Allison and BMW Rolls-Royce contenders. It is very unusual for an airframer to pick a non-existent engine in preference to two which are already on test. Moreover, in this case a British airframer has eliminated both the British, or part-British, engine offerings.

The other finalist was the CF34-8C, an up-graded growth version of the civil derivative of the TF34. This simple turbofan was built in the 9-k class to power the USAF's A-10 attack aircraft and the Navy S-3 anti-submarine aircraft. The derived civil versions power the Challenger bizjets. The CF34-8C is aerodynamically redesigned throughout, and also has a completely new combustor similar to that of the F414 (page 176). It will be rated at 13 k, with growth potential to 18 k.

The proliferation of rivals has made it less likely for the established Rolls-Royce Tay and the later BMW Rolls-Royce engines to sell in their thousands to re-engine older jet aircraft meeting FAR.36 Stage 2 noise legislation. Without modification all Stage 2 aircraft registered in Europe or the USA must be grounded in April or December 2002. Many operators have elected to fit heavy and complicated noise-reduction systems, which are much cheaper than new engines, even though they increase fuel consumption whereas new engines would dramatically reduce it.

One competition which BMW Rolls-Royce did win was propulsion of the BAe Nimrod 2000, the choice to replace existing Nimrod aircraft of the RAF. The selected BR 710-48M was initially to be rated at 14,900 lb, but by June 1997 had grown to 15,500 lb, which compares with just over 11,000 lb for the Spey engines previously fitted and 5,000 lb for the original Comet 1. Originally planned over 50 years ago, the installation of engines inside the wing roots would not be repeated in a new design today, but in any case the wings of the Nimrod 2000 will be new from tip to tip. The airflow through the BR 710 is exactly five times as large as through the Comet 1's engines, but the engine's diameter is actually one inch less! Whereas the Comet 1 was designed for a take-off weight up to 105,000 lb, the Nimrod 2000 will weigh up to 231,165 lb.

This is an unusual case of a modern military aircraft fitted with engines originally designed for the civil market. Another example is the Boeing KC-135 tanker, E-3 AWACS and other types related to or derived from the 707 airliner, which today are re-engined with the F108, the USAF version of the civil CFM56. Most military aircraft have small-diameter engines designed for performance rather than environmental factors or fuel economy. By 1997 the F119 was coming into full production for the F-22, the F100 had been further developed into the Dash-229IPE (improved-performance engine) with F119 features, the Swedish Gripen with the Volvo/GE RM12 engine was in squadron service, and the proven record of the RRTI Adour had held off an attempt by AlliedSignal (Garrett) to displace it as the engine of the Hawk/Goshawk trainers. The two European fighter engines, the Eurojet 200 and SNECMA M88, were fully developed and ready for production, but handicapped by repeated politically motivated delays and post-ponements to the Eurofighter and Rafale.

Surprisingly, despite these repeated and very costly delays, neither aircraft has been designed or modified to have vectored engine nozzles. This technology was pioneered by Britain, initially with the Pegasus and RB.153 (page 170). It is difficult to comprehend how with this pioneering background British designers should, at least until 1997, have been content to leave it to others, notably the Russians and Americans, to put this important technology into practice. When he wrote this book the author chose not to 'rock the boat' by commenting on the fact that, throughout the first ten years of Eurofighter development any suggestion that the nozzles should be capable of vectoring was sure to stir up a hornet's nest. At last, one spokesman for Eurofighter, caught by the author in 1996 just after witnessing an amazing display by the Su-37, said "Maybe we'll get vectored nozzles at the mid-life update".

Eurojet's Spanish engine partner, ITP, is responsible for the nozzle, and in 1997, assisted by Daimler-Benz Aerospace, was preparing to test a non-flying demonstrator nozzle in 1998, with flight testing tentatively scheduled for 2000. It is difficult to understand why, with four partners involved on the aircraft and four on the engine, nobody seems to have been able to persuade the programme managers that plain fixed nozzles were no longer a good idea.

Even the Americans did outstanding work with vectored nozzles and then let the idea gather dust

Pratt & Whitney's F119-PW-100 engine for the F-22 Air Superiority Fighter, on test with the PYBBN nozzle.

Machining Pratt & Whitney's Integrally Bladed Rotor for the F119 derivative engine for the Joint Strike Fighter.

for several years. The pioneer vectored fighter engine in the USA was the Pratt & Whitney F100, though it achieved this merely by happening to be the powerplant of the F-15. In 1984 the USAF ordered an F-15 to be modified as the S/MTD aircraft (STOL/Maneuvering Technology Demonstrator), with various modifications including 2-dimensional rectangular nozzles able to point the jets 20° up or down. This was clever enough, with high-power engines fitted with afterburners, but the nozzles also incorporated reversers to slow the landing. The objective was not only improved combat manoeuvrability but also the ability to operate from a 1,500-ft runway, the (perhaps optimistic) belief being that in a war the F-15 squadrons would not be destroyed where they were parked and would find a 1,500-ft run between enemy bomb craters. The S/MTD began testing its nozzles in May 1989, and from March 1990 demonstrated not only vectoring but also in-flight thrust reversal.

The 1984 order for the S/MTD naturally attracted the attention of the Russians, and in fact Sukhoi and Lyul'ka managed to fly a somewhat similar nozzle first. When this book was written, nothing was publicly known about Russian vectored nozzles, apart from those for the Soyuz engines of the Yak-38 and Yak-141 naval V/STOLs. In fact, even at that time, Aleksandr Sarkisov's team at Klimov Corporation in Leningrad was working on a vectored nozzle for the RD-33 engine, while Dr Viktor Chepkin at Moscow-based Lyul'ka Saturn was scheming vectored nozzles for the AL-31F (the non-vectoring forms of both engines are mentioned on page 180).

As the engine of the Su-27, the AL-31F is larger and much more powerful than the RD-33, though it is similarly a two-shaft turbofan of low (0.6) BPR with a four-stage LP compressor and nine-stage HP. With the Su-27 in service, Chepkin uprated the engine to the AL-31FM, and then used this as the basis for the AL-35 with a vectored nozzle. The first nozzle to be test-flown was a clumsy two-dimensional (rectangular) type generally similar to that of the F-15S/MTD. Even more dramatic is the fact that on 31 March 1989 Sukhoi pilot Oleg Tsoi began testing the Su-27LL-PS with this nozzle fitted to the *left engine only!*

Like most things in the former Soviet Union, the development and application of vectoring fighter engines in Russia has been slowed down and made much more uncertain by crippling lack of money. In 1990 the Sukhoi bureau went ahead with a vectored Su-27M, using engines derived from the AL-31F, and the result is the prototype

Su-37, powered by twin AL-37FU engines, which has been on flight test since 2 April 1996. Each engine has a rating of 18,739 lb dry and 31,966 lb with maximum augmentation. As the normal takeoff weight is 57,320 lb, the performance and manoeuvrability of the Su-37 can fairly be described as fantastic.

The aircraft has a digital flight-control system, which is fully integrated with the FADEC controlling the engines. The nozzles are circular, fully variable in profile and area and with propulsive efficiency in all regimes equal to that of the best fixed nozzles. Each engine weighs 3,660 lb, compared with 3,351 lb for the Su-27's AL-31F and about 3,500 lb for an AL-37F with a fixed nozzle. It is intended by 1998 to reduce weight to about 3,580 lb by making the machined forging on which the nozzle is mounted of titanium alloy instead of refractory steel.

This ring reacts the forces from two pairs of hydraulic actuators, which in a production nozzle would probably be driven by fuel. These drive the nozzle in the vertical plane over limits of ±15°. In a single-engined aircraft the same actuators could point the nozzle to this angle in any direction, but in the Su-37 lateral movement is limited by the proximity of the tailplanes and long rear-fuselage extension. In any case, according to Dr Chepkin, "Differential operation on a twin-engined aircraft has the same effect as a 3-D multi-axis nozzle."

Anatoliy Andreyev, chief designer of the AL-37FU, said in September 1996 that the most difficult design problem had been sealing the joint between the jetpipe and nozzle. In maximum augmentation at sea level the internal gas pressure reaches 7 ata (103 lb/sq in) at a temperature of 2,000°C (3,632°F). According to a Sukhoi chief designer, Vladimir Konokhov, the entire thrust-vectoring system can be retrofitted to existing Su-27 fighters, though probably without digital control. The 40 Su-30MK fighters being delivered to India are expected to be retrofitted with vectored nozzles from 1999.

Dr Chepkin has also stated that "A package of measures has been developed to reduce infrared signature in the non-afterburning mode. These could be integrated into serial production engines at a customer's wish".

Today the 'penny has dropped' in the USA, and, because the Russian rate of progress has slowed dramatically, both GE and Pratt & Whitney are now ahead with vectored fighter engines. GE has developed a nozzle very similar to that of the AL-37FU but with higher performance. Called AVEN (axi-symmetric vectoring engine nozzle), it has internal and external flaps pivoted to the main mounting ring

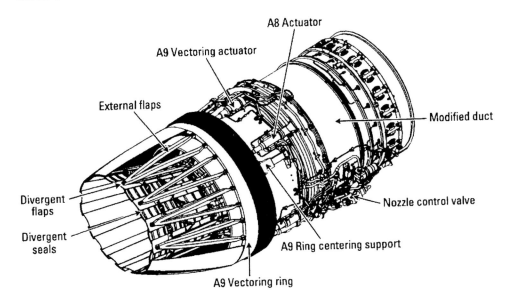

A8 Actuator

A9 Vectoring actuator

External flaps

Modified duct

Divergent
flaps

Divergent
seals

Nozzle control valve

A9 Ring centering support

A9 Vectoring ring

Almost every maker of engines for combat aircraft is urgently developing vectoring nozzles. This is GE's AVEN. The electronic control is not shown.

on which are pivoted the vectoring actuators. These move the nozzle up to 20° in any direction at a rate of 60° per second, or roughly double the pivoting rate of the AL-37FU nozzle. Maximum lateral force is 6,500 lb.

This nozzle was first tested on an F110-129 IPE in 1990. On 2 July 1993 a further refined AVEN nozzle began outstanding flight trials on an F-16. The actual nozzle was exhibited (attached to a YF120 JSF engine, described later) at the 1996 Farnborough airshow.

Rival Pratt & Whitney has released details of two generations of vectoring nozzle. First came the PYBBN, standing for pitch/yaw balanced-beam nozzle. This is basically a regular F100-229 nozzle with three additional moving parts, plus actuators. These vector the nozzle 20° in any direction at the remarkable rate of 120°/sec. Its 'balanced beam' design minimises actuation loads and reduces the need for heavy surrounding structure, though lateral force can reach 4,000 lb. In early 1996 flight testing this nozzle began on an F-15 with F100-229 engines.

This has now led to the SCFN (spherical convergent flap nozzle), which is coming into production for the F119 engines of the Lockheed Martin F-22 Raptor. While the Russians began with a 2-D rectangular nozzle and progressed to axi-symmetric (circular) designs, Pratt & Whitney has gone the other way. The SCFN is approximately square in section, and not only

provides control of area and profile for each flight regime but also ±20° pitch vectoring, ±20° yaw vectoring, inflight thrust reversal, and signature control for enhanced stealth. Almost the entire structure is made of organic-matrix and ceramic-matrix composites, with special coatings. The nozzle is driven by dual pneumatic actuators, and uses brush seals to minimise leakage.

Both the F119 and the rival General Electric F120 are competing to power the Joint Strike Fighter. This programme, which did not exist when the original book was written, is the largest combat-aircraft development effort in the world. It aims to provide a multirole tactical aircraft for the US armed forces (except the Army) and friendly nations. It is being developed in three forms, a CTOL (conventional runway takeoff and landing) version, a naval version for operation from carriers, and an ASTOVL (advanced short takeoff and vertical landing) version.

Originally the ASTOVL was regarded as a slightly undesirable oddball purely to satisfy the demands of the Marine Corps and Royal Navy. Increasingly it has been found that it has more to offer than originally recognised, whilst suffering smaller penalties in (for example) range and performance. Accordingly, a proportion of the Air Force (and possibly Navy) versions may be of this type. This is a monumental, and very welcome, change in attitude.

In early 1997 planned production for the launch

customers was envisaged as 2,036 for the Air Force (mainly replacing the F-16, but also the F-111F and F-15), 642 for the Navy (replacing the F-14 and A-6), 300 for the Marines (replacing the AV-8B and F/A-18) and 60 for the Royal Navy (replacing the Sea Harrier). The RAF has not so far taken part, studying an FOA (future offensive aircraft), but common sense suggests that it is likely through shortage of funds to buy a JSF version instead.

In 1996 McDonnell Douglas/Northrop Grumman/British Aerospace was eliminated, leaving two competing finalists, Boeing and Lockheed Martin. Both will now fly prototypes called the X-32 and X-35 concept demonstrator aircraft. These will be powered by versions of the Pratt & Whitney F119 as main engine.

In June 1997 British Aerospace decided to join with Lockheed Martin's bid to win the whole programme. This is an interesting choice, especially in view of Lockheed Martin's existing possession of the F-22 and derivatives and Boeing's acquisition of McDonnell Douglas and establishment of a new military airplanes division at St Louis. The British firm would have a 12 per cent share.

For the ASTOVL variant, Boeing will use the SE614, with a larger fan (BPR about 0.6 instead of 0.45) and a central 2-D SCFN nozzle, with an integral valve to divert the whole flow to two retractable lateral nozzles for jet lift. Lockheed Martin will use the SE611, with higher temperatures and an uprated LP turbine in order to transmit up to 25,000 shp to an Allison lift fan under the aircraft nose. The single main-engine nozzle will incorporate rotary segments provided by Rolls-Royce to vector thrust through 110°. Both finalists may also use a derivative of the GE F120, called the F120-FX, as an alternative, with or without a lift fan. This was officially accepted as part of the programme in June 1996, to be developed in partnership with Allison Advanced Development Co. and Rolls-Royce.

So far all the emphasis on the vectoring of military engines has been in the vertical plane. As explained above, some engines have multi-axis vectoring, yet are limited to the vertical plane by the proximity of parts of the fixed airframe alongside. This is almost certain to change, as 'stealth' technology forces designers to eliminate vertical tails. Even when these are mounted at a sloping angle, or above the wing (as in the F-22) they are exceedingly undesirable to the designer of any aircraft likely to enter the search area of a hostile radar. The losing JSF contender had no vertical or quasi-vertical tail, and the ongoing Boeing and Lockheed Martin designs are quite likely to lose theirs in the fullness of time. Of course, this also saves weight and drag.

The purpose of a vertical tail is to generate lateral thrust, so it can be replaced by a multi-axis nozzle. Such a nozzle must be driven by actuators with sufficiently high bandwidth to respond not only to pilot inputs but also to sideslip and other stability-augmentation signals from the flight-control system. Lateral thrust to control the aircraft under all possible flight conditions must always be available, even (for a STOVL aircraft) in the hover. If it is impossible to provide sufficient control power at idle thrust, an alternative would be to retain cruise power and vector the engines in opposition, left and right, thereafter modulating the angles to control the aircraft.

This brief overview should be enough to show that aircraft gas turbines and jet propulsion is becoming an ever more diverse subject, and ever more exciting. What the author does not understand is how, with computer power doing the work of vast armies of designers in split seconds, producing even a quite simple new engine takes as many years as in 1944 it took weeks!

Jet and Turbine Aero Engines –
A third update

Throughout its first 100 years the technology of aerospace has moved ahead every single day. Thus, an update written in 1997 becomes sadly out of date by 2002. Of course, the advances are across the board, and indeed it is difficult to think of a single class of engine where there is not much to say about the past four years. The following round-up of news is arranged not by manufacturer but by subject, very loosely in the way the author views the importance of the changes.

Starting with the biggest, the truly remarkable development is that, having for 35 years ruled the world from a monopoly position with the 747, Boeing has seemingly let Airbus

take over. I will return to this presently. For the moment, the interesting and totally unexpected development is what Boeing might do instead.

For more than ten years Boeing had been studying the 747–500/–600, 747–400X and 747–400X Stretch. For 'long thin routes' (thin meaning not much traffic) the company had been developing the 767–400ERX. These aircraft were planned to provide severe competition for the A340–500/–600 and projected A3XX. When the latter went ahead as the A380 in late 2000, Boeing redoubled its efforts to find customers to launch the bigger ultra-long-range 747s. On 29 March 2001 Boeing announced that it was abandoning plans to build a stretched 747,

This is one of the first artworks to be published showing an idea for a Boeing Sonic Cruiser. One thing we can be sure of is that, if a Mach 0.98 aircraft is built, it will not look exactly like this.

having failed to secure a single order, and that it was also cutting back future development of the 767.

This was as astonishing an announcement as it would be if General Motors said it was giving up the manufacture of cars. Yet it was almost ignored by the world's media. This was because Boeing cunningly issued at the same time an image of a possible future passenger aircraft called the Sonic Cruiser. This image appeared in media in virtually every country in the world. The exciting new shape totally overshadowed the real news.

According to Boeing, they had not intended to reveal it, 'but our hand was forced when word leaked out following briefings to airlines'. In my experience, Boeing plan major announcements with care. There is no doubt in my mind that they saw that to take what had been known as the 20XX and name it 'the Sonic Cruiser' was a marvellous way to distract the media from the real news.

At the same time, the dramatic new shape is not pure fiction but a serious study for a future civil transport. Many of the thousands of publications that gave prominence to the Boeing artwork thought the idea was for an SST, especially as its dramatic shape appeared to have more in common with Concorde than with a 747. In fact the Sonic Cruiser is an idea for an aircraft to cruise at Mach 0.95, or about 627 mph. This is only slightly faster than the venerable VC10, which went into airline service in 1963 cleared to 0.94. Perhaps recognising this, by the time of the Paris airshow in June 2001 Boeing were saying '0.98'. On the other hand, Mach 0.98 is very different from the cruising speed of today's jet transports, which are typically around Mach 0.84, or 554 mph. This is because, in this region, the plot of drag coefficient has a very steep slope: a little more speed means a lot more drag. Indeed 0.98 is close to 'the optimum worst', though the peak can be pushed further into the supersonic regime by acute sweepback.

All this has a profound influence on the propulsion system. The most obvious fact is that, for aircraft of any given size, to cruise at Mach 0.98 needs a lot more thrust than at Mach 0.84. To a first order of magnitude, if an aircraft cruises at 0.84 with two engines each giving 18,000 lb thrust at that speed at 35,000 ft, it would need engines putting out 25,000 lb thrust to cruise at the same height at 0.98. In fact, the Sonic Cruiser would cruise higher, at 43,000–48,000 ft, but that does not alter the comparison. At higher altitude the engine thrust falls off in the same ratio as aircraft drag.

Thus, the Sonic Cruiser will need extremely powerful engines. Today's engines giving cruise thrust of 18,000 lb at 35,000 ft give 90,000 lb at takeoff. If the Sonic Cruiser had engines of similar bypass ratio they would give 125,000 lb on takeoff, but we are comparing apples and oranges. Today's engines are enormous, with fans of 10 ft diameter and a bypass ratio of around 8. A Mach 0.98 aircraft would need a much lower bypass ratio, possibly in the region of 3, and so the huge single-stage fan might be replaced by a (for example) three-stage compressor with a diameter of about 5 ft.

Such an engine would be rather like a greatly enlarged version of the engines of modern fighters, though of course without an afterburner. It would look smaller than the monster fans of today's wide-bodies, but in fact it would have a bigger core, be potentially more powerful and *burn fuel faster*. This means that for a Sonic Cruiser to fly 9,000 nautical miles (10,364 miles), which is what Boeing are aiming at, it will have to have room for even more fuel than for an aircraft cruising at Mach 0.84.

It is also a law of nature that to cruise at Mach 0.98 demands engines with a higher jet velocity than one tailored to propulsion at 0.84. At 0.98 the propulsion system is verging on the point at which the propulsive jets must be energized even further by afterburning. If your engines are extremely powerful, as they are in modern fighters, you can supercruise in dry thrust, which the latest fighters can demonstrate. Passenger airliners have a totally different thrust/weight ratio, and to cruise in dry thrust at 0.98 is a very big challenge.

A side issue is that, to minimize transonic drag, the 0.98 aircraft needs to conform to the Area Rule. Broadly, this says that the plot of total cross-sectional area from nose to tail should be a smooth curve, with no sharp bumps. Thus, to compensate for the wing, you should waist (narrow) the fuselage. This is almost a non-starter for passenger cabins, hence the funny shape of the Sonic Cruiser in which the fuselage tapers away along the huge wing glove (the sharply tapered inboard section of wing). As the wing tapers off in cross-section, the engines emerge to make up the difference.

In my opinion, underwing inlets with ducts curving up to engines installed on long mountings stretching back behind the rear spar are not a happy solution. I would not be surprised to see a waisted fuselage behind the

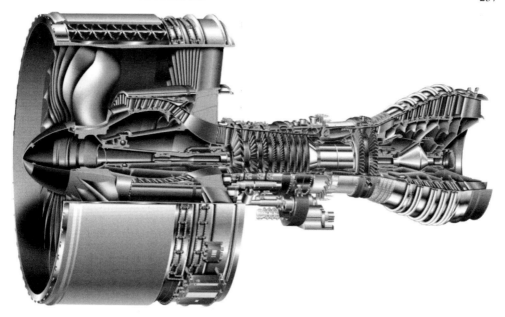

In contrast, we can be sure that, even though it will not run until April 2004, the Engine Alliance GP7200 will look exactly like this. Note the isogrid structure to contain the curvaceous fan blades, and the swept-back booster blades.

rear pressure bulkhead at the rear of the passenger cabin. This could then stick out behind the wing and carry the engines and a single fin. As it is, the heavy engines are trying to twist the wing torsion box and have their inlets nicely placed to collect items thrown up by the landing gears.

Not least, jet noise varies powerfully with the shearing at the boundary between the jet and the surrounding atmosphere, and the higher the jet velocity the greater the noise. The actual noise from a real engine is a complex mix of noises from different sources, but to a rough approximation an engine for a Sonic Cruiser would emit ten times the jet-shearing noise of a turbofan of 8 bypass ratio. The author was astonished to read Sir Richard Branson's comment that such an aircraft 'will be quieter'! He had been misled by Boeing's probably correct belief that a Sonic Cruiser could be quieter *when landing*.

Turning to the closer future, as this was written in mid–2001 the Engine Alliance partners (General Electric and Pratt & Whitney) had still not begun testing a complete GP7000 engine (p. 225), but they had made major changes to its design. Termination of the 747X and 767–400 derivatives means that, at present, the A380 is the only application, the GP7100 family being shelved. The GP7200 now has a complete core section, comprising HP compressor, combustion chamber and HP turbine, scaled directly from that of the enormous GE90. In mid-2000 the Build 5 core test validated several new features, such as the 'swept' first-stage aerofoils and the Stage-4 vane (stator) ring with the blades fixed, instead of being mounted in pivots and linked to the three stators rows in front. At least as important is elimination of the final compressor stage, whilst at the same time holding OPR (overall engine pressure ratio) at 46 by adding an LP stage.

Discarding the original core in favour of a scaled GE90 eliminated the dual-annular combustor, which had been a particular feature of the original GP7000 design. A further fundamental change is introduction of a completely new fan, designed by Pratt & Whitney and driven by a six-stage LP turbine closely related to that of the more powerful members of the PW4000 family (which have either five or seven stages). The new fan has swept wide-chord blades, and a diameter increased from the original 110 in to 116 in. This, combined with the addition of a fifth stage on the LP compressor, explains the 46:1 pressure ratio, and increases bypass ratio to 9 at takeoff and 8.7 in cruise, the highest of any turbofan so far. In early 2001 Pratt & Whitney

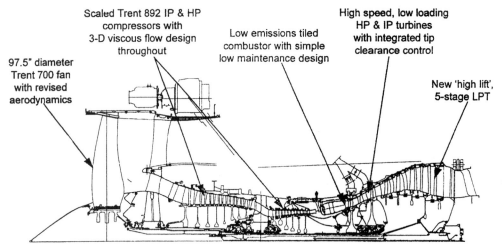

Scaled Trent 892 IP & HP compressors with 3-D viscous flow design throughout

Low emissions tiled combustor with simple low maintenance design

High speed, low loading HP & IP turbines with integrated tip clearance control

97.5" diameter Trent 700 fan with revised aerodynamics

New 'high lift', 5-stage LPT

The Trent 500 is in high-rate production to power the Airbus A340–500/ –600. The debate over two-shaft versus three-shaft seems to get hotter, not colder! Both sides are utterly convincing, but some are more convincing than others.

Canada at Longeuil (across the St Lawrence from Montreal) tested a 42 per cent scale of the fan.

Thus, the whole history of the GP7200 has been one of growth. This reflects the aim of the Alliance to produce a baseline engine which will not need significant further upgrading. Thus, while the original engine was offered at

67,000 lb and 75,000 lb thrust, the GP7200 will be offered at up to 81,500 lb to be ready for future heavier A380s.

Having watched all the early A380 customers pick the competitor engine (Trent 900), the Alliance partners were relieved when on 29 May 2001 Air France ordered ten aircraft with the GP7200. Including spare engines

As if the original GE90 was not big enough, the GE90–115B is easily the biggest aero engine of all time. On take-off it will swallow over 1.8 tons of air each second.

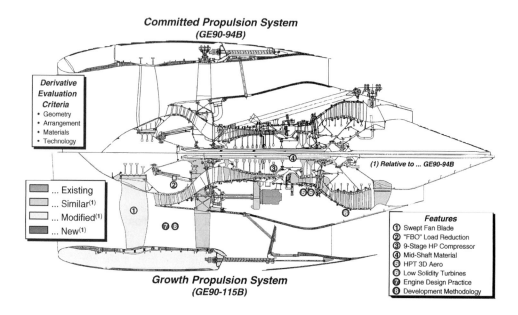

Committed Propulsion System
(GE90-94B)

Derivative Evaluation Criteria
• Geometry
• Arrangement
• Materials
• Technology

(1) Relative to ... GE90-94B

■ ... Existing
□ ... Similar[1]
□ ... Modified[1]
■ ... New[1]

Growth Propulsion System
(GE90-115B)

Features
① Swept Fan Blade
② "FBO" Load Reduction
③ 9-Stage HP Compressor
④ Mid-Shaft Material
⑤ HPT 3D Aero
⑥ Low Solidity Turbines
⑦ Engine Design Practice
⑧ Development Methodology

(probably two), the order was priced at US$900 million. The first Full Engine Test is due in 2005, certification is due in mid-2005, the first GP-engined A380 is to fly in early 2006, with EIS (entry into service) due late that year. Snecma of France and MTU of Germany are partners in the programme.

The rival Rolls-Royce Trent 900 has a fan equally advanced aerodynamically, but with the 110-in diameter of the original GP7200. Other figures so far published are also below those of the rival US engine: bypass ratio 7.8 and OPR 41. At the same time, the author would be surprised if the British engine did not come out significantly lighter. In 2000 published figures were 12,900 lb for the Trent 900 and 12,906 lb for the GP7200, but that was before the Alliance enlarged their fan and added a fifth stage to the LPC.

Meanwhile, the Trent 500, for the impressive A340 growth versions, has had a predictably troublefree development. In a nutshell, this matches a fan of Trent 700 size (97.5 in), with advanced 3–D blades, with a core similar to an upgraded Trent 800 on a 0.8 scale. A single flight-test engine hung under the Airbus A340–300 testbed completed a 60-hour flight programme in the third quarter of 2000. On offer as the Trent 553 (the last two digits indicate thrust, as in the case of the GP7200) for the ultra-long-range A340-500 and as the Trent 556 for the higher-capacity A340-600, this engine was certificated two weeks ahead of schedule on 15 December 2000 at 60,000 lb, allowing a large margin for future growth. At that time 124 of the bigger A340s had been sold, representing almost US$6 billion in engine business. Flight testing the new aircraft began with the A340-600 on 23 April 2001. Service entry of both the stretched A340s is due in 2002.

At the very top end of the engine size scale, the GE90 has grown even bigger, to meet the demands of the heavyweight 777–200X and –300X. That GE were going for this market with an uprated GE90 was known at the time of the last update. What came as a shock to a lot of people is that the Cincinnati company sewed up an exclusivity deal with Boeing which ensures that every future heavyweight 777 will be powered by an engine called GE90-115B. Rolls-Royce and Pratt & Whitney are excluded. The future GE90 is no more powerful than the Trent 8115 was in April 1977, but it is very much bigger and heavier. In size and weight it is more reminiscent of propulsion machinery of ships!

Almost all the 115B core is new, and it is an enlarged version of that being developed for the GP7200. But most of the thrust is generated by the enormous fan, with a diameter of 128.2 in and wide so-called 'swept' blades. Airflow at take-off is increased from the 3,269 lb/second of the GE90-94B to 3,617 lb (almost two US tons)/second. A photograph compares the fan blade with that of previous GE90s. The latter turned the scales at a basic 16,664 lb, but the 115B is likely to weigh at least 18,000 lb, as much as a Second World War medium bomber (current 115B brochures give no weight data).

After a shaky start, which saw British Airways switch from the GE90 in their first fleet of 777s to the Trent 800 in the next batch (and for a customer to change from one engine to another is exceedingly rare), the huge GE engine now appears to enjoy a monopoly position on future 777s, except for customers who buy the original smaller models. GE90 flight time passed the 2,000,000-hour mark in early 2001.

If it is rare for a customer to change engines after putting a fleet into service, it is at least as uncommon for an engine company to announce a new engine and then decide not to build it. On p. 228 appears a section drawing of the SPW14, an apparently attractive turbofan for regional aircraft and large bizjets. In January 1998 the Joint Venture partners, Pratt & Whitney Canada and Snecma, decided not to go ahead with this engine. As it was a completely new design, it ought in theory to have been superior to any older-technology engine or derivative, but what the market likes is to have one winner and then buy that by the thousand. Customers, especially in the civil market, are far happier with old engines that are in use all over the world than with supposedly superior new ones.

Thus, no matter how good it might have been, the SPW14 marketing team would have had an uphill struggle. In contrast, when GE decided to derive a civil engine from the TF34 they could point to more than a million hours in harsh conditions powering A–10 tankbusters and S–3 Vikings being catapulted off carriers. Instead of saying 'But that engine was designed in 1965', the market said 'Great, just what we need!' The result is that this basically ancient turbofan in the 8,600 lb thrust class has gone from strength to strength, got newer and newer, bigger and bigger, and is now on offer at thrusts approaching 20,000 lb.

GE simply can't make this basically very ordinary engine fast enough. The original commercial version, the CF34-3, was picked by

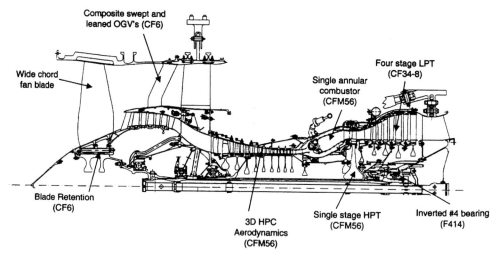

Features of the CF34–10, the most powerful model so far in the best-selling CF34 family. Annotation shows the derivation of some of its technology (mostly from much larger engines).

Bombardier for an upgraded version of the Challenger bizjet. Since then the Challenger has proliferated in different versions, and the CF34-3 has already sold 1,600 firm and 1,300 option engines to 37 customers. As I write, Air Wisconsin have just ordered 75 firm plus 75 option CRJ200s, meaning about another US$600 million in CF34-3 business. As for the much more powerful CF34-8 and CF34-10, these will power whole new families of Bombardiers, Embraers and Dorniers. I am prepared to bet that by 2020 GE will have delivered 12,000 CF34 engines, yet I recall that in the 1960s GE told me 'It's only a small program – the Navy only want 400 at the most'.

In the course of time I would not be surprised to see CF34s hung on Russian airframes. Having all their lives had a captive market in the Soviet Union's mighty aircraft industry, Russian and Ukrainian engine firms are finding life hard. Not only is there now a harsh economic climate, in which each new aircraft programme has to fight for existence, but some of the few aircraft actually in production are being sold with imported Western engines. Among these are the Rolls-Royce 535 and BR710, and a wide range of engines from Pratt & Whitney Canada.

This is tough, because Russian and Ukrainian engine designers are as good as those from anywhere else. Indeed, in some areas, such as advanced fighter engines with vectoring nozzles, they are probably better. How many Western engines have been qualified over a range of inlet conditions which include angles of attack exceeding 90° and airspeeds of *minus* 200 km/h (124 mph), i.e. going backwards?

I am increasingly astonished at attitudes in military procurement. To my simplistic mind it ought not to be difficult, from time to time, to decide what is needed, examine the available technology and fairly quickly place a contract, if necessary for an engine as well as for the new aircraft. We in Europe no longer appear to be able to do this.

The outstanding – or outrageous – example is the need for a new airlifter to replace the C-130 and Transall. This need was self-evident in the mid–1970s, and in 1979 the UK, France and Germany began a formal study of the numerical values for such things as payload/range and the cross-section dimensions of the cargo hold. This effort crystallized in 1982 into what some thought a firm project, called FIMA. Instead of simply getting on with it, the various governments and air forces have so far managed to waste 22 years, and have now reached the point where nine nations have said they would like to buy a combined total of 288, but are determined to keep on delaying a go-ahead.

For example, ten years ago I was assured the engines would be turbofans, but seven years ago it was announced they would be turboprops. The only suitable engine is a product of Ukraine, which apparently is politically unacceptable, so Snecma produced a design with other partners, called M138, while Rolls-Royce Deutschland GmbH in Germany

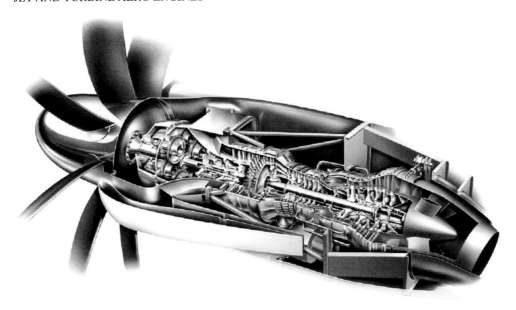

Artwork showing an installed APA TP400. The core is derived from that of the Snecma M88 fighter engine. Messier-Dowty or Ratier-Figeac may collaborate on the 17ft eight-blade propeller, transmitting up to 13,000 hp.

produced a rival, the BR700-TP. Instead of picking one, it was decided that the two proposals could be merged into something called the TP400. At last, in mid-2001 a multinational company, Aero Propulsion Alliance, had been registered in Germany, and precise workshares agreed. Final go-ahead is due in September 2001, but Germany is still dithering, so the latest total for the launch partners is no longer 288 but 'about 200'.

The airlifter itself is now called the Airbus A400M, and this may get into service later in this century. Meanwhile, air forces are buying the C-130K Hercules, and the RAF has leased four Boeing C-17 Globemaster IIIs (it can't afford to buy any), without replacing the Pratt & Whitney F117s with Rolls-Royce 535s.

Similar failure to take decisions is afflicting the UK nationally. For many years the Mid-Life Update of Tornado has tottered from one crisis to another, until the resulting GR.4 has assumed the stature of a national scandal. We are now well on the way to doing even better in creating scandals with the intended next generation, the Future Offensive Air System. This is supposed, or assumed (in absence of knowledge), to be likely to comprise manned aircraft, unmanned combat aircraft and cruise missiles, but persistent refusal to take decisions is crippling

industry's wish to try to write a business case. This book could go through further editions before I shall be able to discuss the RAF's future engines.

In 2001 the only thing I can report is that Rolls-Royce and BAE Systems have expressed concern at their inability to plan ahead on future military engines, and have called upon the Ministry of Defence to boost R&D (research and development) funding for the kind of engines likely to be needed. As things are in 2001, the UK is steadily proceeding into a situation where it will no longer have the capability of providing any future military engines, let alone its own military aircraft.

The only exception to this could be the Joint Strike Fighter in which the UK has a share. The background was covered on p. 234. By 2001 all the prototypes had flown, and a choice between Lockheed Martin and Boeing was due in September 2001. Here I will merely offer a table of the different versions and images of the two contrasting propulsion systems for the ASTOVL versions, the JSF119 SE611S for the Lockheed Martin X-35B and the SE614S for the Boeing X-32B.

In the prevailing situation, in which European governments appear to be incapable of taking any decision, let alone a quick

Pratt & Whitney JSF119 versions

For Boeing X-32:

> X-32A, CTOL version for UAF/USN, configuration 374A, SE614C engine, single vectored nozzle, about 41,000 lb thrust.
> X-32B, ASTOVL version for USMC/RN/RAF, configuration 374B, SE614S engine, main vectored nozzle about 18,000 lb, subsidiary nozzles about 15,000 lb, reaction-control jets 4,000 lb.

For Lockheed Martin X-35:

> X-35A, CTOL version for USAF, configuration 230A, and X-35C for USN, configuration 230C, both SE611C engine, vectored nozzle about 41,000 lb thrust.
> X-35B for ASTOVL version for USMC/RN/RAF, configuration 230B, vectored nozzle about 15,000 lb thrust, shaft-drive fan about 18,000 lb, reaction-control jets 4,000 lb.

decision, engine manufacturers are trying to guess what might eventually be needed. This is doubly essential in view of the fact that engines historically have always taken longer to develop than aircraft. Moreover, future military engines are likely to be all-new, in the sense that they cannot be derivatives of those in use today.

Not knowing how else to move forward, the two principal European *motoristes*, Rolls-Royce and Snecma, decided at the end of 2000 to form a 50/50 joint-venture company. Almost 40 years earlier the previous Rolls-Royce company had begun to collaborate with Snecma to provide the engine of Concorde (p. 190), but at that time there was no attempt to form a joint company. Now Rolls-Royce Snecma Ltd exists, formed in February 2001 with an office in London under the management of Guillaume Giscard d'Estaing. He was previously Sales Director of Turbomeca, which was purchased by Snecma in 2000. The purchase of Turbomeca automatically linked Rolls and Snecma in such engines as the Adour and RTM322.

The immediate task facing the joint venture is to manage a collaborative effort called AMET (Advanced Military Engine[s] Technology, some documents making Engine plural). Broadly, this is to decide what kind of fighter engine will follow the EJ200 and M88. Fortunately for the two air forces and navies, the JSF programme, mentioned earlier, was in 2001 continuing on course under the Bush

Computer graphics showing the Pratt & Whitney (plus partners) propulsion systems for ASTOVL versions of JSF. Upper, SE611S for the Lockheed Martin X-35B. Lower, SE614S, for the Boeing X-32B.

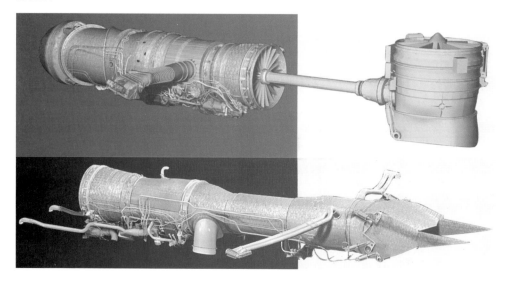

The 'Alternate' (i.e., alternative) engine for CTOL versions of JSF, the General Electric JSF F120 with vectoring nozzle.

Presidency. The JSF appears likely to take the pressure off the need eventually to replace Eurofighter and Rafale.

However, it is only prudent common sense to study what technologies and materials will become available in the next 20 years, and the author is profoundly relieved to note that such a research effort is now being 'funded by the French and UK defence ministries' (though one may be certain that the funding is not generous, and is bitterly opposed by the British Treasury). Even in the absence of any firm idea of what kind of Future Offensive Air System the British MoD and RAF have in mind, a great deal can be done to advance the underlying technologies.

To a considerable degree, this work will benefit from the keen competition to produce better civil engines. I have commented earlier in this book that the technology of aero engines has been driven by different motivating factors at different times. Until 1931 the spur had been air racing, notably for the Schneider Trophy. Throughout the next 25 years the demand had been overwhelmingly military.

Even after the Second World War, and the replacement of piston engines by turbojets, the driving force continued to be military. Thus, in the 1950s Aeroflot put the Tu-104 into service powered by engines developed for the Tu-16 bomber, BOAC put the Comet 4 into service powered by engines based on those developed for the Valiant and Lightning, and PanAm put the 707 into service powered by engines based on those of the B-52 and many other USAF and USN aircraft.

Then it all changed. GE developed airline engines derived from the excellent military J79,

and found few customers. Rolls, having developed the Conway for the Victor B.2 bomber, struggled to sell any to the airlines as supposed superior alternatives to the JT3D and JT4A. In contrast, the all-new Rolls-Royce Spey and Pratt & Whitney JT8D sold in large numbers for many types of aircraft (and, in the reverse of what used to happen, were later fitted with afterburners and developed as engines for supersonic fighters).

In the late 1960s the Lockheed C-5 Galaxy and Boeing 747 wide-body transports gave rise to a totally new species of engine, much larger than anything seen previously. So far these monster turbofan engines have been hung exclusively on large transport aircraft. This is despite the fact that 25 years ago Rolls-Royce made a convincing case for removing the eight J57 or TF33 engines from each remaining B-52 and fitting instead four Rolls-Royce 535E4 turbofans. The transformation would be astonishing – something like 40 per cent greater radius of action, better availability and a $2 billion saving in cost – but the USAF was not interested. A USAF general said to the author in 2000 'If Rolls had bought Allison a bit sooner I think we'd have gone along with it, because the gains were enormous'. The problem was that the USAF will not 'buy foreign', whereas the RAF makes a point of it.

We appear now to have reached a stage where what might be termed 'combat aircraft' never use the same engines as transport aircraft. On the other hand, there are still a few examples of engines forming a link between the two groups. The outstanding example is the CFM56 (p. 188), which was created by matching a new Snecma fan (from a stillborn engine called the M56) with the HP compressor of the F101, the GE engine of the B-1 bomber.

By 2001 this had become the best-selling civil engine of all time. The very last Pratt & Whitney JT8D-200 came off the assembly line at Hartford in 1999, bringing the total of that version to 2,856, and the combined total of all JT8D versions to 14,701. To beat this, orders and options for the CFM56 were then already at that level, and by mid- 2001 the total had gone past 17,000 and was heading for '20 grand'. At a unit price of about $6 million, this shows how important aero engines can be in a nation's economy.

With an income like this, the partners have been able to keep on improving the different versions. One major update, introduced on the CFM56-5B2 (one of the high-thrust versions, for the A321), is a double annular combustor (DAC). As this is part of GE's area of responsibility, it might have been expected that this would just be a scaled version of the DAC of the GE90 (p. 38). In fact, the combustion chambers of the GE90 and CFM56-5B2 are not only very different in size but they have little similarity in design. Moreover, while the GE90 combustor was designed primarily to reduce unburned hydrocarbons (HC) and carbon monoxide (CO), the DAC of the later CFM56 versions was designed principally to cut down oxides of nitrogen (in shorthand, written as NO_x). So far (2001) the DAC has remained a customer option, on the later Dash-5 and Dash-7 engines.

By no means all customers for these engines have thought the DAC worth having. It is a matter of equating a small extra capital cost, and possibly a miniscule extra recurring cost, with environmental friendliness, or 'greenness'.

DAC (double annular combustor) of some CFM56 versions. It is quite different from the much larger DAC of the GE90 (page 38).

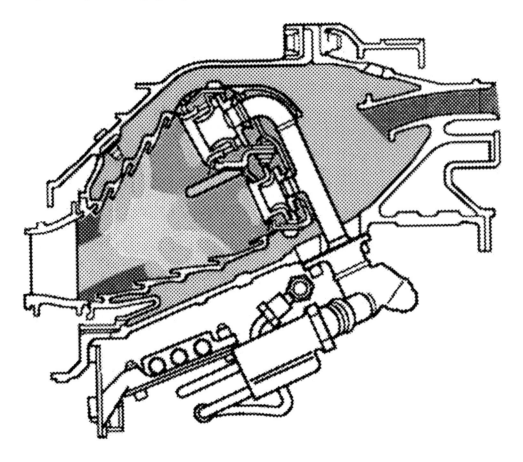

Like all humans, airlines certainly like greenness in principle, but not if it comes with even a small extra cost. I have never seen an airline saying 'Fly with us, we are greener'. On the other hand, CFM have made a big feature of it in their advertising. This is interesting, because the best way to cut down harmful emissions is to burn less fuel, and here CFM's competitor has always had the edge.

There are 'lies, damn lies and statistics', but numbers are the only way comparisons can be made. On the basis of published figures, the cruise specific fuel consumption at Mach 0.8 at 35,000 ft for the CFM56-5B is 0.600, and for the CFM56-7 (the new engine for the 737) 0.603, while the figure for the IAE V2500 is 0.575. Moreover, the CFM figure is for the *uninstalled* engine, while that for the competitor is for the engine installed in its nacelle, with all the losses due to the inlet, reverser and other factors taken into account. CFM admit this, former President Gérard Laviec saying 'We can never be as efficient, because the V2500 has a two-stage HP turbine. . . .'

So why does the CFM56 sell like the proverbial hot cakes? One answer is that, if you want to buy a 737, it comes with CFM56 engines – no argument. This is fine by such people as Southwest Airways, who will soon have 400 aircraft, every one a 737. But if you want any of the single-aisle Airbus family you have a choice, and here CFM have to fight hard, saying 'OK, the competition burn less fuel, but we really have the better engine.'

The Editor would not dream of commenting on this, except to note that, with so much money in the bank, CFM are now in a position to keep on updating their complete engine family. Unlike IAE, whose V2500 engines are all basically the same (the customer just adjusts the fuel control to get the thrust he wants, from 22,000 lb to 33,000 lb), the CFM56 comes in a range of versions with major differences. For example, the CFM56-2, used in Sentries, KC-135s and re-engined DC-8s, has a fan of 68.3-in diameter with 44 slender blades. The Dash-5C, used in A340s, has a fan diameter of 72.3 in, while the latest 737 engine, the Dash-7, has a more modern 61-in fan with 24 wide-chord blades (but still made in solid titanium).

Thus, to produce a better CFM56, where does one start? The most externally obvious change is a new fan, with so-called 'swept-forward' blades made of hollow titanium alloy. Though this is a Snecma creation, GE has the best test facility at Peebles, Ohio, and in spring 2001 this was testing the new fan with a diameter of 1,735 mm (so-called 68 in). Perhaps even more significant, GE are at last moving on from the old F101 HPC (high-pressure compressor). By mid-2001 they were getting excellent results with a totally new HPC with forward-swept blisk rotors (blades and disks in one piece) with only three variable stators instead of four, and only 968 blades instead of 1,518, yet operating at a pressure ratio 29 per cent higher! Not least, by 2001 outstanding results were being achieved with a contra-rotating HPT/LPT (high- and low-pressure turbines) rig, with a dramatic reduction in numbers of blades, yet higher performance, and with a 22 per cent reduction in cooling airflow (airflow which is largely lost to thrust).

Whenever a product is improved, one can never know how well it would have gone on selling without change. Certainly, the ancient features of the CFM56 were no handicap in the market place. Many airlines, told about the major changes in the offing, will feel 'I wish they'd leave it alone'. However, competition is a great driver towards better engines. With the CFM56 we have the chance to study how an engine first run in June 1974, incorporating a core which was already mature at that time, can be progressively transformed into an engine that will still look modern 50 years after the first run.

To be even-handed, I must report that, from the outset, the rival IAE V2500 beat the CFM56 on almost every count. Certainly it is more fuel-efficient, quieter and greener. By 2001 it had everything going for it except one: it is still excluded from the 737.

One could set up an interesting public debate 'That this house considers repeated updating of an engine over many decades to be a mistake'. I cannot help musing over such imponderables as what will eventually happen, probably many years hence, when an updated GE90 is finally seen as basically too old to continue as the exclusive engine of the 777 – always assuming that the 777 itself has not been swept away by Sonic Cruisers! Will GE's exclusivity extend into a totally new-generation GE90 successor?

Going to the opposite end of the size scale, it is now more than 42 years since I visited Indianapolis and saw Allison starting to test the first T63 helicopter turboshaft. It was to give 250 hp, so the civil version was called the Allison 250. Today, with almost 30,000 delivered, the Rolls-Royce 250 is still in production. Of course, today's 250 is a very different beast from the T63 prototype, and it can give more than three times the power, but can it go on for ever?

Over the years I have been assured by the engineers at Indianapolis that they were near 'the end of the road' with the 250. In the mid-1980s they became fired up over the prospect of collaborating with colleagues who were working on the Allison World Engine. This was to be a truly low-cost gas turbine for general industrial and even domestic use. In particular, a market had been identified to take care of power cuts, which in California have become a serious problem to hospitals, laboratories and production plants.

I recalled that 40 years ago, when I was Technology Editor of *Science Journal,* I commissioned a Ford Motor engineer to write about future engines. He predicted that they would be almost 'dirt cheap' gas turbines, 'because the vital blading will be silicon dioxide, which is found on the beaches of the world'. This is the thinking behind the World Engine, intended to be the first gas turbine to be made in batches of 10,000 at a time. Unfortunately, the underlying technology was found to be incompatible with aerospace, where more severe levels of certification are demanded. Whether or not these qualification criteria might be eased is another subject for impassioned debate.

Be that as it may, the Indianapolis engineers, who are now part of Rolls-Royce, have had to go back to square one and decide where to go with the previously evergreen Model 250. Tommy Thomason, V-P of Customer Operations, does not beat about the bush. In 1999 he said 'The perfection of the 250 is nigh, and a replacement is required'. Such matters are market-driven. When customers, in this case mainly the builders and operators of helicopters, are no longer eager to buy, and start eyeing competitors, it is time either to upgrade the product (again) or start designing a replacement. After so much brilliantly successful development, Rolls-Royce Corporation (the new name at Indianapolis) is now at last beavering away on a completely new engine, this time starting at about 500 hp.

While on the subject of helicopters, by 2001 large numbers of Rolls-Royce Turbomeca RTM322 engines were coming off the two assembly lines for the British Apache, the Anglo-Italian EH101 and Heliliner, and the NH90 for France, Germany and the Netherlands. These helicopters are certain to find many customers around the world. Their excellent engine is available at powers up to

3,122 shp, and could admirably fit the requirement issued by the US Army for a CEP (Common Engine Program) to replace the T700 in the Black Hawk/Seahawk family and Apache. However, the US Army does not buy foreign, so it has framed the requirements to exclude the RTM322, for example by demanding a 60 per cent improvement in operation and support costs, and announcing that the European engine cannot achieve this. There is no way the Rolls-Royce Turbomeca team can prove that they can. Instead, the US Army is having a completely new CEP engine developed jointly by General Electric and Pratt & Whitney. The work split is likely to be 60:40 in GE's favour. Imagine the outcry there would be if Europeans framed their requirements to keep out US companies.

By 2001, despite many severe difficulties, the MVZ (Moscow helicopter factory) named for M.L.Mil had managed to deliver nearly 300 Mi-26 helicopters to customers in 'about 20' countries. With a 105-ft eight-blade main rotor, driven by two 11,400-shp D-136 engines, the Mi-26 is by an enormous margin the most capable helicopter in the world. To the author, it is almost beyond belief that no Western manufacturer has thought it worth while making a helicopter more capable than the Chinook and Sea Stallion, both of which go back in design 40 years. Compared with these, the Mi-26 is more than twice as capable, and I am astonished that it does not appear to have been noticed by many Western customers. Be that as it may, the Mil designers are now looking further ahead to helicopters able to lift 100 tons.

The stumbling block is the main drive gearbox. Even the relatively modest Mi-6, with two D-25V engines of 5,500 shp, needed a gearbox with a drive ratio of 69:1, measuring more than 5 ft wide by over 6 ft long and almost 10 ft high. Called the R-7, this gearbox weighed (ignoring oil) 7,054 lb, or heavier than the two D-25V engines combined! The gearbox of the Mi-26 is half as heavy again, and the weight and design problems of a gearbox to transmit 50,000 shp would be truly awesome.

There is thus enormous pressure to avoid shaft drive altogether. There are many ways of driving a helicopter rotor by providing thrust at the tips of the blades, or even along their length. Small machines have flown with tip-mounted ramjets or pulsejets, but these are deafeningly noisy, cannot start from rest and (this is the killer) generally offer very poor specific fuel consumption.

Several companies, notably Fairey Aviation

in the UK, developed pressure-jet systems in which tip-mounted combustion chambers are fed with compressed air ducted through each blade. With a fraction more foresight and common sense, the Fairey Rotodyne could have been a great success, escaping from the need for a giant drive gearbox. In the typical British bungling manner, just as it was solving its last problems, which centred around noise, the factory was closed down and the Rotodyne handed to Westland. That company's management suffered from NIH (not invented here), and smartly terminated the entire effort. An 80-seat Rotodyne cruising at 250 mph would be quite useful even today; 40 years later, the best we have is 30 passengers at 161 mph.

In fact, Westland themselves had enthusiastically publicised tip-drive helicopters in 1950 (see p. 219). They came to the conclusion that hanging turbojets on the tips of a helicopter rotor was simply not possible, and in 1954 Technical Director Hollis Williams told me he was 'personally embarrassed that such nonsense should have been published'. There the matter rested until 1999, when Gennadi Lazarev, a leading designer of the Mil MVZ, revealed that Russia had identified a clear need for helicopters even more capable than the Mi-26, and that the Mil team were convinced that their proposed super-large design should have *tip-mounted turbojets*!

The Mil designers tested this concept in 1959 with a small helicopter designated V-7, from V*intokryl,* screw-wing (i.e., helicopter). It is such an attractive machine a refined version might even go into production, in which case it would have an Mi designation. Though Mil is a Russian company, its partner working on the engine is ZMKB Progress, of Ukraine. Team leader on the V-7 propulsion system was S.I. Slobodkin. He has discussed the problems which were encountered with the two kinds of engine flown on the V-7.

One engine, the MD-3, featured starting by compressed air fed from the fuselage, while the AI-7 had electric starting. The main problems were caused by the centrifugal loads of up to 200 g. One concerned the main shaft bearings, and the lubrication system. Oil temperature rise was found to be double that of the same engine running on a stationary testbed, and while the oil tank surrounded the engine inlet the radiator formed the leading edge of the rotor blade. Another problem was obtaining uniform combustion throughout the combustion chamber. To prevent excessive rotor drag it was necessary to limit the engine angle of attack to 6π, even when hovering at maximum weight. This was the most difficult situation, because

Three-view of one of the monster Russian helicopter projects being studied by Mil, made possible by driving the main rotor from the tips. Dimensions are metres.

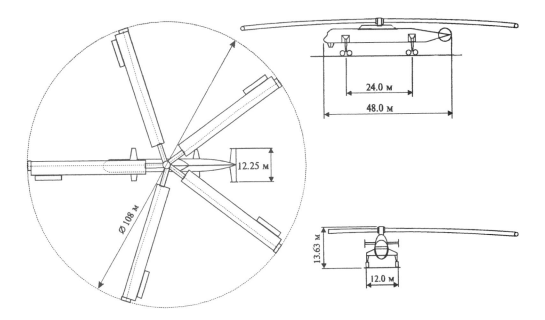

the engine axis was in the tip-path plane. A future giant helicopter with six blades would have the peripheries of the hot propulsive jets (defined by 70°C excess temperature) only 1.7 m (67 in) apart.

Calculations and rig tests suggest that a tip-drive helicopter with a four-blade 50-m (164-ft) rotor, with a 40-ton payload, should have tip engines with a unit thrust of 3,527 lb. Engine specific weight would be about 0.18, and sfc about 1.22, centrifugal forces having added 3 per cent over that of a static engine. Noise is said to be no serious problem: 'In the V-7 cabin we could talk without using intercom'. According to Slobodkin, 'The economical effectiveness of the future large helicopter will be close to that of the transport aeroplane, and indeed to that of a cross-country surface vehicle'.

Dating from 1995, the biggest firm project at Mil is the Mi-32, with a take-off weight of 150 tonnes (330,688 lb), but for the more distant future the tip-drive monsters include sizes up to a rotor diameter of more than 100 m. One shown here is a crane helicopter, to lift loads of up to 250 tonnes (551,146 lb). It would have a rotor with a diameter of 108 m (354 ft 4 in), with five constant-profile blades. The rotor would drive shafting to twin tail rotors to provide positive control of the fuselage axis. I believe that this is the largest design currently being studied, in partnership with the Progress engine KB.

Going back to the opposite end of the size scale, the new century has seen unprecedented interest in very small engines. Wren Turbines, a three-man company in Yorkshire, England, has developed both turboprop and jet engines with cores that could easily fit in a typical trouser pocket. Their customers are mainly enthusiasts, but also include Rolls-Royce, Los Alamos Scientific Laboratory and NASA.

One can go much smaller. Among a large and rapidly growing range of MEMS (micro-electromechanical systems) are several species of micro-turbojet. For example MIT (Massachusetts Institute of Technology) has tested a turbojet with overall dimensions of 12 mm (0.47 in) diameter and 3 mm (0.12 in) long, and a weight of 1 g (0.035 oz). Though it rotates at 2,400,000 rpm, it has driven an alternator, giving 50 W (0.067 hp).

MIT is one of many laboratories – others include B.F. Goodrich, Honeywell, Northrop Grumman and Caltech – working on MEMS applications to full-size engines and their inlet systems. Among several promising developments are dramatic improvements in flow control throughout the engine aerodynamics. Initially these will be applied to small engines for UCAVs (unmanned combat aerial vehicle) and cruise missiles, but there is no doubt they will also progressively improve the performance of the largest aircraft gas turbines.

In conclusion, in 1936 Whittle showed mathematically that some turbofans would be improved by inserting a reduction gear, so that a small high-speed turbine could drive a large fan. This is actually a feature of a few engines, such as the TFE731 and ALF502, but the proposed PW8000 has yet to be run. Imagine my surprise when Pratt & Whitney Canada announced the geared-fan PW800. Since 17 March 2001 testing has been in progress with an ATFI (advanced-technology fan integrator) using a PW308 core, and flight test is due in 2002. Rated at '10,000-20,000 lb thrust, for regional aircraft and bizjets', the PW800 could be delivered 36 months from go-ahead. One Pratt leader went so far as to agree with my belief that we shall see gearboxes on the biggest future airline engines.

Jet and Turbine Aero Engines – A fourth update

With amazing slowness, engine designers are at last approaching the point where they will have closed the gap between the turbojet and the turboprop, with various species of turbofan and propfan. From the first turbofans, which were just turbojets with a cool outer casing, it has taken 50 years to reach really effective turbofans with a BPR (bypass ratio) of 11. Going much beyond that point will call for insertion of a reduction gear, which will add an extra source of weight and possible failure.

For the moment, the latest airline engines, with BPR of around 11, are the rival General Electric GEnx and Rolls-Royce Trent 1000. Both were originally conceived to power the Boeing 787. The former is derived from the GE90–94B, and is thus a two-shaft engine with a pressure ratio of from 36 (in the 2B67 version, which is the exclusive engine of the Boeing 747–8) through 40 (for the 787) to 45 (in the 1A72 version for the Airbus A350). In the 787 the GEnx is competing against the Trent

The GE90 is a family of high by-pass turbofan engines. The GE90-115B is the world's most powerful jet engine at an unprecedented 115,000lb of thrust. GE

1000, which continues the unique three-shaft formula to new heights of efficiency, with a pressure ratio of no less than 50. This means the air is essentially *red-hot* as it enters the combustion chamber, before any fuel has been burned!

Perhaps surprisingly, Airbus launched the A350 on the basis of the GEnx 1A72 as exclusive powerplant. Substantially larger than most 787 versions, the A350 was at first little more than an A330 with new engines, but Airbus wisely thought again. By the autumn of 2005 the A350 had become an almost totally new aircraft, and it is already proving a formidable rival to the 787. Launched in late 2005 when Boeing had already taken deposits on over 200 787s, the A350 then proceeded to notch up a succession of cutomers, and its appeal was enhanced by admitting a new version powered by an advanced Trent version called the 1700. Whereas 787 engines have thrust ratings from 63,000 to 69,800 lb, the A350 engines begin life at 72,000 lb.

Coming down the scale of size and thrust, the CFM56 and V2500 battle it out in the 22,000–34,000 lb thrust bracket, the latter continuing to be handicapped by the former's exclusive position on the 737. At the bottom end of this class, at 22,000 lb, Pratt & Whitney have had a dreadful experience with the PW6000. This was the engine meant to succeed the globally best-selling JT8D, but trouble with the compressor caused serious delays. Ultimately it was decided that MTU, the German company responsible for the low-pressure turbine, should also replace the American compressor with one of its own design. This has delayed certification from 2000 to 2006, and when the PW6122 at last entered service it was with LAN of Chile, one of a small handful of customers. Probably the Pratt & Whitney engine's simplicity will steadily increase its position in the marketplace, ultimately as preferred engine of the A318..

In the category just below 20,000 lb a newcomer with healthy prospects is Powerjet's SaM 146 Created jointly by Snecma of France and Saturn of Russia, this all-new turbofan was from the start tailored to the RRJ (Russian Regional Jet), yet another passenger aircraft with two underwing engines. Rated at two levels, 15,645 and 17,209 lb, the basically simple SaM 146 hopes to get customers in the West as well as in Russia and neighbouring countries.

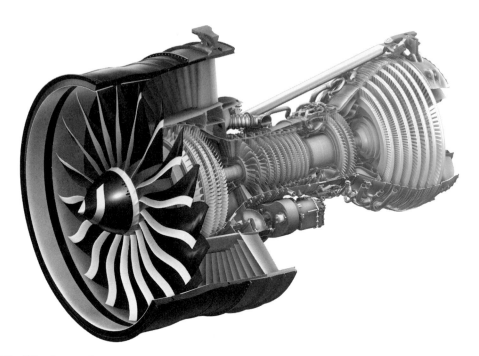

The GEnx is one of two engines selected by Boeing to power the 787, and is the exclusive engine of the 747–8. GE

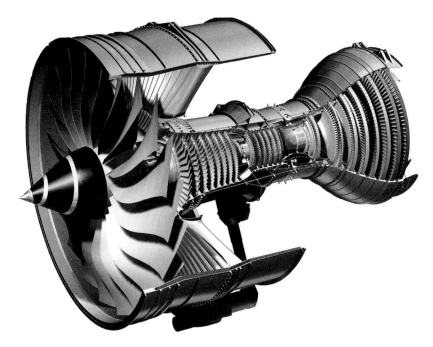

The Rolls-Royce Trent 1000 has been chosen by Boeing as the lead engine for its 787 Dreamliner.
Rolls-Royce PLC

By aiming at the bracket 17,000–23,000 lb thrust, Pratt & Whitney Canada have been trying to expand their global dominance into a new field of power well above that of the company's existing engines. The SPW14 featured in the 1997 update was abandoned when General Electric claimed it encroached on the CFM56 market. To replace it, the Montreal firm had been hoping neighbour Bombardier would launch the C-Series, a new jetliner tailored around the projected Canadian engine. In January 2006 Bombardier finally gave up and abandoned the C-Series, but Pratt & Whitney continued work on geared-fan engines in this general thrust class.

General Electric's CF34 has astonished its builders by becoming a global best-seller for transports and bizjets by Embraer of Brazil, Bombardier of Canada and AVIC of China. Starting out at the 9,000 lb thrust level, this simple two-shaft turbofan is now in production at over double this thrust in AVIC's ARJ121 and Embraer's EMB–190/195 family, in which the engines are moved to under the wings (unlike earlier Embraer passenger jets).

With impressively methodical lack of haste, in about 1990 Honda of Japan began studying small turbofan engines. Eventually, after flight testing various development engines, this led to the HF118-2, which from 2004 has been a joint project with GE. Rated in the 1,700 lb class, this remarkably simple engine is being flight tested in unusual overwing pods on a specially designed Honda twin-jet, but Honda have not said that this aircraft will be built in quantity.

In a previous (1996) edition of this book I commented on the fact that Williams, previously the mass-producer of cruise-missile engines, was replacing this dwindling market by moving into the field of manned aircraft. Assisted initially by Rolls-Royce, their FJ44 had by the new century carved out an impressive market for twins and even singles needing engines in the 1,500 to 3,000 lb thrust class. Now, without needing Rolls-Royce, the Walled Lake, Michigan, firm have moved further down with the FJ33, in the 1,500–1,700 lb bracket. In the 1990s there was virtually no market for really small jet transports, but now Williams is cashing in on their strategic vision with a dozen eager customers.

The one customer where Williams failed was in providing engines for the Eclipse E500. This baby jet flew with FJX engines, of 700 lb

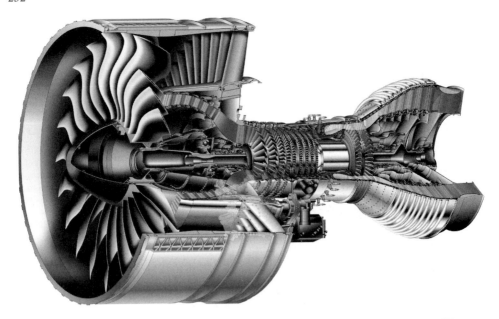

Cutaway of the GP7200, a joint venture between General Electric and Pratt & Whitney. GP

thrust, but Eclipse very publicly said the engines were not sufficiently powerful. The Eclipse is now in production powered by Pratt & Whitney Canada PW610F engines of 900 lb thrust.

Among a wealth of newcomers to the global gas-turbine engine market, one making a particularly simple engine is Agilis, based in Florida. Formed in 1993 to make small jet engines, Agilis Engineering is now producing a range of very simple two-shaft engines all using a basically similar core. The low-pressure (fan and LP turbine) section is small in the TF1000, and gets progressively bigger in the TF1200, 1400 and 1500, the designations giving the thrust in lb.

At least as simple are the turboprops of the Pennsylvania-based Innodyn company. Again, these feature a basically standard core around which are peripherals and other features (including twin jetpipes), with a propeller reduction gear on the front. The Innodyn TE (turboprop engine) range at present comprise the 165TE, 185TE, 205TE and 255TE, these designations giving the horsepower.

Pratt & Whitney may have taken hard knocks in the airline market, but in the military sphere they are in a good position. The 35,000-lb-thrust F119-PW-100 is now in full service as the engine of the F-22A Raptor, and the derived F135, in the 40,000-lb class, is coming into production for the first four batches of Lockheed Martin F-35 Joint Strike Fighters. The F135-PW-600 Stovl (Short-takeoff, vertical landing) version drives an extension shaft geared to the amazing LiftFan produced by Rolls-Royce which converts 29,000 shp into a vertical thrust of 21,000 lb. The jet of pure air can be rapidly vectored through up to 60° by the Vane-Box Nozzle which is the structural basis of the F-35B, and the first part to be assembled.

Via an engine called the YF120, General Electric and Rolls-Royce have developed the F136. This turbofan has been planned as a later and rather more powerful rival to the F135. It is (in early 2006) planned to be available from the 6th production batch of F-35 aircraft. The intention has been that from that point customers will have a free choice of either engine, and that F-35 pilots would be unable to tell which engine was fitted. Unfortunately, in late 2005 the US Navy was so short of money that it announced that it would terminate funding of the F136. In May 2006 funding was restored by US Congress, but the battles are not necessarily over, and 'buy American' is a powerful force affecting strategic decisions.

The only other super-power fighter engine is Lyul'ka Saturn's AL-31, which powers the

Honda's fuel efficient HF118 is a small turbofan engine for the light business jet market. Honda

brilliant Su-27 in all its many versions. In partnership with production plant Salyut, this engine is one of the few in Russia to have achieved such important export sales that new versions are fully funded. Today versions are in production with thrust increased and the nozzle able to vector under computer control in any direction. In theory this could enable later Su fighters to dispense with a tail.

This is in contrast to the EJ200, engine of the Eurofighter Typhoon. This outstanding aircraft was by 2006 in service in large numbers with the original partners (p.179), but progress towards getting a vectoring-nozzle version continues to be slow. The good news is that export customers for the Typhoon are making up for the handicap of the cash-strapped United Kingdom.

Turning to shaft-drive engines, the big news is selection of the European multinational Europrop TP400-D6 as the engine of the Airbus A400M, the definitive name of what in the 1997 update was called the FLA. Selection

of the Europrop engine caused a storm of protest from Pratt & Whitney Canada, which claimed that their proposal (based on twin PW150 power sections driving a single propeller gearbox) was in every way superior. The TP400-D6 first ran on the testbed at near its 11,000-shp design rating in late 2005, and before this new edition appears it will have run with the 17ft 6in single-rotation eight-blade Ratier propeller. Like the de Havilland Hornet of World War 2, the engines will be handed, the inboards rotating in opposition to the outboards. The A400M shows every sign of being a successor to the C-130 as a globally adopted multirole airlifter.

By 2006 the Bell/Boeing V-22 Osprey was at last established in service, though there was no sign of funding for proposed derivatives with four instead of two swivelling Rolls-Royce T406 engines. Three of these 6,000-shp engines appear to be the likely choice to power the US Marine Corps' Sikorsky CH-53K Super Stallion successor.

Not least, Turbomeca, as a business unit of the mighty Safran Group (another is Snecma), continues to be the only company to offer real competition to Pratt & Whitney Canada in the small-turbine category. Their product range for the immediate future has been completed with the Ardiden, a simple and attractive helicopter engine in the 1,500-shp class. It is rapidly finding customers, some of the most important being Indian helicopters for which Ardiden engines will be jointly developed and produced by Hindustan Aeronautics.

Glossary

absolute, see temperature.

AOA Angle of attack.

blisk A rotor comprising disc (US=disk) and blades formed from one piece of material.

BPR Bypass ratio.

bypass ratio Ratio of mass flow through the fan or bypass duct to that through the core.

CFRP Carbon-fibre reinforced plastics.

CG Centre of gravity.

chord Breadth of a blade from leading to trailing edge.

core The central power-producing part of a two- or three-shaft engine, comprising compressor, combustion chamber and turbines.

DB Diffusion bonding.

dry rating Not using water injection.

dry weight The usual weight quoted for an engine, with no fluids (oil, fuel or water).

DS Directional solidification, molecular structure arranged in lines along the axis of load.

EGT Exhaust gas temperature, generally synonymous with JPT.

ehp Equivalent (shaft) horsepower, equal to shp+P/2.6 (in the USA, P/2.5), where P is the residual jetpipe thrust.

FADEC Full-authority digital electronic (or engine control).

FOD Foreign-object damage.

FPR Fan p.r.

gas generator The power-producing part of an engine; generally synonymous with 'core', though a gas generator can be run by itself to feed gas to a turbine or helo rotor.

GFRP Glass-fibre reinforced plastics.

HP High-pressure.

hp Horsepower.

IGVs Inlet guide vanes.

IP Intermediate pressure.

JPT Jetpipe gas temperature.

LP Low-pressure.

mass flow Mass of air passing through the fan or engine per second.

NGV Nozzle guide vane.

normal shock A shockwave at 90° to the air-flow.

OEI One engine inoperative.

power section Gas producer; term normally used where two share a common gearbox.

power turbine Turbine not driving the compressor but an output shaft, usually to a gearbox.

p.r. Pressure ratio, ratio of pressure at the compressor delivery to that at the inlet.

rating Certificated power of a particular engine model, qualified according to operating regime: take-off, max climb, max continuous, flight idle, OEI contingency etc.

rpm Revolutions per minute.

SC Single-crystal.

sfc Specific fuel consumption, rate at which fuel is burned divided by power produced.

shp Shaft horsepower, power transmitted via a rotating shaft.

speed In the context of engines, rpm.

SPF Superplastic forming.

temperature In engines, measured either by °C or by °K (Kelvin), an absolute reading in which 0° is absolute zero; thus 0°C is 273°K. The USA still often uses °F; 0°F is –17.77°C =255.22°K.

TET Turbine entry temperature.

TGT Turbine gas temperature, generally the same as TET.

V/STOL Vertical or short take-off and landing.

VSVs Variable stator vanes.

wet Rating using water or water/alcohol injection.

zero stage A stage of blading added at the front of an axial compressor.

Index